An Introduction to
Numerical Methods
in C++

An Introduction to
Numerical Methods
in C++

REVISED EDITION

B. H. FLOWERS

Fellow and former Rector of
Imperial College
University of London

OXFORD
UNIVERSITY PRESS

ΧFORD

UNIVERSITY PRESS

Great Clarendon Street, Oxford OX2 6DP

Oxford University Press is a department of the University of Oxford.
It furthers the University's objective of excellence in research, scholarship,
and education by publishing worldwide in

Oxford New York

Athens Auckland Bangkok Bogota Buenos Aires Calcutta
Cape Town Chennai Dar es Salaam Delhi Florence Hong Kong Istanbul
Karachi Kuala Lumpur Madrid Melbourne Mexico City Mumbai
Nairobi Paris São Paulo Singapore Taipei Tokyo Toronto Warsaw

with associated companies in Berlin Ibadan

Oxford is a registered trade mark of Oxford University Press
in the UK and in certain other countries

Published in the United States
by Oxford University Press Inc., New York

A catalogue record for this book is available from the British Library

Library of Congress Cataloging in Publication Data

ISBN   0 19 850693 7

Typeset by AMA DataSet Ltd, Preston, Lancs
Printed in Great Britain
on acid-free paper by
Biddles Ltd, Guildford & King's Lynn

For Peter

LECTOR: Why on earth did you write this book?
AUCTOR: For my amusement.
LECTOR: And why do you suppose I got it?
AUCTOR: I cannot conceive . . .

— Hilaire Belloc: *The Path to Rome*

Preface to Revised Edition

The review I conducted in 1996 of the first edition to prepare for its reprinting convinced me that some quite modest revisions and additions would improve the book significantly. In this revised edition I have tried to clarify some of the material by minor programming changes and explanatory material. The routine `quicksort` in 14.4.1, for example, although correct, lacked elegance and has been improved, and a list version has been provided in 20.2 for comparison. The function `gauss` in 16.4 was wrong and has been corrected.

There are a few additional topics briefly supplementing those already treated: the chief ones being a fast recursive function for computing integer powers of a number in 3.11, the Brent method for finding a minimum of a function by fitting a local parabola to it in 4.11, the method of steepest descent for finding a root of a system of non-linear equations in 12.6, and a remark about multiple integrals in 16.9.

In the preface to the first edition, I noted with regret that I had not made use of the template facility with which C++ is endowed. It is in fact a useful tool, and I have introduced both function templates and class templates in the revision. The adoption of a list template has powerfully clarified the treatment of lists in chapters 20 and 21.

By far the most substantial addition to the book, however, is the addendum—which tries to describe, and to some extent explain, how to go about programming in Windows as an alternative vehicle to chapter 15 for producing the graphical representation of the results of numerical computation. The standard textbooks on the subject occupy 700 pages or more, and I have attempted to achieve all that is necessary for this book within a few percent of that. I have presented it as an addendum instead of an additional chapter simply for that reason: the level, and to some extent the style, of treatment are different. Nevertheless, I hope that readers wishing to work in Windows will find it helpful.

Virtual functions play an important part in programming in Windows. No use was found for them in the first edition, so no account was given of their properties. They are explained in the addendum, I hope sufficiently for the purpose. Together with the introduction of templates mentioned above, more of the C++ language has now been expounded in the particular context of numerical computation.

I hope that these modifications will be found to have added interest, utility, and enjoyment to my book.

London   B.H.F.

Preface

This book is exactly what its title describes it to be. It is an introductory text on numerical computing, presented through the medium of the C++ language. I hope that it will prove helpful to undergraduate students of science and engineering who are seriously studying numerical methods for the first time, and perhaps to graduate students too. I also hope that it may be of interest to computing scientists who wish to see how C++ can be used in earnest for numerical computation.

Mathematically, the book pre-supposes a working knowledge, at undergraduate level, of elementary calculus, of linear algebra, and of differential equations. Many of the classical algorithms of numerical computation are explored as befits an introductory text, together with the mathematical analysis which underlies their utility and behaviour in computing programs. I have tried to provide sufficient mathematical background for that purpose, but it falls short of a systematic textbook on numerical analysis. It would have been rewarding to extend the treatment of non-linear systems in order to witness the onset of chaos; the essential techniques are provided, but the temptation was resisted; it would have taken me too far into analytical dynamics. Likewise, I have entered into the vast field of statistical inference only so far as to understand and test the most common random sequence generators. By contrast, the book ends with treatments of some classical problems in terms of lists, which are not, to my knowledge, to be found in the standard textbooks on numerical methods.

The book also assumes familiarity, at about the same level, with computing methods. Some elementary experience of programming in Basic, Fortran or Pascal is essential, and a sound working knowledge of C would undoubtedly prove an advantage. No prior knowledge of C++ as such is assumed, however. The language is developed *ab initio*, but as far as possible in step with mathematical need. There are two consequences. First, the language is developed from the point of view of its applications to numerical computation; the development does not necessarily follow the logical path quite properly trodden by an expert computing scientist. Exposition of the language, to an extent, is therefore piecemeal, the style of presentation inductive, at times even narrative. (It was even more so in earlier drafts.) The same could be said, however, of the early chapters of the classic presentation of the C language by Kernighan and Ritchie. Secondly, not all of the C++ language is expounded, only those parts for which I have so far found an application in numerical computation. One of the strengths of C++ is the ease with which it may be moulded to suit the user's purpose. It is a relatively modest sub-set of C++ that is presented in this book, adapted so as to reflect ordinary mathematical usage as far as possible.

The Preface, written with hindsight by the author and supposedly read without foresight by the reader, is not the place for technicalities. Let me then confine myself to saying, without further explanation, that I have found no need in the chosen context for variables explicitly declared to be of the storage class register, nor to declare variables or class members volatile. The crucial concept of a class is presented in some detail, and the derivation of classes from a base class; but I have found no need for multiple inheritance of classes, one of the richest features of the language, nor therefore for the concept of virtual functions and classes to which it gives rise. The C++ language also provides systematic techniques for exception handling, which enable programs to be written that are relatively tolerant of fault conditions. In long and complex computations these are invaluable. In an introductory text, however, in which the programs are mostly short and relatively uncomplicated, I thought fit to handle errors in a more straightforward fashion.

One feature, which with hindsight I rather wish I had included, is the template, although a relatively recent addition to the language. That would have made it possible, for example, to define a class of vectors of arbitrary type, instead merely of floating point numbers of double precision, as I have done. There are a few places in the book where that additional flexibility would have helped, because it would have made the treatment of certain operations more uniform. The serious student of C++ will no doubt wish to provide that facility in person.

Those are the major omissions from my limited exposition of the C++ language. What remains, however, is a powerful framework for numerical computation, and more than enough for an introductory text.

It is usual in a Preface to disclose how the book came to be written. I belong to that generation of theoretical physicists who could always rely upon willing assistants to satisfy their computing requirements. During my active research life I had therefore never learned how to compute for myself, except in the most rudimentary fashion with paper and pencil. The work which eventually culminated in this book began about 10 years ago when I was deeply involved in the higher strata of academic administration. It was an enjoyable hobby, and immensely relaxing during interminable committee meetings, to write snippets of programs which could later be tried out at home, and it was less visible to one's colleagues than most other portable pastimes, such as wood carving or taking snuff. Some of the earlier routines presented in this book I first wrote in Basic, later in Pascal. A few routines I wrote in Assembly Language, because I had a horror of "black boxes" and wanted to know as closely as I could what went on inside the machine. It was some time before I discovered C, which first seemed to me to have been designed deliberately to disguise its potential for serious numerical work. For the sake of discipline, however, I set myself the limited task of translating into C the book by Atkinson and Harley, *An Introduction to Numerical Methods with Pascal*.

Nevertheless, it was not long before my new interest was detected. A well-wisher generously presented me with the three volumes of Donald E. Knuth's great work *The Art of Computer Programming*. Then I chanced upon a copy of Stroustrup's *The C++ Programming Language*, and realized that C++ contained the facilities I had been seeking but failing to find in C. Finally, another encouraging colleague suggested I consult Kendall E. Atkinson's *An Introduction to Numerical Analysis* if I was really serious about numerical computing. Between them, these three books transformed my modest objective into something much more ambitious. I now had to understand what I was doing. I read many books, but these are the three upon which I came to rely most. I have not provided detailed references. A list of all the books I have used is given in the Bibliography. I do not wish to suggest that it includes all the books that might have been useful. Occasionally I have indicated in the text where I have followed a particular treatment rather closely. In particular, I have adapted a few routines from the book *Numerical Recipes in C*, by Press *et al.*, as examples of more advanced programs.

I began to work in C++ soon after I retired from academic life in 1990, and the first draft of this book was written at home during 1993. There was an immediate difficulty. There was not, and there still is not, an agreed ANSI standard for the C++ language, although one is expected soon. (We have a rather strong indication of what it will contain in *The Annotated C++ Reference Manual* provided as an ANSI base document by Ellis and Stroustrup.) Moreover, many features are, and apparently will continue to be, dependent upon the particular implementation used. Among them are the input and output mechanisms, which in any case are not themselves part of the language. Obviously, one would wish one's work as far as possible to be portable, but without adequate facilities for input and output, in both text mode and graphics, life is limited to say the least. On all sides I have been advised to confine myself to such matters as may be said to be independent of implementation, or—since we do not yet have a standard—thought to be so. That is precisely where I have met the greatest difficulty in understanding how actual programs are run in C++, especially if formatted input and output, or graphical output, are desired—simply because textbook writers by and large avoid discussing such implementation dependent details (although Stroustrup gives some helpful indications of what might be involved). To know what one is doing in running a complete program in *some* implementation, it seemed to me, was better than to offer snippets of programs, no matter how ingenious, without providing comprehensive facilities for using them.

And so I wrote my own schematic account of input and output in C++, as an illustration of the class concept, which led me on to the discussion of formatting that I have used throughout the book; and I wrote a graphics package adequate for my immediate purposes. I am well aware that some may disapprove, and others will have access to other means of handling input and output, through the system software, for example. Those who share my dislike of black boxes, however, may nevertheless be interested to see how these matters *can* be handled directly.

After some experimentation, I selected Borland's TurboC++ as the implementation I would use with my home based PC running on DOS. As far as possible I have avoided features peculiar to that implementation; but for input and output, and for graphics, I have freely used the facilities thereby made available to me, identifying and explaining, as far as seemed reasonable, what was involved. The particular library functions provided by TurboC++ can in any case be readily translated into other implementations, if so desired.

All the routines presented in this book have been run and tested in a variety of contexts. Nevertheless, they are intended as illustrations of the translation of numerical algorithms into the C++ language rather than as foolproof packages. In any case, even standard packages are used blindfold at the user's peril.

There is also the vexed question of programming style, as hotly debated as any theological question. Some computing experts insist on writing statements sequentially from top to bottom, believing that to be the only acceptable presentation of a program. We are all, however, from the earliest age, taught to read from left to right (or some of us from right to left), as well as from top to bottom. Among many others, I find it more natural to do that than to insist upon a strict top to bottom style. It should be a matter solely of clarity. When clarity is enhanced by writing from left to right I have not hesitated to do so. When, on the contrary, there is a need for comment, to explain what is going on behind the computer's back, so to speak, then the top to bottom style is to be preferred. It should not be a matter of dogma, only of clarity.

Since this book has been written as a personal exploration of a fascinating field, largely as a diversion from the weighty matters for which I was being paid to be responsible, there are few acknowledgements I have to make, except of the general tolerance of my colleagues. However, I am particularly grateful to Professor O. P. Buneman, of the University of Pennsylvania, who not only presented me with Knuth's three volumes at a particularly critical moment in my life, but responded in a stimulating manner to innumerable questions about the intricacies of computing science, mostly while we were both engaged (myself as his assistant) upon various constructional projects in the Scottish Highlands. It was he who suggested that I illustrate my graphics package with Bezier curves; he, moreover, who proposed that I implement the Fast Fourier Transform in terms of lists, the subject matter of the last chapter of my book, and the incentive of much that comes before. He is in no way responsible for the outcome, however. I also gladly acknowledge the contribution of the referees provided by the Oxford University Press, as well as the staff of the Press itself. Their criticisms were as penetrating as their encouragement was warm, and both I and the book have benefitted greatly as a result of my responding as well as I could to their suggestions.

Above all, however, I have to thank my wife. She has suffered from having a preoccupied husband huddled solitary, incommunicable and round-shouldered over a computer for far too long, her contribution to our joint welfare gladly but sadly restricted to the provision of life's essentials. She too should enjoy my retirement. The time has come. She shall.

London   B. H. F.
September 1994

In this reprint I have taken the opportunity to correct a number of minor errors and infelicities. There was also one serious error in the function decompose on page 215, which was very kindly corrected for me by Mr Benjamin Keeping of Imperial College, to whom I am most grateful.

London   B. H. F.
July 1996

Contents

2 Expressions, statements and functions   20

3 Errors, theorems and speed

4 Roots of non-linear equations

13 Matrix eigenvalue problems   250

14 Interpolation and data fitting   280

15 Graphics   305

Addendum Programming in Windows 481

Bibliography 533

1 Preliminaries

We begin with an elementary introduction to input and output in the C++ language. We go on to describe the basic data types, both integral and floating point, and the elementary types derived from these, including pointers. We end the chapter with a brief discussion of the more common preprocessor directives, and with notes on keywords and identifiers.

1.1 Elementary input and output

Throughout this book we shall need to be able to deal with input and output of character strings and numerical data. In later chapters we shall have to consider more elaborate procedures for dealing with formatted data in tabular form, and with user-defined types.

1.1.1 Strings

Every program, if it is to be used, must contain some output instructions, otherwise there would be no means of knowing what it had achieved. In C++, the most elementary program with output would be something like this:

```
#include <iostream.h>
main()
{
        cout << "This is me!\n";
}
```

When compiled and run, a C++ program executes sequentially the instructions contained between braces { ... } in the function main. We shall introduce the concept of a C++ function more fully in the next chapter. In this case, the program will print on the screen:

```
This is me!
```

The identifier cout represents the standard output stream, usually directed to the terminal screen, and << ("insert" or "put to") represents the output operator. The program simply instructs the computer to print the character string "This is me!" to the screen. The character

'\n' represents a new line (usually written *newline* in text), so the output of the program terminates neatly at the beginning of the next line. Note that a character string always appears between double quotation marks.

However, the compiler does not itself recognize the identifier cout. Input and output are not part of the C++ language. They are supplied by the user, or in certain standard forms by the compiler writer in compiled modules. Definitions of identifiers used by standard modules are conventionally provided in *headers*, which have to be included in the program before the identifiers concerned are used. In TurboC++, for example, standard input and output conventions are defined in the header file iostream.h, and accordingly our program is prefaced with the compiler instruction #include <iostream.h>. In other implementations the name of the file and the precise conventions might be different, but they would serve essentially the same purposes.

Most programs also need some input data. The identifier cin represents the standard input stream, typically provided from the terminal keyboard. Input is then obtained with the operator >> ("extract" or "get from"). If name represents a character string, the statement cin >> name; causes the computer to pause while characters are entered from the keyboard and stored in memory at a location identified by name and conventionally terminated by the character '\0' when the return key is pressed. The following program,

```
#include <iostream.h>
char name[10];
main()
{     cout << "Who are you?\nEnter name: ";
      cin  >> name;
      cout << "You are " << name << "\n";
}
```

requests that a name be entered—"Peter", say—whereupon it responds to its own question "Who are you?" with the statement "You are Peter", followed by a newline. On the screen we shall see:

```
Who are you?
Enter name: Peter
You are Peter
```

Here, name has been declared as an array of 10 characters by the global declaration char name[10];. It will accept an input string of up to 9 characters, leaving room for the terminator '\0'. The conventional chaining property of the << operator has been used to concatenate the two strings "You are " and name, and to append a newline. Note that the >> operator automatically follows the entered string with a newline. Note also that if the entry includes "white space" (spaces, tabs, newlines or comments), only the characters before the first such space are included. Thus, if instead of "Peter" we enter "Peter Bloggs", the program will still respond with "You are Peter". If we expect to receive two names, we should write instead,

```
#include <iostream.h>
typedef char array[10];
main()
{      array name, surname;
       cout << "Enter name and surname:\n";
       cin  >> name >> surname;
       cout << "Your name is "
                 << name << " "
                 << surname << "\n";
}
```

On the screen we shall see:

```
Enter name and surname:
Peter Bloggs
Your name is Peter Bloggs
```

Note that with the keyword `typedef` we have defined the identifier `array` to be a type name for an array of 10 characters. The declaration,

```
array name, surname;
```

then declares two arrays of 10 characters each. This time we have declared them locally within the function `main` because there is no advantage in their being global variables. Note that the succession of `<<` operators can be spread over several lines.

1.1.2 Numbers

The input and output operators are "overloaded": that is, they can be used with inputs and outputs of different types. If we declare n to be an integer and assign to it the value 10 (`int n = 10;`), then `cout << n;` will recognize n as an integer and print it accordingly. If, on the other hand, we define x to be a floating point number (`float x = 3.141593`), it will be printed in floating point fashion. The input operator `>>` behaves similarly.

The following program requests that an integer be entered, then prints the result to screen:

```
#include <iostream.h>
main()
{      int n;
       cout << "Enter an Integer: ";
       cin  >> n;
       cout << "The Integer is " << n << "\n";
}
```

On entering 10, say, the screen will show:

```
Enter an Integer: 10
The Integer is 10
```

Notice that the output operator concatenates quantities of different types, in this case an integer and two strings. The input operator behaves similarly. If n is an integer and x is a float, we may write the following program:

```
#include <iostream.h>
main()
{       int n;
        float x;
        cout << "Enter an Integer "
             << "followed by a Float\n";
        cin  >> n >> x;
        cout << "n = " << n << ", x = " << x << "\n";
        cout << "n times x = " << n * x << "\n";
}
```

Here we have divided the first string because it is too long for the page. Entering 2 and 3.141593 separated by a space, the screen displays:

```
Enter an Integer followed by a Float
2 3.141593
n = 2, x = 3.141593
n times x = 6.283186
```

Finally, note that the output operator can manage an arithmetic expression, correctly printing out the product n * x of the two variables n and x.

1.1.3 Standard error output

There are two further standard output streams: cerr, which is used for printing error messages (usually to screen) and is flushed after each insertion; and clog, which is a fully buffered version of cerr, and which may be used for storing error messages. The following program fragment illustrates the usefulness of cerr:

```
int m, n;
........
if (n == 0)
{       cerr << "Cannot divide by zero!";
        exit(1);
}
```

```
int k = m/n;
........
```

The standard function `exit`, when invoked, causes the program to terminate gracefully, in this case leaving on the screen the message:

```
Cannot divide by zero!
```

Notice that in C++, as in C, the boolean equality operator is denoted by == to distinguish it from the assignment operator =.

1.2 Comments

In order to be comprehensible to human beings, most programs require additional explanatory material which, although incorporated in the program text, is ignored by the compiler. In C++ there are two ways of writing comments. The first kind, also found in C, is:

```
/* This is a comment */
```

It may spread over several lines, and may even be distributed according to some pleasing pattern:

```
/* This is a comment
 * which spreads over
 * several lines
 * of source code
 */
```

The second kind, not found in C, is:

```
// This is another comment
```

which tells the compiler to ignore the remainder of the line following `//`. It cannot spread to succeeding lines. In the case of simple comments the choice between the two is often made on aesthetic grounds.

Comments of the same kind may not be nested, although a comment of one kind may be used to "comment out" the other kind. For example, a program may contain a fragment such as

```
........
int n;        // store integer
........
```

If the programmer wishes to test the program with this fragment temporarily removed, it can simply be converted into a comment as follows:

```
/*
........
int n;        // store integer
........
*/
```

But if one were to write

```
/*......../*........*/........*/
```

the compiler would protest with an error message.

1.3 Basic data types

Every object or expression in a program has a *type* that determines its memory storage requirements and the operations that may be performed on it. There are two groups of basic types, the *integer* types and the *floating point* types, and a wide range of types that may be derived from them. The precise nature of these types, however, may be machine dependent.

1.3.1 Integer types

In C++ the integer types are denoted by the *type specifiers* char, short, int and long. The number of 8-bit bytes of storage required for each of them is given by the operator sizeof. In general:

```
1 = sizeof(char) ≤ sizeof(short) ≤ sizeof(int) ≤ sizeof(long).
```

These types may be signed or unsigned, of which one is chosen as the default when the compiler is installed. If the default is signed then signed int is synonymous with int, but unsigned int is distinct, having a different range of possible values. Similarly with char, short and long. In this book, unsigned char is sometimes abbreviated to byte, but the compiler must in that case be told so, and similarly with unsigned int and unsigned long:

```
typedef unsigned char byte;
typedef unsigned int word;
typedef unsigned long longword;
```

These types may not all be distinct. In TurboC++, where `int` and `short` each occupy two bytes, while `long` occupies four bytes, the various synonyms and the corresponding range of values of the integral types are given in the following table, in which it is assumed that the default is `signed`, and the types `byte`, `word` and `longword` have been properly defined as above:

char, signed char	−128 to 127
unsigned char, byte	0 to 255
int, signed int, short, signed short	−32768 to 32767
unsigned, unsigned int, word, unsigned short	0 to 65535
long, long int, signed long, signed long int	−2147483648 to 2147483647
unsigned long, unsigned long int, longword	0 to 4294967295

On mainframe machines, however, the ranges may be considerably wider.

In this book, we shall confine ourselves to `char`, `int` and `long`, in their signed and unsigned versions, occasionally using the synonyms `byte`, `word`, and `longword`. Note that the declarations: `int i = 2; long l = 2;` although apparently defining the same number, are in fact defining different representations of it, the first (in the TurboC++ convention) in two bytes, the second in four bytes. The compiler converts automatically from one to the other, so that by writing, for example,

```
int ni = 2;   long nl = ni;
```

nl is set equal to a long 2, sometimes written 2L. Likewise, if one writes `char c = 'A';` `int n = c;` n is assigned the ASCII value of `'A'`, namely 65. Conversely, if one writes `int n = 321; char c = n;` n will first be converted to n $\%$ 256, namely 65, where $\%$ is the C++ symbol for *modulo*, and c is thereby assigned the character `'A'`.

Sometimes it is necessary to force type conversion explicitly. There are two ways of doing this. If i is an `int` which we wish to express as a `long`, we may use the traditional C cast notation `(long)i`, or the functional notation `long(i)`. The latter is perhaps preferable because it better conveys the idea of a conversion process, but it is a matter of taste. Similarly, one may convert a `long` to an `int`—with the obvious danger that the `long` we start with will be truncated (from the left!) if its value lies outside the range of an `int`.

1.3.2 Floating point types

Likewise, there may be several floating point types. In standard C++ there are two, the `float` and the `double`. They differ both in their range of possible values and in their attainable precision, and these attributes are machine dependent. In TurboC++, for example, they occupy 4 bytes and 8 bytes, respectively, and a 10-byte `long double` is also provided. The TurboC++ precisions and ranges are shown in the following table:

float	7-digit	-3.4×10^{-38} to 3.4×10^{38}
double	15-digit	-1.7×10^{-308} to 1.7×10^{308}
long double	19-digit	-3.4×10^{-4932} to 1.1×10^{4932}

In this case, sufficient numerical accuracy in lengthy arithmetic manipulations is unlikely to be maintained if numbers are represented in `float` and 5- or 6-digit accuracy is required in the result. In such arithmetic `double` is normally used in spite of the extra demand on memory, and we shall confine ourselves for the most part to working to that precision. We shall discuss precision more fully later.

Again, automatic conversion takes place between `float` and `double`, and between the integer types and the floating point types. In the latter case, truncation will take place on conversion to an integer type. One may write, for example:

```
double d = 3.14;
int n = 2;
d = n + d;
n = n + d;
```

The resulting values will be 5.14 for d, and 7 for n. However, the direction of truncation of a negative floating point number on conversion to an integer is machine dependent.

Note that floating point numbers may be written in the normal decimal notation, or in floating point notation. For example, `123.45` may also be written `1.2345E+2`, and `0.012345` written `1.2345E-2`. By default, these values are taken to be double. If floats are required, `f` or `F` may be added as a suffix. Thus, `1.2345F` is the float representation of the number `1.2345`. Naturally, when conversion takes place from `double` to `float` there is in general a loss of precision. As with integers, conversion may be forced between `float` and `double`, and between the integer types and the floating point types.

There is an authoritative discussion of the hazards of type conversion in the book by Stroustrup.

1.3.3 Void type

For formal purposes it is also convenient to define a "type" which is in fact no type at all. It is denoted by the keyword `void`, which stands syntactically wherever a normal type identifier might stand. It is useful for declaring functions which return no value (sometimes called *procedures*), or which take no argument. We discuss functions in 2.5. It is also useful for declaring pointers to objects of undetermined type. We discuss pointers in 1.5.

1.4 Derived types

From these basic types others may be created with the aid of special keywords, as follows.

1.4.1 Constants

The simplest extension to the basic types is obtained by preceding any of them (except void) with the keyword const in an initializing statement. This defines the object concerned once and for all; it is an error later to attempt to assign to it another value. For the same reason, it is an error to forget to initialize a constant when it is declared. Examples of typed constants are:

```
const char tab   = '\t';
const int quad   = 4;
const float pi   = 3.141593;
```

Because they are less likely to lead to errors of interpretation, the introduction of typed constants is much preferable, for most purposes, to the traditional C custom of introducing constants in untyped fashion by means of macros, such as:

```
#define QUAD   4
```

Nevertheless, we sometimes make use of #define when it appears safe to do so.

1.4.2 Enumerations

A derived integer type may be obtained by *enumeration*, using the keyword enum. The enumeration

```
enum { sun, mon, tue, wed, thu, fri, sat };
```

by default defines sun = 0, mon = 1, tue = 2, and so on, although other values may be set if desired. The basic type is int. Enumerations can also be named, thus creating a distinct type. For example:

```
enum day { sun, mon, ... };   // defining type day
day thisday;                  // variable of type day
thisday = mon;                // initializing variable
thisday = 1;                  // ILLEGAL!
mon = 2;                      // ILLEGAL!
```

Note that thisday should not be set equal to 1 even though the integer value of mon is 1, because the compiler should recognize day as a type formally distinct from int. Nor can the values assigned in enumeration be changed: they are *constants*.

Similarly, the statement enum bool { FALSE, TRUE }; defines the boolean type bool which, although unnecessary in C++, sometimes adds considerably to the clarity of code, and we shall therefore use it in this book. For example:

```
bool wrong;
. . . . . . . .
if (wrong) dosomething;
. . . . . . . .
```

The integer values assigned in the above enumerations are by default 0, 1, . . . They may also be set explicitly:

```
enum { mar=3, jul=7, jan=1,... };
```

where this time we have chosen not to name the enumeration, preferring to leave it *anonymous*.

1.4.3 Structures

In C++, as in C and many other languages, structured types may be defined, the *members* of which are of known types. This is done with the aid of the keyword `struct`. For example, if we wish to deal with points in the euclidean plane, each identified by a letter—$A(x,y)$, say—we can write:

```
struct point {
      char ch;     // name of point
      float x;     // x coordinate
      float y;     // y coordinate
};
```

This defines the type `point` as a structure with three members. The labelling character which identifies the point is stored in the first member, while the cartesian coordinates are stored in the second and third members. Note that in the declaration of a structure the keyword `struct` *must* come first, that the declaration of the members of a structure *must* all be contained within the braces which follow the name of the structure, and that the definition of a structure type *must* end with a semi-colon. However, the placing of the material on the page is otherwise arbitrary. We have chosen this particular arrangement because it is easily recognizable. It would have been perfectly correct to write instead,

```
struct point { char ch;   float x, y; };
```

but this might have seemed more difficult to read, and it would certainly have been more difficult to include comments. Although we prefer the former style of writing, there are occasions when the latter may be acceptable.

Objects of structured type might be expected to occupy in memory just sufficient space to accommodate all their members. A `point`, for instance, might be expected to occupy `sizeof(char) + 2*sizeof(float)` bytes. However, the memory allocated to a structure is implementation dependent. Some implementations prefer to "pad" the memory allocated to an object of structured type so that it always occupies an even number of bytes, or so that it meets some other architectural feature of the hardware. The actual space required is given by `sizeof(point)` no matter what implementation is used.

Members of a structure may be accessed by means of the . (dot) operator. For example,

```
point a;
a.ch =   'A';
a.x  =   1.234F;
a.y  =  -3.21F;
```

The contents of the point a can then be printed out in the usual way:

```
cout << a.ch << "(" << a.x << ", " << a.y << ")\n";
```

whereupon there will appear on the screen

```
A(1.234, -3.21)
```

The concept of a structure type will be massively developed in later chapters in the context of *class*. There we shall see, for example, how to obtain the same output by writing simply `cout << a;`.

1.5 Pointers

Pointers are declared with the aid of the unary * operator (the "dereference" or "indirection" operator, which must not be confused with the binary multiplication operator). Thus, `int* ip; double* dp;` and so on. They may be written equivalently as `int *ip; double *dp;` etc, which relies upon the fact that the value of an object pointed to by a pointer p is by convention written *p. Thus, we may speak of the "pointer type" *type**, where *type* stands for any given type, by which we indicate pointers which point to objects of type *type*. We may even give the pointer type a name; for example,

```
typedef type* tptr;
```

in terms of which we may proceed to declare particular instances of that type:

```
tptr thisptr, thatptr;
```

A pointer is most easily thought of as the address in memory of the object pointed to, and should indeed be initialized accordingly. If we write:

```
int i = 10;   int* ip = &i;   int j = *ip;
```

where & is the unary "address-of" operator (not to be confused with the binary & operator standing for bitwise AND which we shall meet later), then the value held in i is 10, ip holds the address of i, and the value of *ip assigned to j is 10.

A typed pointer may be assigned the integer value 0. This by convention guarantees that it points to no object whatever. It is then known as a *null pointer*. The zero value of a null pointer is sometimes denoted by a token such as NULL, predefined by #define NULL 0, so that double *dp = NULL; is the same statement as double *dp = 0;. However, this initialization statement must not be confused with, say, the assignment statement *dp = 0;, which defines the value of the (non-null) object pointed to by the pointer dp to be zero. No object at all is pointed to by a null pointer, so that no value can be associated with it, not even zero.

Pointer conversions can be achieved by means of casts, but they are implementation dependent, ambiguous and in general dangerous. Consider the following fragment, in which one is attempting to access the integer value of a floating point number by means of a pointer:

```
float f = 3.141593F;
float *pf = &f;
int *pi1 = (int*)pf;      // DON'T!
int i = (int)f;
int *pi2 = &i;            // CORRECT
```

The value of *pi1 is unpredictable (computer scientists like to say that it is "garbage"). This is not really surprising. What is the object that pi1 is supposed to be pointing to? The value of *pi2, however, is 3 as desired. Pointer conversions should be confined to conversion to pointers to void (*cf* 1.5.1), and between derived and base classes (*cf* 6.1).

The number of bytes occupied by a pointer in memory is machine and implementation dependent, but if required may always be revealed as sizeof(*type**), where *type* is any type. It is important, however, not to confuse this with the number of bytes occupied by the object pointed to, which is sizeof(*type*), and which varies from type to type.

One further word of explanation may be helpful, especially to PC users. In machines with segmented memories, two bytes are necessary to denote the location of a given object within a segment, and a further two bytes to denote the location of the segment within memory, a total of four bytes altogether. For small programs it can be arranged that all the work takes place within fixed segments, and in that case only the first two bytes are needed. We may then speak of *near pointers*, and it may well be that the implementation offers these as the default option. If it is necessary to step outside the segment, because the memory requirements of the program cannot otherwise be met, then the full four byte *far pointers* must be used. This is the case in TurboC++, for example, where a far pointer to a double would be written double far *fdp; (unless it is on the contrary the default option), and this is capable of addressing many different segments. This book confines itself to small programs for which near pointers are adequate, and where we may therefore forget about the distinction between "near" and "far". The main exception is the work on graphics where, as we shall see in chapter 15, the TurboC++ implementation uses far pointers, and we have chosen to follow that convention even though it is not really necessary to do so.

1.5.1 Pointer to void

It is also useful to be able to define a pointer unrelated to type. This may be done using the keyword void. The statement void* vp; declares a pointer to void, that is to say, a pointer to an object of unknown type, for which, therefore, the "value" *vp is undefined. Any pointer may be converted to a pointer to void, and we shall see how to do so in chapter 20. Void pointers are useful in certain circumstances for defining functions which do not need to know the type of object involved in order to fulfil their purpose. In the meantime, it is well to understand that a void pointer and a null pointer are two quite different concepts, the first pointing to an object of unknown type, the second a typed pointer pointing to no object at all.

1.5.2 Constants and pointers

Pointers can be defined to constant objects, and can themselves be defined to be constant. A pointer to a constant integer is written const int *p, whereas a constant pointer to an integer is written int *const q. A constant pointer to a constant integer would then be written const int *const r.

The value of a constant cannot be changed by use of a pointer to it, but there is some subtlety involved. If, for example, int m = 10; int n = 20;, consider the following statements:

```
int *o = &m;
const int *p = &m;
int *const q = &m;
const int *const r = &m;
```

Each defines a pointer to m, the value of which is 10, but here m itself is not defined as constant. Then the following statements are allowed:

```
m =  m + 1;          // increase m by 1
*o = *o + 1;          // increase *o by 1
 p = &n;              // change p to point to n
*q = *q + 1;          // increase *q by 1
```

but these are forbidden because they attempt to vary a constant quantity:

```
*p = *p + 1;          // ILLEGAL!
 q = &n;              // ILLEGAL!
*r = *r + 1;          // ILLEGAL!
 r = &n;              // ILLEGAL!
```

Similar considerations apply to the other basic types. For example, it is sometimes useful in output routines to define the *separator* "," by a constant pointer to a constant string:

```
const char *const sep = ", ";.
```

Note in particular that it is an error to define a pointer to a constant without declaring the constancy of the number pointed to, because otherwise the pointer could be used to modify the constant:

```
const int num = 123;        // constant defined
int *pnum = &num;           // ILLEGAL!
const int *pnum = &num;     // OK
```

1.5.3 Pointers to structures

A pointer may be defined to an object of structured type. In that case the type members may be accessed using the operator ->. Note that the symbol -> consists of *two* contiguous characters and will not be understood by the compiler otherwise. Thus, if p is a pointer to point a as defined in 1.4.3:

```
point a;
. . . . . . . .
point *p = &a;              // initialize p
. . . . . . . .
char  name = p->ch;         // access member ch
float xpos = p->x;          // access member x
float ypos = p->y;          // access member y
. . . . . . . .
```

Note that a.ch, etc, give exactly the same results as p->ch, etc., since p = &a.

1.5.4 Arrays

Arrays are particularly simple structured objects denoted by square brackets, as in

```
char chrs[10];
```

representing an array named chrs of, in this case, 10 characters; or, in general,

```
type name[num];
```

where *type* is any basic type and num is a *constant* integer representing the number of elements of the array. The ith element of the array is given by name[i], where $0 <= i < $ num.

In C++, when used in an expression, the name of an array behaves as if it were the address of the initial element of the array: `name == &name[0]`, and therefore `*name == name[0]`. One is sometimes tempted to write `type *name;` in order to declare the array, but that declares only a single element. We need to allow for the total storage requirements of the array, which in this case must be just sufficient to store num objects of type `type`. In C++ this storage requirement may be taken care of by means of the operator new, in terms of which one may write:

```
type *name = new type [num];
```

The pointer name is then equivalent to the array name, except that num need no longer be a constant. In other words, storage may be *dynamically* allocated according to the needs of the moment, using the operator new; whereas the array declaration `type name[num];` allocates fixed storage. In both cases, however, the *i*th element of the array is given by `name[i]` or by `*(name + i)` according to taste and convenience. Note that this implies

```
name + i == &name[i]
```

whatever the size of the objects of type `type`. On the left-hand side, an integer added to a pointer is automatically multiplied by `sizeof(type)`. Pointer arithmetic differs from ordinary arithmetic in this and other ways, and should not be lightly engaged in.

In terms of pointers, dynamic storage for a character string of exactly n characters is also provided by means of the operator new:

```
char *str = new char [n + 1];
```

where n must be incremented by 1 to allow for the terminal character `'\0'`.

Any "aggregate", namely an array, a structure or a class object, may be initialized by listing its elements or members in order within braces. In the case of an array of integers, for example, the statement

```
int a[] = { 1, 2, 3 };
```

defines an array a (or *vector*) of three elements. Note that the size of the array, 3, is here taken by default to be the number of elements in the listing. This definition is precisely equivalent to the sequence of statements:

```
int a[3];
a[0] = 1;
a[1] = 2;
a[2] = 3;
```

On the other hand, the statement

```
int b[5] = { 1, 2, 3 };
```

defines b to be an array of 5 elements, the first three of which are those listed, the remaining two being initialized to zero by default.

We may also define multi-dimensional arrays. The statement,

```
int c[4][3] = {
      { 1, 2, 3 },
      { 4, 5, 6 },
      { 7, 8, 9 }
};
```

defines an array (or *matrix*) of 4 vectors of 3 elements each, of which only the first 3 vectors are initialized to non-zero values. The same array can equally well be defined by the alternative statement

```
int c[4][3] = { 1, 2, 3, 4, 5, 6, 7, 8, 9 };
```

but the matrix nature of the array is then not so clear.

1.6 References

The unary & operator may also be used for a *reference*, which is simply an alternative name for a given object. A reference to an object t of type *type* is written *type*& r = t;, or equivalently, *type* &r = t;. Notice that the reference must be initialized: any operation carried out on r is then an operation on t. For example:

```
int i = 1;        // i = 1
int& r = i;       // r refers to i
r = 2;            // now i = 2
```

A reference may most obviously be implemented by a *constant* pointer which is implicitly de-referenced whenever it is used:

```
int *const p = &r;    // pointer implementation
*p = 5;               // now i = 5
```

References are very useful when passing variables to functions, as we shall see in 2.5.3.

1.7 Preprocessor directives

It is convenient to describe some of the more useful directives which can be included anywhere in the program and which will from there to the end of the program cause the compiler to carry out certain actions, including actions that are conditional upon the value of some defined quantity. They are all prefaced by the symbol #. We have already met #define in 1.4.1 and #include in 1.1.1. There is also the trivial directive consisting of a line containing the character # standing alone; it is always ignored, but may be useful for aesthetic purposes, or to emphasize blank lines in the source code.

1.7.1 Token replacement

The #define directive defines a macro, which will expand wherever it occurs into the prescribed sequence of tokens:

```
#define macro_name <token_sequence>
```

Thus, #define SIZE 256 on compilation will cause the preprocessor first to replace every occurrence of SIZE by 256. Likewise,

```
#define STRING "This is a string"
```

will replace every occurrence of STRING by "This is a string". Notice that it is not usual to terminate a macro with a semi-colon because we do not know the context in which it might be used. If a semi-colon is required to terminate a statement which involves a macro it should be provided explicitly, as in cout << STRING;, for example. It is also possible to treat program segments in this way, although it can be a dangerous practice. For example,

```
#define INPUT int n; cout << "enter n: "; cin >> n
```

will replace INPUT on every occurrence by the sequence which follows (although it is usually preferable to use an inline function (*cf* 2.5) in such circumstances).

Macros may also be defined which take parameters, although great care must then be exercised to ensure that there are no type or syntax errors. The following macro is quite useful:

```
#define sign(x) ((x)<0 ? -1 : 1)
```

where the *ternary expression* in parentheses has the value -1 if $x < 0$, 1 otherwise. The type of the result is the type of x, assuming that x is a type for which the operator < has been defined. The reason that (x) appears in the boolean expression (x)<0 instead of simply x is that we might wish to know the sign of some arithmetic expression, all of which has to replace x. It is, however, redefined as an inline function template in 2.5.7.

A macro, once defined, can later be undefined. If `macro` has been defined using `#define`, it becomes undefined as soon as the directive `#undef macro` is met. The bare statement `#define IDENTIFIER` by default defines `IDENTIFIER` to be 1. If it has not been so defined, or if it has been undefined, its value is taken to be zero.

Although the use of macros can sometimes be useful, and indeed we use them in this book, it can also be dangerous unless they have been very carefully constructed to convey the intended meaning in every circumstance in which they are likely to be used. Typed constants and inline functions are usually to be preferred.

1.7.2 Conditional directives

A directive of the form `#if` *expression* checks whether the (constant) expression evaluates to non-zero. There are similar directives `#else`, `#elif` (short for `else if`) and `#endif`, which may be used in the usual conditional chains. There are also the directives `#ifdef` and `#ifndef`; the directive `#ifdef identifier` checks whether identifier has been and remains defined; `#ifndef identifier` checks whether identifier is still undefined. All three directives `#if`, `#ifdef` and `#ifndef` may be followed by an arbitrary number of lines, possibly including lines beginning `#elif or #else`, but the sequence must end with the directive `#endif`. If the checked expression is true, any lines between `#elif` or `#else` and `#endif` are ignored. If it is false, any lines between the check and the first `#elif` or `#else`, or in their absence, `#endif`, are ignored.

1.7.3 File inclusion

A convenient and useful example of the use of conditional directives arises from the problem of avoiding multiple declarations. If the same object or token is defined in more than one file there is a danger that the various definitions will not be identical; in that case, if the files are combined in some way the compiler will signal an error. Standard definitions are therefore put into header files which may be included in any source file which requires them, thus greatly reducing the risk of inconsistency. The preprocessor simply copies the header file into the source file at the point indicated before compilation begins. However, if the header file `head1.h` itself contains a line `#include <head2.h>`, there is a danger of chaining, and even of endless repetition. A well designed header file, `head.h`, say, will therefore take the form:

```
#ifndef HEAD
#define HEAD
........
........
#endif
```

where the dotted lines represent the body of the header file itself. Hence, if the file head.h has already been included by the preprocessor, HEAD will have been defined and the body of the file will not be included more than once.

1.8 Keywords and identifiers

Words such as the standard type specifiers int and double, and the preprocessor directives define and include, are *keywords*, or *reserved words*. They must not be used for any purpose other than those for which they have been defined; they are part of the language.

The names given to variables, type specifiers, functions, and so on, are called *identifiers*. In C++ they may contain any number of characters, although the compiler may place a limit on the number of characters it recognizes. TurboC++ discards all identifier characters after the first 32. Upper case and lower case letters are treated as distinct and may be used at will, as may digits and the underscore character '_'. An identifier *must* begin with a letter or an underscore, however. Permissible identifiers are:

```
myident      MYIDENT      MyIdent      my_ident
ident2       x            _12345       _my_ident_
this_extremely_long_but_permissible_identifier
```

although the last may be truncated. The following, however,

```
12345           3ident        $ident        id[]ent
```

and many others, are all ILLEGAL! A keyword must *never* be used as an identifier.

2 Expressions, statements and functions

The following program implements a simple version of Newton's algorithm for extracting the square root of 10, starting with 3 as first approximation:

```
#include <iostream.h>
#include <math.h>
const double tolerance = 1.0E-10;
double x = 3;
main()
{       while (fabs(x*x - 10) > tolerance)
                x = (x + 10/x)/2;
        cout << "Square root of 10 is "
                << x << "\n";
}
```

When compiled and run, the program displays on the screen:

```
Square root of 10 is 3.162278
```

The result is correct to the default precision of the output operator, although the program is unsatisfactory in several ways. It is sufficient, however, for immediate purposes of exposition. Later we shall refine it, thereby introducing a number of helpful practices.

The small quantity `tolerance`, initialized to the value 10^{-10}, is optionally declared `const` because it is not to be (and is not!) changed by the program. The successive approximations to the square root, starting with 3, are assigned to `x`. The core of the program takes the form of a *conditional statement* which gives rise to an *iterative loop*:

```
while (condition) statement
```

where *condition* is a boolean expression which evaluates to true or false. Assuming the condition is initially true the statement is repeatedly executed as long as the condition remains true; but as soon as it fails, the `while` loop terminates. Better conditional expressions can be given (*cf* 4.7), but in this case it is taken to be

```
fabs(x*x - 10) > tolerance
```

where in general fabs(t) is the absolute value of the floating point variable t, and binary *
is the multiplication operator. We assume that the function fabs is declared, together with a
library of other useful mathematical functions provided with the compiler, in a header file
called math.h (as it is in TurboC++), which is included in the program. If not, we show how
to define such a function in 2.5.

The statement in this case is the assignment statement x = (x + 10/x)/2, which
replaces the current value of x by the improved value (x + 10/x)/2. The condition is
therefore that the absolute value of the difference between 10 and the square of the current
approximate square root, x, exceeds tolerance: if it does, the next approximation is calculated; if it does not, the loop terminates and the current value is sent to screen as the desired
result.

In this chapter we shall develop the concept of an expression, especially in the context of
arithmetical operations; expand the idea of a statement, including the conditional statement;
introduce the general features of a function; and see how to define a pointer to a function. We
end the chapter with a brief discussion of the function main.

2.1 Expressions

An *expression* is composed of one or more *operations*. If we define int m = 10;
int n = 3; then m + n is an expression having *operands* m and n and *operator* +, the
value of which is 13. The binary operator + is the addition operator. Operators may be *unary*
or *binary*. A unary operation induced by an operator @, say, may be written @x, or sometimes
x@. A binary operation is written x @ y, or sometimes @(x, y). For completeness, a single
identifier of declared type may also be said to constitute an expression. Thus m and n are
themselves expressions.

2.1.1 Arithmetic expressions

The standard *arithmetical* binary operations for addition, subtraction, multiplication and division are all defined, and the corresponding operators are denoted by +, −, *, and /, respectively. There are unary plus and minus, also denoted by + and −. These operators are defined
for all integer and floating point types, and there are appropriate automatic conversions
between different types as discussed in 1.3.1 and 1.3.2, and conventions regarding overflow.
Note, however, that the result of division of two integers is an integer: if int p = 14;
int q = 4; then int i = p/q has the value 3. If the floating point value is sought
instead, it is necessary to convert p to floating point before performing the division:
double d = (double)p/q;, or equivalently double d = double(p)/q;, gives
d the value 4.666 . . . This is an example of where explicit type conversion is necessary. Take
care, however: the value of double(p/q) is 3.0, because the integer division is carried out
before the conversion to floating point. In the case of integer division, the *remainder* is given
by p % q (where % stands for *modulo*) which in this case equals 2.

If these arithmetic operations are used in succession, more complex expressions such as

```
a + (b - c*d)/(e + f/g)
```

can be built up, using parentheses where necessary to override the normal precedence of operators (*cf* 2.1.7), or to improve clarity.

2.1.2 Relational expressions

There are the *relational* operators >, <, >=, <=, == and !=, the last representing "not equal". In general, ! is the *negation* operator. The result of evaluating a relational operation is boolean: it is either true or false; *ie*, it has the integer value 1 or 0. They may be applied to floating point numbers as well as integers, and indeed to other types when their definitions are suitably extended. Care, however, must be taken when testing for equality between floating point numbers owing to the finite precision with which the latter may be represented. One usually must be satisfied, for example, to know that the absolute value of the difference of two floating point numbers is less than some small fixed number, or less than some small fixed proportion of either of them, depending on the context.

2.1.3 Logical expressions

There are also the *logical* operators AND, written &&, and OR, written ||, with the aid of which more complex boolean expressions may be built up, such as:

```
i >= 0 && i < n
```

which tests whether the integer i lies in the range $0 \ldots (n-1)$. Note that there is no need for parentheses in this expression (except perhaps for aesthetic reasons) because the relational have higher precedence than the logical operators: they are evaluated first. We have already introduced the logical NOT, written ! in C and C++.

2.1.4 Bitwise expressions

There are the binary *bitwise* operators AND, XOR and OR, written &, ^ and |, respectively. To see their effect, one must first express the operands in binary notation. Thus, the integers 13 and 7 in binary notation are $0 \ldots 01101$ and $0 \ldots 0111$, respectively. Bitwise operations are obtained by comparing these two numbers bit by bit. If two corresponding bits are both 0 the result of any bitwise operation, so far as those bits are concerned, is 0. If they are both 1, the AND operation results in 1, the XOR operation in 0, and the OR operation in 1. If one bit is 1 and the other is 0, AND results in 0, while XOR and OR both result in 1. Applying these operators to each bit in turn the final results are

```
13 & 7 = 5,        13 ^ 7 = 10,        13 | 7 = 15.
```

The binary & operator will not be confused with the unary address operator & because the context is different.

There are also the binary *shift* operators << (shift left) and >> (shift right). These will not be confused with the identically written output and input operators because, again, the context will be different. If n and s are integers, n << s shifts the pattern of bits in the bit representation of n, s places to the left, thus multiplying n by 2^s (provided none of the leftmost s bits is 1). Similarly, n >> s shifts the pattern s places to the right, thus dividing n by 2^s (integer fashion). Thus, the value of 16 << 2 is 64, and the value of 16 >> 2 is 4. (If it seems difficult to remember which operator is which, << may be thought of as "increase" and >> as "decrease".)

There is also the the bitwise *complement* operator (or NOT) denoted by ~. If n is an integer, the value of ~n is the "ones complement" of n, which exchanges 0s and 1s in the binary representation of n. Thus, the value of ~0 is −1, and the value of ~123 is −124.

2.1.5 Comma expressions

Finally, there is the *comma* operator, used in comma expressions. An expression such as expr1, expr2, ..., exprn, where the various terms are themselves expressions, evaluates left to right, discarding the previous term as it goes. Its final value is the value of the last expression. This enables the writing of very concise multiple statements. Thus, if i is an integer, the value of the comma expression i = 3, i + 2 is 5. (It may sometimes be necessary to enclose a comma expression in parentheses to avoid confusion with the more usual use of the comma as a punctuator.)

2.1.6 Other expressions

Expressions are by no means only arithmetical. For example, cout << str, where str is a character string, is an output stream expression. If the operator + is overloaded (additionally defined) to apply to strings, the expression str1 + str2 might mean the concatenation of the two strings str1 and str2.

2.1.7 Operator precedence

As we have remarked, parentheses may always be used to clarify or override the normal precedence of operators in expressions. However, a program overburdened with parentheses is difficult to read. It is therefore best to rely upon the natural precedence as much as possible. The following table, taken from the book by Lippman, summarizes the order of precedence of all the operators used in C++. Those with higher precedence are evaluated before those with lower precedence. Those having equal precedence are evaluated from left to right if marked L the table, right to left if marked R.

Level	Operator	Function
17R	: :	unary global scope
17L	: :	binary class scope
16L	->* .*	member pointer selectors
16L	[ ]	array index
16L	( )	function call
16L	( )	type construction
16L	sizeof	size in bytes
15R	++ --	increment, decrement
15R	~	bitwise NOT
15R	!	logical NOT
15R	+ -	unary plus, minus
15R	* &	dereference, address-of
15R	( )	type conversion
15R	new delete	free store operators
14L	-> .	member selectors
13L	* / %	multiply, divide, modulo
12L	+ -	add, subtract
11L	<< >>	bitwise shift
10L	< <= > >=	relational operators
9L	== !=	equality, inequality
8L	&	bitwise AND
7L	^	bitwixe XOR
6L	|	bitwise OR
5L	&&	logical AND
4L	||	logical OR
3L	?:	ternary operator
2R	= *= /=	assignment
2R	%= += -= <<=	assignment
2R	>>= &= |= ^=	assignment
1L	,	comma

Strict use of this table should make it possible to write code with relatively few parentheses. Additional parentheses should always be used, however, where otherwise the reader might not know what was intended. This, of course, is a matter of judgement. Consider, for example, the fragment:

```
int i = 10;
int* p = &i;
int j = 2**p;
```

which defines j to hold the value 20 because the dereferencing operator has higher precedence than the multiplication operator, and so *p will be evaluated before being multiplied by
two. That takes sufficient puzzling out to warrent parentheses in an introductory text, and we
therefore prefer to write the offending statement in the unmistakeable fashion

```
int j = 2*(*p);
```

even though it is unnecessary to do so. A simple arithmetical expression like
(x + 10/x)/2 needs no further parentheses because it is so easy to remember that division
carries higher precedence than addition. Of course, parentheses must always be used to override precedence when otherwise the compiler would be misled. For example, a + b/c + d
is not the same as a + b/(c + d). A very common error is to confuse, say, a/b*c, which
is the same as (a/b)*c, with a/(b*c), which is not. For an authoritative account of operator precedence, see the book by Stroustrup.

2.2 Statements

A C++ program proceeds by executing *statements* sequentially, or according to *jumps*, which
may be predetermined or conditional upon the value of some expression. Statements are of
various kinds, of which we shall here deal with declarations, assignments, compound statements, conditional statements, iteration statements (loops), jump statements, and selection
statements. Later we shall encounter return statements in function calls. All statements end
with a semi-colon. The *null statement* is written simply ; and is useful when the syntax
requires a statement to be provided even though no action is desired, as illustrated in 2.2.4.
More generally, an *expression statement* is written expression;.

2.2.1 Declaration and initialization

The *declaration* int n; declares the variable n to be an integer, and reserves for it
sizeof(int) bytes of addressable memory, the contents of which are in general undefined. A later statement n = 10; then assigns to n the value 10, and we say that the variable
has been *initialized*. Declaration and initialization may be combined in the single statement
int n = 10; and the variable n may then be said to have been *defined*. All variables must
be declared before they are used. Several declarations of variables of the same type may be
combined in one line of code. For example,

```
int i, j, array[10];
```

has exactly the same effect as

```
int i;              // first integer
int j;              // second integer
int array[10];      // integer array
```

but in the second form it may be easier to include comments.

The particular block of memory reserved for a variable is sometimes called an "object"; if we define int n = 10; the *name* of this particular object is n. There are therefore *two* numbers associated with a variable: its value, in this case written n, and the location or address of the object in memory, written &n. Similar considerations apply to variables of other types. There is a certain ambiguity about the identifier n: it is the name of the object, but it also represents the value currently stored in that object. This ambiguity is apparent in the symbolism of an assignment operation.

2.2.2 Assignments

In C++, as in C, assignment is regarded as an expression which includes the assignment operator =. Assignment mixes freely with other expressions. This has already been encountered in the case of comma expressions such as i = 3, i + 2. If a, b and c are integers, the *assignment statement*

```
a = b + c;
```

tells the compiler that it must add together the values it finds in the objects named b and c, and put their sum into the object named a which is to be found at the address &a. The identifiers b and c on the right-hand side are being interpreted in a different sense from the identifier a on the left-hand side. When it is important to make the distinction one speaks of *rvalues* and *lvalues*. A convenient mnemonic is to recall that an rvalue is obtained by *reading* the contents of an object. An lvalue, by contrast, is associated with *writing*: that is, with assigning the value of the expression on the right-hand side to the object named on the left-hand side. It is clear, therefore, that in C++, as in C, a statement such as a + b = c; is meaningless: it is in fact an error.

2.2.3 Compound assignments

The assignment operator may be combined with any of the binary arithmetic or bitwise operators to form a compound assignment operator. If @ is one of these, x @= y means x = x @ y. The device lends itself to compact code. Thus, we may write x += 1, meaning that the value of x is increased by 1; or, if x is an integer, x &= 5, meaning that to x is applied the bitwise operation x & 5. Notice that there must be no white space between the two operators; each pair is to be regarded as a single operator. The complete set of compound assignment operators is as follows:

```
+=      -=      *=      /=      %=      &=      |=      ^=      <<=      >>=
```

2.2.4 Increments and decrements

A still more compact notation is provided for increasing or decreasing the value of an integer variable by unity. If n is an integer, n++ and ++n both increase the value of n by 1. In the postfix form, n++, the original value of n is used in the expression in which it occurs before its value is changed; in the prefix form, ++n, the value of n is increased before it is used. Similarly, there is a decrement operator --, which may appear in postfix or prefix form. Expressions using these special operators are really assignments; they may be applied only to lvalues, and an expression such as (a + b)++ has no meaning. They have the advantage of being compact, but they are also efficient in that they call the object concerned once only. Increment and decrement operators are especially useful in loops where the number of iterations has to be counted, as in the following fragment:

```
int count = 0;
while(condition)
{       ++count;
        statement
}
```

These operators may also be applied to pointers. However, if p is a pointer, ++p points to the object immediately after that pointed to by p, whereas p++ points to the object pointed to by p, but once used points to the next object. There are similarly interpretations of --p and p--. Incremented and decremented pointers lead to some extremely compact code. For example, if char *q = "........"; defines a zero-terminated character string, then the statement:

```
while (*p++ = *q++) ;
```

copies q into the string p. To see this, note that initially q points to the initial character in the string, that character being given by *q. The assignment *p = *q copies that character into the initial character of p, following which both pointers are incremented and attention passes to the next character. The process continues until the terminal character '\0' has been copied, whereupon the while loop terminates (*cf* 2.4.2). Notice that no additional statement has to be executed: the semi-colon represents the null statement. It is clear from this example why it is useful to have the convention that a string always ends with the null terminator '\0'.

Statements of this very concise kind are the delight of expert C programmers. They are also fraught with danger because of their obscurity. For example, it would have been easy to write *p++ == *q++ instead of *p++ = *q++ (a boolean expressions being expected after while), thereby changing the meaning completely. For this reason, some compilers (TurboC++ included) do not accept such expressions without protest. To avoid the danger, it is better to be explicit:

```
while ((*p++ = *q++) != '\0') ;
```

It may seem less elegant to some eyes, but it is easier to see what is intended, and is therefore less liable to error.

Note that, in either case, the while statement is equivalent to the much less compact statements

```
while (*p != '\0')
{       *q = *p;
        p = p + 1;
        q = q + 1;
}
*q = '\0';
```

in which, moreover, the increment operator is not employed at all.

2.3 Compound statements

The syntactical conventions of C++ often require that only one statement is specified. For example, if (condition) statement, requires a single statement. Nevertheless, a sequence of statements is frequently what is needed. A compromise is reached by enclosing the sequence of statements, each terminated by a semi-colon, within braces:

```
{ statement1 statement2 statement3 ... }
```

It may equivalently be written

```
{
        statement1
        statement2 statement3
        ........
}
```

or in many other ways, because the only significant separator is the semi-colon which terminates each statement. As usual, clarity will be our guide. The result is a *compound statement* or *block*, within which the statements will be executed in sequence, but which for syntactical purposes is a single statement. Notice that there is no need for an additional semi-colon to terminate a block. Using a block, one may now write:

```
if (condition)
{       statement1
        statement2
        ........
}
```

The sequence of statements will be executed if the condition evaluates to true; otherwise the whole block will be ignored.

Since a block of statements is syntactically equivalent to a single statement, it follows that blocks can be nested, and they frequently are:

```
{
        ........
    {
            ........
    }
        ........
}
```

2.3.1 Scope

In C++, unlike C, a declaration is a statement and can occur anywhere another statement can. If it occurs within a block,

```
{
        ........
    double x = 1.0;
        ........
}
```

it is called a *local* variable. The object named by x is accessible only from the point of declaration to the end of the block enclosing it; it has local *scope*, and any attempt to access it outside the block in which it is declared must fail. A variable declared before a block, however, can be accessed within that block, and within all inner blocks, provided it is not "hidden" by another variable defined within such a block and having the same name. Thus, two or more variables may not be given the same name if they are of the same scope. A variable declared outside *all* blocks is said to be a *global* variable; it has *file scope*, and is accessible anywhere from its point of declaration to the end of the file in which it is declared. However, if there is a variable within a block having the same name as a global variable, the global variable is again "hidden" within the block. In C++, however, it can still be accessed with the aid of the scope operator : :. The following fragment makes the point somewhat starkly:

```
int x = 10;                  // global declaration
main()
{       int x = 20;          // x = 20
        ........
    {       int x = x;       // NONSENSE!
            int y = x;       // y = 20
            int x = ::x;     // x = 10
            ........
    }
        ........
}
```

Likewise, a member variable declared within a structured object, such as the object a of type `point` defined in 1.4.3, is local to the structure. As we have seen, it may nevertheless be accessed with the aid of the dot operator, as in `a.name`, say, or in the case of a pointer to the object, `p->name`.

2.3.2 Lifetime

Unless otherwise specified, the "time" a variable has addressable memory associated with it coincides with its scope. The variable is "created" at the point of declaration and "destroyed" at the end of the block in which it is declared. Such a variable is said to be *automatic*. If necessary this property can be made explicit with the aid of the keyword `auto`, as in `auto float x = 1.0;`, but this is usually unnecessary because a local variable, unless otherwise specified, is automatic by default. A global variable, on the other hand, "lives" from its point of declaration to the end of the file in which it is declared.

There is another possibility, however. A variable declared within a block, and therefore local to that block, can be declared to be *static*:

```
static int count;
```

In that case, its lifetime is that of the file, and it will retain its value even when it is temporarily out of scope.

Finally, a variable may be declared *external*:

```
extern double x;
```

In this case its definition may be given in another file rather than in the file in which it is declared. An `extern` declaration does not of itself create a new object; it merely informs the compiler of the type of the object referred to.

2.4 Conditional statements

Although program statements are by default executed sequentially, only the simplest problems can be programmed so that this is the case in practice. It is usually necessary that the order of execution depends on the value of some parameter, or on the result of executing some earlier statement; or that statements are repeated a fixed number of times, or until some condition is satisfied. We have already encountered the `if` statement and the `while` loop. We examine these and other conditional statements more closely now.

2.4.1 if-else statements

The simplest conditional statement uses the keyword `if`:

```
if (condition) statement
```

The statement is evaluated and executed if and only if the expression *condition* is first found to be true (*ie*, evaluates to 1); if the condition is false (*ie*, evaluates to zero), the statement is not evaluated at all (*cf* 2.1.2). Thus, a condition is evaluated against the boolean test, *condition* != 0. If **statement** is a simple statement, that statement will be passed over if *condition* evaluates to false; if **statement** is a compound statement, the whole block of statements which comprise it will be passed over, execution being resumed at the first statement following the block. Notice that the condition *must* be enclosed within parentheses.

A conditional statement such as

```
if (n > 0) statement
```

executes if and only if n > 0 is true. Similarly,

```
if (n == 0) statement
```

executes if and only if n is zero. On the other hand,

```
if (n != 0) statement
```

executes if and only if n is *not* zero. This last conditional statement may alternatively and more concisely be written,

```
if (n) statement
```

because the condition n != 0 succeeds if and only if n is non-zero. Correspondingly,

```
if (!n) statement
```

succeeds if and only if n is zero.

Depending upon the evaluation of some condition, program execution may take alternative paths. This may be brought about by means of the keyword **else**:

```
if (condition) statement1
else statement2
```

If the condition evaluates to true, *statement1* is evaluated and executed, while *statement2* is ignored; if it is false, *statement1* is ignored, but *statement2* is evaluated and executed instead. Both statements may, of course, be compound, so that a program containing such a conditional statement is capable of major branching depending on the evaluation of the condition.

For example, in the program for computing the square root of 10 using Newton's algorithm, given at the beginning of this chapter, there is in principle the possibility that the current value of x becomes so small that a new value cannot be calculated because it would cause overflow in the evaluation of the expression x + 10/x. Clearly, very small values of x should be treated separately. If `toosmall` is a very small constant number, then we might modify the `while` loop to read:

```
while (fabs(x*x - 10) > tolerance)
{       if (fabs(x) < toosmall)
        {       x = -1;
                break;
        }
        else x = (x + 10/x)/2;
}
```

If the danger of a zero divisor arises, the loop assigns the impossible, and therefore easily recognized value −1 to x before the `break` statement (*cf* 2.4.5) causes the loop to terminate prematurely. If this condition does not arise, the loop behaves normally.

Multiple chaining is also possible. In the statement

```
if (condition1) statement1
else if (condition2) statement2
else statement3
```

`statement1` is executed if and only if *condition1* is true; if it is, the rest of the compound statement is then ignored. If *condition1* is false, *statement1* is ignored and *statement2* or `statement3` is executed depending on whether *condition2* is true or false. That is the default interpretation; it may be written more explicitly in block form:

```
if (condition1) statement1
else {
        if (condition2) statement2
        else statement3
}
```

Ambiguities creep in with deeper chaining. In that case block form may be essential, both for clarity and to avoid error. For example,

```
if (c1) {
        if (c2) s1
        else s2
}
else if (c3) s3
else s4
```

is not the same as

```
if (c1) s1
else if (c2) s2
else if (c3) s3
else s4
```

The default interpretation is that each `else` is paired with the preceeding unpaired `if`, as in the last example, unless the presence of blocks demands otherwise.

We have already met the *ternary conditional operator* `? :`, the expression `a ? x : y` meaning "if *a* then *x* else *y*". This enables very concise conditional expressions and statements to be written. For example, if `a` and `b` are defined integers, the statement

```
int i = a > b ? a : b;
```

defines `i` to be equal to the greater of them. It is, however, exactly equivalent to the `if-else` statement

```
if (a > b)
        int i = a;
else
        int i = b;
```

which some may prefer on the grounds of clarity.

2.4.2 while statements

We have already met the `while` statement several times:

```
while (expression) statement
```

where the (possibly arithmetic) expression is treated as a boolean condition. A non-boolean expression `expr` is automatically treated as if it were the boolean condition, `expr != 0`. Any variables in terms of which `expression` is written must already have been declared and initialized. Execution of the statement is repeated only so long as the expression evaluates to true; if it is not true initially, the statement will not execute at all.

2.4.3 do statements

The do statement resembles the `while` statement, but has the form:

```
do statement while (expression);
```

Execution of the statement continues until the expression becomes false; it is always executed once.

2.4.4 **for** statements

The for statement provides the most general loop. An example is:

```
int i;
for (i = 0; i < 10; i++)
        cout << i << "\n";
```

which prints out each integer from 0 to 9 on a separate line. It is precisely equivalent to

```
int i = 0;
while (i < 10)
        cout << i++ << "\n";
```

However, in C++, unlike C, the declaration of the "control variable" (which need not be an integer) may take place without change of scope in the initialization statement of a for loop. After execution of the statement

```
for (int i = 0; i < 10; i++)
        cout << i << "\n";
```

the variable i is still in scope and has the value 10.

In any case, the general form of a for loop is

```
for (initialize; condition; expression)
        statement
```

where *initialize* represents statements initializing the control variable or variables (for there may be more than one), *condition* provides loop control, and *expression*, which is evaluated after each iteration of the loop, indicates how the variables initialized in *initialize* are to be modified as the iteration proceeds. The loop proceeds until *condition* evaluates to false. If the first evaluation of *condition* is false neither *expression* nor *statement* is evaluated at all.

2.4.5 **break** and **continue** statements

A particular statement which can be used in a conditional statement is provided by the keyword break. If the statement

```
if (condition) break;
```

occurs within a block, and the condition is true, execution passes to the first statement after the block; if it is false, the block is executed normally. It is useful especially in breaking out of a loop if some condition is fulfilled. The fragment:

```
int n = 0;
while (++n < 10)
{      cout << n << "\n";
       if (n == 6) break;
}
cout << "finished";
```

prints out the integers from 1 to 6, and then prints "finished", having exited the block.

It is also possible to omit a particular iteration in a loop using the keyword continue, which skips over the remainder of the block. The similar fragment:

```
int n = 0;
while (++n < 10)
{      if (n == 6) continue;
       cout << n << "\n";
}
```

prints out the integers from 1 to 9, omitting 6.

It is sometimes useful to use a break statement to control a for loop. The following fragment may seem a bit extreme, but it works:

```
int n = 0;
for (;;)
{      cout << ++n << "\n";
       if (n == 6) break;
}
```

The variable n is initialized before the loop is entered; its increment is provided for within the output statement; and a break statement terminates the loop at n == 6.

2.4.6 goto statements

It is sometimes necessary to make an explicit jump from the point of current execution to a possibly distant point which has been labelled with an identifier:

```
........
if (condition) goto label;
........
........
```

```
label: statement
........
```

The labelled statement can occur before or after the `goto` jump. The iterative statements we have just introduced can readily be expressed in terms of a jump instruction. For example, the `do-while` statement

```
do statement while (condition);
```

can be written

```
loop: statement
if (condition) goto loop;
```

Similarly, a `continue` statement (*cf* 2.4.5) within a `while` loop can be replaced by

```
while (expression)
{       ........
        if (condition) goto endblock;
        statement
        endblock:
}
```

A `goto` statement is sometimes useful if a jump has to take place out of nested blocks. However, it should be used sparingly. A succession of `goto` statements which cause execution to leap around in the program is very difficult to follow, and can lead to loss of control in unexpected places.

2.4.7 switch statements

The `switch` statement is useful for multiple choice situations, and avoids the ambiguities of chained `if-else` statements. We shall use it mainly in the context of an enumerated type, although it is of quite general application. It uses the special keywords `case` and, optionally, `default`. Suppose we have an enumeration:

```
enum state { ITERATING, SUCCESS, WONTSTOP, NEARZERO, SAMESIGN };
```

in terms of which we may declare and initialize a variable of type `state`, say:

```
state s = ITERATING;
```

This variable may then be used to store the state of an iterative loop: it may be iterating, or it may have terminated having obtained the desired result, or it may have failed to terminate normally after a given number of iterations, or a very small divisor may have occurred causing premature termination, or perhaps two variables which should have been of opposite sign have acquired the same sign, again leading to premature termination. Of course, there may be other possibilities. Then in printing out the outcome of the loop it may be desired that the final state of the iteration be known. The `switch` statement can be used to achieve this. For example:

```
switch (s) {
        case SUCCESS    :   cout << result;
                            break;
        case WONTSTOP   :   cerr << "Too many iterations!";
                            break;
        case NEARZERO   :   cerr << "Cannot divide by zero!";
                            break;
        case SAMESIGN   :   cerr << "Variables have same sign!";
                            break;
        default         :   cerr << "That's very peculiar!";
}
```

If success is achieved, the program prints out the result; otherwise, it prints out a suitable error message depending on the final state of the loop. If the final state is one that does not appear in the list of case labels, the optional `default` statement at the end is executed. Notice the appearance of the `break` statement after each outcome except the default: without it, every following statement in the `switch` statement would be executed. There are occasions when it may be useful to allow this to happen:

```
switch (ch) {
        case 'A' :
        case 'a' :   statement
                     break;
        ........
}
```

The statement is executed whether ch has the value `'A'` or `'a'`. In general, the switch expression may be of any type for which integral conversion is provided, and the case expressions must be of the same type.

2.5 Functions

In C++, a function is a block of statements which may be invoked by name whenever required; after execution of the block, program control reverts to the statement immediately following the function call (control being exercised by means of the program's run-time stack). When executed, the function may return a single value of given type. It may also require argument values before it can be evaluated, and any such arguments must be of declared types. A general function which takes as argument a variable of type *arg_type* and returns a value of type *ret_type* may be declared:

```
ret-type func(arg-type ...);
```

where the ellipsis **. . .** here merely indicates that the function may take more than one argument, not necessarily of the same type, separated by commas. If the return type is not explicitly stated, it is assumed to be `int` by default. In a declaration it is not necessary to give names to the arguments, and we have not done so. It is only the types that matter. Declared in this manner, the function `func` is said to have a *prototype*.

Of course, a function *definition* also serves as its declaration. When the function is defined, the names of its arguments must also be given, and the block of statements that comprises its *body* must be added. For example:

```
int    twice(int i)
{      int t = 2 * i;
       return t;
}
```

When called, the function `twice` takes a copy of the actual value of the formal argument variable `i`, doubles it, stores it in the local variable `t`, and returns the resulting value, which is itself an integer. It does not change the actual value (*cf* 2.5.1). The purpose of the return statement (`return` is a keyword) is to substitute for the function the value obtained by executing its body, given the actual value of its argument. Whenever it is desired to double an integer, one can now call this function:

```
int a = 3;        // a = 3
int b = twice(a); // b = 6
```

In a function call, each argument type, and the return type, is checked against its declaration, and it is an error if they do not all coincide.

A function call is an expression, in which the function name, here `twice`, is immediately followed by the *function call operator* `( )`, the parentheses containing the actual value of the function's argument, if any; or, if there is more than one argument, by a sequence of values of the correct types in the correct order, separated by commas. Since a function call is an expression it may mix freely with other expressions, provided its return type is appropriate. For example, if the function f returns an `int`, and if a is an integer, then we are entitled to write, for example:

```
int i = a + f(...);
```

Clearly, a function call is an rvalue. Whereas we may be able to write `y = f(x);` (if f returns the same type as `y`), in no circumstances (except in a comment) may we write

f(x) = y. The correspondence with ordinary mathematical function notation is unidirec-
tional, because the operator = is the assignment operator. However, this fact enables us to
make reassignment statements of the form x = f(x);. In such a statement, the present value
of x is passed to the function f as the actual value of its argument, and the value f(x) is then
assigned to the variable x. The statement is shorthand for the more explicit sequence *type*
t = f(x);   x = t;, and the compiler is expected to recognize it as such, automatically
creating a temporary variable t of appropriate type in order to do so.

A function can return the value of an expression, rather than of a single variable only.
The general form of a return statement is return   *expression*;, but the type of the
expression must be the return type of the function, using type conversion if necessary,
otherwise an error will be reported. Thus, in the function twice there is no need for the
explicit local variable t because the compiler can create it implicitly if it so wishes. The
function can just as well be written:

```
int    twice(int i)
{      return 2 * i;   }
```

The result is the same. Note that there is no need for parentheses around the expression
returned, but they may always be used in the interests of clarity.

This example illustrates the advantage of declaring a function before it is defined, because
its actual definition may not have been decided upon, and yet one wishes to call it formally in
other parts of the program. To call a function before declaring it generates an error message
on compilation. It is good practice to include the declarations of functions of general application in a header file, so that conflicting declarations of what is intended to be the same function do not arise.

Any type may be returned except an array type or a function type (although a pointer may
be returned to either); the default return type is int. However, any variables declared locally
within the body of a function disappear after the function call has been executed. It is therefore an error to attempt to return a pointer or a reference to a local variable, because the object
pointed to may change unpredictably. None of the following functions is allowed:

```
int*   f(...)
{      int i = 1;
       ........
       return &i;          // ERROR!
}


int&   f(...)
{      int i = 1;
       ........
       return i;           // ERROR!
}
```

```
int&   f(...)
{      .........
       return 1;              // ERROR!
}
```

the last because the value returned is treated as a constant local variable.

A more satisfactory version of Newton's algorithm for square roots may now be written with the aid of an auxiliary function:

```
double sqroot(double A, double x)
{      while (fabs(x*x - A) > tolerance)
            x = (x + A/x)/2;
       return x;
}
```

where we have ignored the possibility of a zero divisor. Here we pass to the function `sqroot` not only the approximate value of the root, x, but also the number A of which we require the square root. Omitting the usual headers, this function may now be called in the program function `main`, which might take the form:

```
main()
{      double A, x;
       cout << "Enter positive number: ";
       cin  >> A;
       cout << "Enter approximate root: ";
       cin  >> x;
       x = sqroot(A, x);
       cout << "Square root of " << A
            << " is " << x << "\n";
}
```

We obtain the square root of 10 by entering 10 in response to the first request and 3, say, in response to the second. Do not choose A = 0, however (*cf* 4.8).

Notice that we have here called the function `sqroot` within the function `main`. Since a function call is an expression, this is perfectly allowable. Indeed, it is commonplace for one function to call others within its body. For example, the following function returns the value of $4x^2$ by first calling the previously defined function `twice`:

```
double foursquare(double x)
{      double t = twice(x);
       return t * t;
}
```

This flexibility allows one to break a complicated function down into smaller self-contained pieces, each of which may be separately defined as a function. Increased clarity usually results, which more than compensates for the overhead of additional function calls.

Note that the execution of a return statement, no matter where it occurs, terminates the function; anything following it is ignored. However, premature termination may be desirable after a *conditional* statement, as in:

```
int    f(int i)
{      if (i%2) return i;
       return i/2;
}
```

which returns the value of i if it is odd (% being the modulo operator), i/2 otherwise. An if-else statement would here be superfluous.

Similarly, a null return statement can be used in a function returning no value (*cf* 2.5.3):

```
void   v(...)
{      ........
       if (condition) return;
       statement
       ........
}
```

The statement is executed only if the condition fails.

Finally, we may introduce the promised definition of the function fabs, which returns the absolute value of a floating point variable. It is very simple:

```
double fabs(double x)
{      if (x < 0) return -x;
       return x;
}
```

2.5.1 Inline functions

Function calls are a somewhat wasteful overhead for simple one- or two-line function bodies. The arguments must be copied, machine registers must be saved on stack, and the program must branch to a new location, and back again, no matter how simple the function. It may nevertheless be a great advantage to have some way of calling repeatedly, and perhaps in different circumstances, upon a few given statements, without fear of variations developing. In such cases, it is better to introduce an *inline function* using the keyword inline. For example, the function twice can instead be written:

```
inline int twice(int i) { return 2 * i; }
```

When an inline function is called, it is type checked and the return expression substituted for the function call before compilation. It is, in a sense, a typed version of the macro

```
#define twice(i) 2*(i)
```

but being typed is much safer to use. In some implementations, however, conditional statements are not allowed in inline functions, and recursive functions (*cf* 2.6) should never be defined inline. However, the keyword `inline` is only a recommendation to the compiler, which should be capable of making up its own mind where the advantage lies.

2.5.2 Passing arguments to functions

In the above program, the key statement `x = sqroot(A, x);` takes account of the fact that the value of `x` is unchanged by execution of the function `sqroot`; it is changed only by reassignment. Recall that function arguments are *passed by value*. When a function is called, and its body evaluated, copies of the actual arguments are made in temporary local variables, and it is these copies upon which the function operates, not the argument values themselves. This process can be mimicked by recasting the function `sqroot` as follows:

```
double sqroot(double A, double x)    // UNNECESSARY!
{       double T = A, t = x;
        while (fabs(t*t - T) > tolerance)
                t = (t + T/t)/2;
        return t;
}
```

It is totally unnecessary to do this, however, because a C or C++ compiler does it automatically. In other languages, argument values may not be passed in this manner; they may instead be passed by reference (*cf* 2.5.4).

2.5.3 Functions returning void

It is not necessary that a function should have a return value. In some languages such a function is called a *procedure*: it lacks a return statement, and in C++ its return type is declared to be `void`. It is important to distinguish between a function of return type `void` and a function the return type of which is unstated: the latter is taken to return an `int` value by default, and if it does not include a return statement an error message may result on compilation.

Since in C++ the values of the arguments passed to a function are unchanged, it would at first sight seem that a function returning `void` achieves nothing. This is not so, however, because the function can have *side effects*: it can operate upon global variables. For example, the standard input and output streams `cin` and `cout` are global variables. So the function:

```
void   output(float x)
{      cout << x;   }
```

prints out the value of x as a side effect on the standard output stream; there is no return value.

2.5.4 Reference arguments

It would nevertheless be convenient if changes could sometimes be made to the value of an argument passed to a function. There are two ways of satisfying the spirit, although not the letter, of this wish. The first is to pass instead of the argument itself a reference to the argument (call by reference); that is to say, its lvalue rather than its rvalue. This means (*cf* 1.6) that any changes made to the argument within the function body change the argument itself and not a local copy. In this way we can rewrite the function `twice` so that it becomes of return type `void`:

```
void   twice(int& i)
{      i *= 2;      }
```

The reference itself is unchanged by the function, but the value of the variable it refers to *is* changed:

```
int a = 3;
twice(a);
cout << a;          // a = 6
```

To take a less trivial example, suppose that we wish to exchange the values of two integer variables by means of a suitable function. We might be tempted to write:

```
void   swap(int i, int j)        // WRONG!
{      int t = i;
       i = j;
       j = t;
}
```

This function fails to achieve the desired result because the values of the arguments passed are unaffected by the function call `swap(i,j)`. Instead, we may pass references to the arguments; then anything that happens to the references (*cf* 1.6) happens to the arguments:

```
void   swap(int& i, int& j)      // CORRECT!
{      int t = i;
       i = j;
       j = t;
}
```

which successfully performs the task when swap(i,j) is called.

There is a further reason for passing references. Since function execution begins by making copies of the variables passed to the function, there is an overhead to be paid in a function call. If the arguments passed are large (for instance, if they are long arrays or structures of many members) that overhead may be prohibitive. Passing a reference eliminates the problem. But there is a potential danger. If a variable is passed by reference, execution of the function may modify its value. That was our original reason for introducing the passing of variables by reference! Particular care must be taken if it is desired that the variable concerned should *not* be changed in value during function execution. One way of ensuring that this will be so is to declare that the reference is to a constant, as in the declarations:

```
struct st {.........};
void f(const st& s);
```

Any attempt by the function f to modify the value referred to by s should then cause the compiler to generate an error message. We shall make good use of this device in later chapters.

2.5.5 Pointer arguments

The alternative approach is to pass pointers to the arguments instead of the arguments themselves. This is hardly surprising in view of the close connection between references and pointers noted in 1.6. In the case of the swap function, we can therefore write:

```
void   swap(int *p, int *q)      // ALSO CORRECT!
{      int t = *p;
       *p = *q;
       *q = t;
}
```

If now we call swap(&i, &j) the desired result is again obtained. As before, one can pass a pointer to a constant object if the context warrants it.

2.5.6 Overloading function names

It will have been observed that we used the same name for the three versions of the function swap above. This was deliberate. Each version has a different *signature*: in C++ the compiler automatically distinguishes between functions which differ in the type(s) of formal arguments passed to them. It is, in fact, a convenience to use the same name for functions which perform a very similar task. The common name is said to be *overloaded*. However, functions which differ only in the return type cannot use the same name. Overloading is a commonplace device in the definition of classes (*cf* chapter 5).

A particular form of overloading occurs when functions are passed variables having default values. If we declare

```
double f(int = 0);
```

then on calling `f()` without an argument it will be assumed that the variable passed has the value 0. There is no need to call `f(0)`, although to do so would give the same result.

2.5.7 Function templates

Sometimes it is useful to systematize the overloading of related functions in terms of a *function template*, of which actual functions are particular instances called *template functions*. Using the keyword `template`, and the angle brackets <>, one may, for example, write:

```
template <class T> inline T sign(T x) { x < 0 ? -1 : 1; }
```

Then for all types `T` for which the "less than" operator < has been defined, and for which the numbers 1 and –1 are meaningful, the template function `sign` will return 1 or –1 according to the type and sign of `x`. If one wishes to know the sign of a floating point number, for example,

```
float x = ...;
float s = sign(x);
```

will suffice because the function template `sign` guarantees that the type of the returned value is the same as that of the function argument. The function template as here defined is therefore much safer to use than the macro version introduced in 1.7.1.

We also take the opportunity to define a very useful pair of function templates,

```
template <class T>
inline T max(T x, T y) { return x < y ? y : x; }
template <class T>
inline T min(T x, T y) { return x > y ? y : x; }
```

which we employ to good effect in due course. They return the larger and smaller value, respectively, of the variables `x` and `y` whatever the type of the variables concerned, provided only that it is one for which the relations < and > are defined. Note that the types of the two arguments of these template functions must be identical; it is not enough that there should be automatic conversion. For example, if we were to seek the value of `max(65, 'Z')`, that would be an error since `int` and `char` are different types in spite of the ASCII conversion between them (*cf* 1.3.1).

We shall define a number of further function templates later on, but we note here that it is also possible to define *template classes*, or classes which take a type-name as a parameter (*cf* 5.4.3 and 20.2).

2.6 Recursive functions

A function may call itself, and is then said to be *recursive*. Consider, for example, the definition of the factorial function. This may be written recursively in a very concise way:

```
long  fac(int n)
{     if (n == 0) return 1L;
      return n * fac(n-1);
}
```

The conditional statement `if (n == 0) return 1L;` is a stopping statement. Without it there would be infinite recursion. If possible, a recursive function should only call itself in its last line, otherwise there will remain a stack of residual statements awaiting execution after the last function call has been made, and a consequent demand for time and memory.

Note that this function may be written even more concisely using a ternary expression:

```
long fac(int n) { return (n==0 ? 1L : n * fac(n-1)); }
```

where for clarity we have enclosed the ternary expression in parentheses.

An iterative version, on the other hand, may be written

```
long  ifac(int n)
{     long f = 1;
      for (int i = 2; i <= n; ++i)
            f *= i;
      return f;
}
```

which gives the same result, but is less concise and much less elegant. However, it dispenses with the repeated function calls of the recursive version, and is therefore more efficient. Note that, if we set n = 0 or n = 1, the condition i <= n cannot be satisfied, so the for loop is not executed at all (*cf* 2.4.4) and f remains equal to 1.

As we have seen, it is sometimes possible to extract a value from a void function by means of a reference variable. The void version of ifac might be

```
void  vfac(int n, long &f)
{     int i = 1;
      while (++i <= n) f *= i;
}
```

and this certainly works if f is initialized to 1:

```
int   n = 5;
long  f = 1;
vfac(n, f);
cout << f;          // f = 120
```

Similarly, the recursive version of the void function might be

```
void   vrfac(int n, long &f)
{      if (n == 0) return;
       vrfac(n-1, f*=n);
}
```

where we have chosen to put the successive assignments to f in the recursive function call. Notice that `return` here serves only to terminate an otherwise endless loop; it does not return a value.

A value can also be extracted as a side effect, and this may sometimes be useful even though it cannot then be used in an assignment statement. Here is a `void` recursive function with output as side effect;

```
void   vrsefac(int n, long f)
{      if (n == 0)
       {      cout << "\nfac = " << f << "\n";
              return;
       }
       vrsefac(n-1, f*n);
}
```

It does *not* require f to be passed by reference, although it does still require it to be initialized to 1. However, the elegance of the original recursive function has now been completely lost.

Finally, it is possible to extract an output from a `void` recursive function by means of a global variable:

```
long  f = 1;               // global
void  vgfac(int n)
{      if (n > 0)
       {      f *= n;
              vgfac(n-1);
       }
}
```

This is a particularly useful device if several variables are affected by the function and we wish to extract more than one of them. We use it powerfully in Chapter 8.

However, the manner of execution of these versions of the factorial function differs widely. Consider what happens when we call `fac(4)`. We shall keep track only of multiplications and function calls. In an obvious notation:

```
fac(4) ⇒  4*fac(3)
       ⇒  4*3*fac(2)
       ⇒  4*3*2*fac(1)
       ⇒  4*3*2*1*fac(0)
       ⇒  4*3*2*1*1
       ⇒  4*3*2*1
       ⇒  4*3*2
       ⇒  4*6
       ⇒  24
```

A call to the function `vrfac`, on the other hand, leads in the same notation to:

```
vfrac(4,1) ⇒ vfrac(3,  4 = 4*1)
           ⇒ vfrac(2, 12 = 3*4)
           ⇒ vfrac(1, 24 = 2*12)
           ⇒ vfrac(0, 24 = 1*24)
```

In the first case, the numbers to be multiplied are successively saved on stack until all function calls have been made; in the second, the multiplications take place at once and accumulate in the variable `f`: it is called *tail recursive*. If the value of n is large, the second may be more efficient; but this depends upon the implementation.

The reader may now care to recast the function `sqroot` in one or more recursive functions, and trace its execution.

For the sake of completeness, we remark that two or more functions may be *mutually recursive*. We offer the trivial example of a pair of functions which test whether a given integer is even or odd.

```
bool odd(int);


bool even(int n)
{     if (n == 0) return TRUE;
      return odd(n - 1);
}


bool odd(int n)
{     if (n == 0) return FALSE;
      return even(n - 1);
}
```

Note that since the function odd is called within the function even before it has been defined, it is necessary to begin by declaring its prototype.

If we now wish to know whether 5, say, is even, we call

```
even(5)         ⇒ odd(4)
                ⇒ even(3)
                ⇒ odd(2)
                ⇒ even(1)
                ⇒ odd(0)
                ⇒ FALSE
```

Obviously, in practice it is much simpler just to test the value of (n%2).

2.7 Static local variables

By default, variables defined within the body of a function are automatic; they lose their value after the function call is complete. That property can be overridden by use of the keyword static. A static variable, though still local, persists across invocations; it maintains its value after the function call is complete, and that value may change with each call. Consider, for example, the following implementation of Euclid's famous algorithm for the greatest common divisor of two integers:

```
int     gcd(int m, int n)
{       static int call = 1;
        cout << "call #" << call++ << "\n";
        if (n == 0) return m;
        return gcd(n, m%n);
}


main()
{       int m, n;
        cout << "enter m n: ";
        cin >> m >> n;
        cout << "gcd(" << m << ", " << n << ") = "
             << gcd(m, n) << "\n";
}
```

When the function gcd is called in the function main, the static integer variable call is set equal to 1. It then increases by 1 each time the function gcd calls itself. When run, the screen will show something like this:

```
enter m n: 30 12
call #1
call #2
call #3
gcd(30, 12) = 6
```

Note that, if the function gcd is called by the program more than once, the static variable will continue to be incremented by successive calls. Had call not been declared static, however, its output value would have remained resolutely 1. Of course, the same result could have been achieved by means of a global variable, but it then might be subject to changes due to some other, possibly unintended, feature of the program.

We see here the interplay between two related aspects of a static variable. Whereas, on the one hand, an automatic variable within a function block, for example, is created on the stack (and subsequently destroyed) each time the function is called, a static variable is allocated memory at a fixed address the first time the function is called, and remains until the file has completed execution. On the other hand, a static variable is local in its scope. If defined globally in a file, its scope is that of the file; it cannot be accessed from another file (*cf* 2.3.2); if defined inside a function block it cannot be accessed outside the function. And in 5.3 we shall see that a static class member is limited in scope to its class. These two meanings of the specifier static—one having to do with memory storage, the other with scope—sometimes give rise to confusion.

2.8 Pointers and functions

Functions can involve pointers in three different ways. First, as we have already noted, a function can take a pointer as argument. The function

```
int    strlen(char *s)
{      int c = 0;
       while (*s++) ++c;
       return c;
}
```

calculates the number of characters in a string.

Second, a function can return a pointer. The function

```
char   *strcpy(char *d, char *s)
{      char *t = d;
       while((*d++ = *s++) != '\0') ;
       return t;
}
```

copies the string s into the string d (which must be big enough), and returns a pointer to the resulting string.

Third, we may define a *pointer to a function*. Suppose we want to define a pointer to the function gcd defined above, the declaration of which is

```
int gcd(int, int);
```

A pointer fp to this function must be declared

```
int (*fp)(int, int);
```

using the same prototype in order that its type shall accord with that of the function it is intended to point to. The dereferencing is enclosed within parentheses to avoid confusion with a function that returns a pointer: the function call operator ( ) has higher precedence than the dereferencing operator *, so this is necessary. Its type now determined, the pointer can be initialized by the assignment:

```
fp = &gcd;
```

However, there is no need at this point for the address-of operator &; we may write instead simply

```
fp = gcd;        // preferred
```

showing that fp is just another name for gcd. Recall 1.5.4, where we saw that an array name behaves as if it is a pointer to the initial element of the array. Similarly, a function name behaves as if it is a pointer to a function of its type. Thus, if we call fp(30, 12) we shall get exactly the same result as if we had called gcd(30, 12). Alternatively, we may call (*fp)(30, 12);, the result is the same.

2.9 Function types

We can now define a *function type*. Persisting with the function gcd, we may define a type func:

```
typedef int (*func)(int, int);
```

and then declare and initialize fp:

```
func fp = gcd;
```

A pointer to a function can be passed as an argument to another function like any other pointer. For example, we can write a function which calls another function of which as yet only the type (*ie*, the prototype) has been declared:

```
int    ff(func f, int m, int n)
{      int i = f(m, n);
       return i;
}
```

Note, however, that a function cannot *return* a function (only the *value* of a function); but it can return a pointer to a function like any other pointer. Note also that there are as many function types as there are function signatures and return types. Each function type has to be separately defined. For example, the types `onefunc`, `twofunc`,..., might be defined by:

```
typedef double (*onefunc)(double);
typedef double (*twofunc)(double, double);
.........
```

Finally, note that a function type is not really a new type, only an alternative name for a pointer to a function of given signature and return type.

2.10 The function `main`

So far, we have conspicuously neglected to declare what sort of function `main` is, except to say that every executable program must contain a function of that name. In 2.5, we said that if the return type of a function is not explicitly stated, then by default it is `int`. We also said in 2.5 that the type of the return expression must be the return type of the function. We introduced void functions in 2.5.2 as functions which had no return value. Except in the case of a void function it is an error not to include a return statement which returns the value of a variable or an expression of the return type of the function. (As we saw in 2.6, a void function may contain a return statement, but if so it returns no value.)

At first sight, therefore, the function `main` as we have presented it so far should be declared

```
void main();      // NOT STANDARD!
```

or, if we wish optionally to emphasize that it contains no arguments,

```
void main(void);  // NOT STANDARD!
```

In certain implementations it may indeed be so declared, but, as we have indicated, that is not standard practice and may therefore be frowned upon. The widely accepted convention is that `main` returns an `int`, and so should be declared:

```
int main(void);
```

That implies that it should contain a return statement returning an `int`:

```
int   main()
{       int r = ...;
        .....
        return r;
}
```

When the function `main` has been executed, the program terminates and control reverts to the operating system. Therefore the return value of `main` should be one which informs the operating system of something it would be useful for it to know: for instance that the program was completed without fault, or that it contained a number of non-fatal errors—which it can do by allowing the number of such errors to be accumulated in the variable `r`. Such a facility can evidently be useful, especially if many large programs are being run in a batch. In this book we are concerned only with small programs run singly, and we do not need such a facility. We therefore have two choices. One choice is to return an arbitrary integer constant, as in:

```
int   main()
{       .....
        return 0;
}
```

Some authors follow this practice, but although correct it seems a little artificial. The other choice is to neglect to declare the return type, in which case the function is nevertheless deemed to return an `int`, and then to omit the return statement altogether. The latter is evidently not strictly correct, but it is allowed in the special case of the function `main`, and it causes no error because in fact there is no place left for the program to return to; it has already reached the end. This is the convention we have followed so far, and we shall continue to do so.

There is one more thing to say about the function `main`. So far, we have provided essential data to a program in response to appropriate interactive questions and answers. There is an alternative approach, which is by means of *command line parameters*. Suppose we have a program compiled into an executable file named `PROG.EXE` (using DOS operating system conventions). Normally, we would call the program by entering the DOS command `PROG`. The program would then execute, calling interactively for any data it requires in the process. The alternative is to include the data in the command line.

Suppose, for example, that we want the square root of 10, given the initial approximation 3, and the square root program is to be found in the file PROG. Then it would be convenient if we could simply enter PROG 10 3 to obtain the required result. This can be done if PROG is compiled from the following source code, in which the function sqroot, given in 2.5, is unchanged:

```
main(int argc, char *argv[])
{       double A, x;
        if (argc > 1)
        {       A = atof(argv[1]);
                x = atof(argv[2]);
        }
        else
        {       cout << "Enter positive number: ";
                cin  >> A;
                cout << "Enter approximate square root: ";
                cin  >> x;
        }
        x = sqroot(A, x);
        cout << "Square root of " << A
        << " is " << x << "\n";
}
```

In this version of the program, the standard global variable argc counts the number of strings separated by white space in the command line, in this case 3 because we have called PROG 10 3. The individual strings are stored in the standard global array of strings argv[]. By convention, argv[0] is always the program name including its path, if any; so in our case the value of argv[0] is "PROG". The remaining strings, in this case, store the values of A and x (considered as strings), namely, "10" and "3". The standard library function atof converts a numerical string into a double. In this way, the input data can be retrieved from the command line parameters. After that the program continues as before, and we have also allowed for the fact that it may be preferred *not* to supply the input in this manner.

In this book, we shall make no further reference to input *via* the command line.

3 Errors, theorems and speed

In this chapter we consider the arithmetical errors that arise in computing, certain basic theorems useful for error estimation and algorithm development, and some factors which affect the speed of computation.

3.1 Truncation

Most computations are approximate. One of the most common approximations is the truncation of an iterative statement when some predetermined condition fails. We have seen an example of this in Newton's algorithm for the square root of a number A, where the condition

```
fabs(x*x - A) < tolerance
```

cuts off the iteration as soon as the difference between the square of the current approximate square root coincides with A to within the preset tolerance. To a large extent, the errors arising from truncation are within the user's control, but that does not mean that it is always easy to estimate their magnitude. In this case, however, it is easy enough. If we wish to calculate the square root of 10 to six significant decimal places, the truncation error must be less than $\varepsilon = 5.10^{-7}$ and this implies that the tolerance $\tau < 2\varepsilon x \approx 3.10^{-6}$. The estimation of truncation errors in iteration resembles that of termination of infinite series. There will be many examples later in this book.

3.2 Rounding errors

A quite different source of error results from the fact that floating point numbers are represented in a computer to high but limited precision. Owing to the fixed number of bytes associated with a such a number, be it `float` or `double`, errors occur which may be amplified by arithmetical processes.

The precise conventions according to which floating point numbers are represented in binary form in a computer depend upon the implementation, but it is sufficient in order to understand the nature of the errors involved to imagine that they are written in normalized decimal form. Let us denote by $\mathrm{fl}(x)$ the floating point value of the real number x, and let us work (with the aid of a pocket calculator) to four significant decimal digits; *ie*, we represent any floating point number as $\pm 0.d_1d_2d_3d_4.10^n$, where n is called the exponent. In a representation in which this number might occupy four 8-bit bytes, one bit would represent the sign, seven bits the value of the exponent (positive or negative), and the remaining 24 bits the fractional part, or mantissa, suitably normalized. "Exact" values we shall represent in seven figure precision to sufficient approximation. For example,

$$\text{if } x = 2/3 \ (= 0.6666667) \text{ then } \text{fl}(x) = 0.6667.10^0,$$
$$\text{if } y = 1/7 \ (= 0.1428571) \text{ then } \text{fl}(y) = 0.1429.10^0.$$

The exact values resulting from the basic arithmetic operations are:

$$x + y = 17.0/21 = 0.8095238$$
$$x - y = 11.0/21 = 0.5238095$$
$$x \times y = 2.0/21 = 0.0952381$$
$$x / y = 14.0/3 = 4.6666667$$

In four-figure floating point precision we shall define the arithmetic operations by the relations:

$$x \oplus y = \text{fl}(\text{fl}(x) + \text{fl}(y)) = .8096.10^0$$
$$x \ominus y = \text{fl}(\text{fl}(x) - \text{fl}(y)) = .5238.10^0$$
$$x \otimes y = \text{fl}(\text{fl}(x) \times \text{fl}(y)) = .9527.10^{-1}$$
$$x \oslash y = \text{fl}(\text{fl}(x) / \text{fl}(y)) = .4667.10^1$$

where we have also given their approximate values.

The *absolute error* of an approximation $x*$ to an exact value x is given by the absolute difference $|x - x*|$. The *relative error*, which may be more relevant in practice, is given by $|x - x*|/|x|$. The errors resulting from the four-figure arithmetic above are:

Operation	Absolute	Relative
$x \oplus y$	$7.6.10^{-5}$	$9.4.10^{-5}$
$x \ominus y$	$0.9.10^{-5}$	$1.8.10^{-5}$
$x \otimes y$	$3.2.10^{-5}$	34.10^{-5}
$x \oslash y$	33.10^{-5}	7.10^{-5}

Clearly, care must be taken especially in subtracting two nearly equal numbers, in multiplying by a large number and in dividing by a small number, if the numbers cannot be stated exactly. Even when higher precision is employed, which is usually the case, successive arithmetic operations can easily reduce precision drastically.

3.3 Quadratic equations

The extraction of the real roots of a quadratic equation with real coefficients illustrates the point. The equation

$$ax^2 + bx + c = 0$$

possesses two solutions, α and β, the product of which is $\alpha\beta = c$. The roots are given by the well-known formula

$$(-b \pm \sqrt{(b^2 - 4ac)})/2a.$$

To retain accuracy, the minus sign should be taken in the numerator if b is positive, the negative sign otherwise, to obtain the numerically larger root, α, say. The second root is then given by $\beta = c/\alpha$. In this way, we avoid subtracting two numbers which may be almost equal in value. The following function carries out this algorithm:

```
void   quadroots(double a, double b, double c,
                         double &x1, double &x2)
{      double d = b*b - 4*a*c;
       if (d < 0) error("...");
       double s = -(b + sqrt(d)*sign(b))/2;
       x1 = s/a;
       x2 = c/s;
}
```

where the macro `sign` was defined in 1.7.1, and the function `sqrt` is a standard library function.

Note that we have taken avoiding action in case the discriminant d is negative. In the second statement, a standard error function will be chosen to be

```
void   error(char *errmsg)
{      cerr << "\nRuntime Error ...\n";
       cerr << errmsg;
       cerr << "\n... Quitting Program!\n";
       exit();
}
```

It is a development of the fragment offerred in 1.1.3. If we choose the error message `"Discriminant negative in quadroots(...)"` there will appear the following message on the screen when the function `error` is called:

```
Runtime Error ...
Discriminant negative in quadroots(...)
... Quitting Program!
```

The program will then terminate gracefully thanks to the standard library function `exit` which will have "cleaned up", for example, output streams and variables still in scope, before termination. Note that this error function is of fairly general application. We shall use it freely hereafter; only the error message has to be changed. More sophisticated error handling is available in C++, which may, for instance, indicate more precisely where an error has occurred, or which allows the program to try again before it is terminated, and we refer the reader to the book by Stroustrup for a fuller discussion of the subject.

We shall use the function `quadroots` whenever in this book we need to extract square roots. Note, however, that in the event of the parameter a being zero or very small, prior avoiding action should be taken.

3.4 Floating point equality

Often one wishes to write a conditional statement of the form:

```
if (x == y) statement.
```

If x and y are integers there is no problem. But if x and y are floating point numbers, the evaluation of the boolean expression x == y is liable to error owing to the finite precision with which they may be represented, and the precision to which one may be working in practice. Given an appropriately small number ε, exact equality $x = y$ may be replaced by $|x - y| < \varepsilon$. This is adequate if $|x| \approx 1$, but may be unsatisfactory if $|x| \ll 1$ or $|x| \gg 1$. (Here, of course, we imply "much less than" or "much greater than".) In the latter cases, it may be better to write a relative expression rather than an absolute one, of which the simplest is $|x - y|/|x| < \varepsilon$, which amounts to the same thing if $|x| \approx 1$, except that its evaluation requires an additional multiplication. Of course, the case in which x is very close to zero has to be treated exceptionally. One of many boolean functions which express these considerations may be written:

```
bool  equal(double x, double y, double del, double eps)
{     double fx = fabs(x), fy = fabs(y);
      if (fx < del)                          // x ≈ 0
      {     if (fy < del) return TRUE;    // y ≈ 0 also
            else return FALSE;
      }
      else if (fabs(x - y) < eps * fx) return TRUE;
      else return FALSE;
}
```

where `del` and `eps` are two numbers of very small absolute value. But, in any case, it is always better to avoid testing *exact* equality of floating point numbers.

3.5 Conditioning and stability

A function is said to be *well-conditioned* if small changes in its parameters result in small changes in its value; otherwise, it is *ill-conditioned*. The classical example of an ill-conditioned equation is one having roots at $x = 1, \ldots, 20$, which can be written in terms of the polynomial

$$P(x) \equiv (x - 1)(x - 2) \cdots (x - 20) = 0.$$

If this is expanded in powers of x, and the coefficient of x^{19}, say, changed by approximately 10^{-7}, the roots change drastically. The polynomial now has 5 pairs of complex roots, and one of the real roots becomes -20.847. (We shall see how this comes about in 4.10.2.)

Similarly, a numerical computation is said to be *stable* if small changes in the input data, or small round-off errors, induce only small changes in the computed result; otherwise it is said to be unstable. In a practical computation one wishes the error in the result to remain under control, and to be capable of estimation if possible. In a repetitive process, if the error E_n at the nth iteration (n exceeding some small number) has the form $|E_n| \propto n$, error growth is said to be *linear*. If $|E_n| \propto k^n$, where $k > 1$, the errors are said to grow *exponentially*; if $k < 1$, they decrease exponentially.

Let $\phi = (\sqrt{5} - 1)/2 \approx 0.618034$ (the "golden mean"); it is a solution of the quadratic equation $x^2 + x - 1 = 0$. The integer powers of ϕ satisfy the recurrence relation:

$$x_n = x_{n-2} - x_{n-1},$$

which may easily be written as a recursive function to the precision `float`:

```
float phipow(int n)
{       if (n == 0) return 1.0;
        if (n == 1) return GOLDMEAN;
        return phipow(n-2) - phipow(n-1);
}
```

The results are tabulated below, compared with the "exact" values obtained by reworking the function with `double` instead of `float`. There is, however, another solution to the quadratic equation, namely $\phi_1 = -(\sqrt{5} + 1)/2 = -1.618034$. The recurrence relation, being linear, admits solutions which are of the form $\phi^n + \varepsilon\phi_1^n$, where $\varepsilon \approx 10^{-7}$ represents the precision we have assumed of the `float` representation of floating point numbers. Since $|\phi_1| > 1$, the error grows exponentially:

n	float	double
0	1.0	1.0
1	.618034	.618034
2	.381966	.381966
5	.090170	.090170
10	.008130	.008131
15	.000743	.000733
20	−.000045	.000066
25	.001238	.000002

All is well until powers of about 10 are reached, after which errors build up until the value returned bears no relation to the correct value.

3.6 Local and global errors

Many computations, such as the solution of differential equations, proceed in steps. At each step it may be possible to estimate and control the *local* errors involved, but the results overall may contain cumulative errors which are difficult or impossible to predict or control. Overall, or *global*, errors may be guessed at by computing solutions to model problems, the answers to which are known exactly by analytical means, or by working to higher precision. We shall frequently use such devices in later chapters.

3.7 Basic theorems

Here we recall without proof some of the basic theorems of calculus which are fundamental to the development of numerical algorithms. All are intuitively obvious, but proofs may be found in standard textbooks. We use the usual set notation. If $a < c < b$, we say that c is on the *open* interval (a, b) and write $c \in (a, b)$; but if $a \le c \le b$, we say that c is on the *closed* interval $[a,b]$, and we write $c \in [a, b]$. Sometimes we may write $c \in (a, b]$ or $c \in [a, b)$. Let $f(x)$ be a continuous and differentiable function on $[a, b]$. Then the following theorems are true:

Rolle's theorem
     If $f(a) = f(b) = 0$, there exists a number $c \in (a,b)$, such that $f'(c) = 0$.

Mean value theorem
     There exists a number $c \in (a,b)$ such that $f'(c) = (f(b) - f(a))/(b - a)$.

Extreme value theorem
     There exist $c_1, c_2 \in [a, b]$ such that, for all $x \in [a, b]$, $f(c_1) \le f(x) \le f(c_2)$. The values c_1 and c_2 occur either at the ends of the interval or where $f'(x) = 0$.

Intermediate value theorem

If K is any number such that $f(a) < K < f(b)$, then there exists a number $c \in (a, b)$ such that $f(c) = K$.

Weighted mean value theorem

If the function $g(x)$ is integrable and does not change sign on $[a, b]$, then there exists $c \in (a, b)$ such that

$$\int_a^b f(x)g(x)\,dx = f(c) \int_a^b g(x)\,dx$$

When $g(x) \equiv 1$, we obtain the average value of f over the interval $[a, b]$:

$$<f> = \int_a^b f(x)\,dx/(b - a)$$

Taylor's theorem

Suppose that $f(x)$ is differentiable $(n + 1)$ times on $[a, b]$. Let $x_0 \in [a, b]$. Then for every $x \in [a, b]$, there exists a number $\xi(x)$ between x_0 and x such that

$$f(x) = P_n(x) + R_n(x),$$

where
$$P_n(x) = f(x_0) + f'(x_0)(x - x_0) + (1/2!)\, f''(x_0)(x - x_0)^2$$
$$+ \cdots + (1/n!)\, f^{(n)}(x_0)(x - x_0)^n,$$

and
$$R_n(x) = (1/(n + 1)!)f^{(n+1)}(\xi(x))(x - x_0)^{n+1}.$$

Here $P_n(x)$ is called the *n-th Taylor polynomial* for f about x_0, and $R_n(x)$ is called the *remainder term* or *truncation error* associated with P_n. If $x_0 = 0$, the series is sometimes called a *Maclaurin series*, and the theorem is then referred to as *Maclaurin's theorem*.

Limits can sometimes be placed on the value of the remainder. Thus, for example, we may expand the exponential function about $x = 0$:

$$e^x = 1 + x + x^2/2! + e^\xi x^3/3!, \quad 0 < \xi(x) < x.$$

Setting $x = 1$ then gives us $e = 1 + 1 + (1/2) + R$, where the remainder R lies between $1/6$ and $e/6$. Thus, $2.67 < e < 2.95$ to this approximation, since $e \approx 2.72$.

It is convenient at this point to introduce *divided differences*. These may be defined inductively:

$$f[x_0] = f(x_0)$$
$$f[x_0, x_1] = (f[x_1] - f[x_0])/(x_1 - x_0)$$
$$f[x_0, x_1, x_2] = (f[x_1, x_2] - f[x_0, x_1])/(x_2 - x_0)$$
$$\cdots\cdots\cdots\cdots\cdots\cdots\cdots\cdots\cdots\cdots\cdots\cdots\cdots\cdots\cdots$$
$$f[x_0, \ldots x_{k+1}] = (f[x_1, \ldots, x_{k+1}] - f[x_0, \ldots, x_k])/(x_{k+1} - x_0)$$

Then Taylor's theorem gives: $f[x_0, \ldots, x_k] = (1/k!)f^{(k)}(\xi),$

where ξ is a number in the range

$$\text{Min}(x_0, \ldots, x_k) \leq \xi \leq \text{Max}(x_0, \ldots, x_k)$$

and Min and Max are the lowest and highest values of the numbers x_i, respectively. We shall conventionally write this: $\xi \in I[x_0, \ldots, x_k]$.

Note that the value of $f[\ldots]$ is independent of the order of the variables. We may use this fact to obtain an expression for the differential coefficient:

$$\begin{aligned}
(d/dx)f[x_0, \ldots, x_n, x] &= \lim_{h\to 0}(f[x_0, \ldots, x_n, x+h] - f[x_0, \ldots, x_n, x])/h \\
&= \lim_{h\to 0}(f[x_0, \ldots, x_n, x+h] - f[x, x_0, \ldots, x_n])/h \\
&= \lim_{h\to 0}f[x, x_0, \ldots, x_n, x+h] \\
&= f[x, x_0, \ldots, x_n, x] = f[x_0, \ldots, x_n, x, x].
\end{aligned}$$

We shall use this result when we consider numerical integration.

Finally, it is convenient to introduce the *Cauchy–Schwartz Inequality*:

If $\mathbf{x} = (x_1, x_2, \ldots, x_n)$ and $\mathbf{y} = (y_1, y_2, \ldots, y_n)$ are two sequences of n numbers (vectors), then

$$\left|\sum_i x_i y_i\right| \leq \left(\sum_i x_i^2\right)^{1/2}\left(\sum_i y_i^2\right)^{1/2}.$$

3.8 Rates of convergence

Since many computations calculate a sequence of successive approximations, it is important to be able to express the rate at which a sequence converges. This will determine the number of steps required to achieve given precision. Suppose $\{x_n\}$, where $n = 1, 2, \ldots, \infty$, is a sequence that converges to a number x. We say that $\{x_n\}$ converges to x with *rate of convergence* $O(y_n)$, where $\{y_n\}$ is another sequence with $y_n \neq 0$, $n = 1, 2, \ldots, \infty$, if

$$|x_n - x|/|y_n| \leq K \text{ for sufficiently large } n,$$

where K is a constant independent of n. Then we write

$$x_n = x + O(y_n).$$

Consider, for example, the sequence $\{x_n\}$ with $x_n = (n-1)/n$. We have

$$x_n = 1 - (1/n) \to x = 1.$$

If $\{y_n\}$ is the sequence $y_n = 1/n$, then

$$|x_n - x|/|y_n| = (1/n)/(1/n) = K = 1,$$

and $x_n = 1 + O(1/n)$.

On the other hand, the sequence $\{x'_n\}$ with $x'_n = n(n+2)/(n+1)^2$ satisfies

$$x'_n = 1 - 1/(n+1)^2 \to x' = 1;$$

and if $\{y'_n\}$ is the sequence $y'_n = 1/n^2$, then

$$|x'_n - x'|/|y'_n| = n^2/(n+1)^2 \le K = 1,$$

and $x'_n = 1 + O(1/n^2)$. Both sequences converge to unity, the first linearly, the second quadratically, in $1/n$. The following table shows the first few terms in each sequence:

n	=	1	2	3	4	5	6	7
x_n	=	0	.500	.667	.750	.800	.833	.857
x'_n	=	.500	.889	.938	.960	.972	.980	.984

Obviously, we shall wish to compute sequences of approximations that converge as fast as possible—although sometimes there may be little choice!

The concept of the rate of convergence may also be applied to a function. If f is a function with the property $f(x) \to L$ as $x \to 0$, and $g(x)$ is another function such that

$$|f(x) - L|/|g(x)| \le K \text{ for sufficiently small } |x| > 0,$$

then we write $f(x) = L + O(g(x))$.

For example, using Taylor's theorem to expand $\cos x$ about $x = 0$,

$$\cos x = 1 - (1/2!)x^2 + (1/4!)x^4 \cos \xi(x),$$

where $0 \le \xi(x) \le x$. Thus:

$$\cos x + (1/2)x^2 = 1 + O(x^4)$$

since $\quad |(\cos x + (1/2)x^2) - 1|/|x^4| = |(1/24)\cos \xi(x)| \le K = 1/24.$

3.9 Reciprocals without division

We shall now present the complete convergence analysis for a well-known simple example of an iterative computation. Suppose we wish to find the reciprocal of a positive floating point number a without using the built-in division operation. (Early computers could not do division directly, and this was the process they used.) We seek the root of the function $f(x) = a - (1/x)$

by an iterative method derived from Newton's algorithm. Let x_0 be a first approximation to the root. Draw the tangent to $f(x)$ at $x = x_0$, which crosses the x-axis at x_1, where $f'(x_0) = f(x_0)/(x_0 - x_1) = 1/x_0^2$. We assert that x_1 is a better approximation to the reciprocal of a than is x_0, and that the sequence

$$x_{n+1} = x_n(2 - ax_n), \ n \geq 0$$

converges to $1/a$.

If we define the residue $r_n = 1 - ax_n$, then $x_{n+1} = x_n(1 + r_n)$, and the error at each step is given by

$$E_n \equiv (1/a) - x_n = r_n/a.$$

But $\qquad\qquad r_{n+1} = 1 - ax_{n+1} = 1 - ax_n(1 + r_n) = 1 - (1 - r_n)(1 + r_n) = r_n^2.$

By induction, therefore,

$$r_n = r_0^m, \text{ where } m = 2^n,$$

and $r_n \to 0$ as $n \to \infty$ if and only if $|r_0| < 1$, or equivalently $0 < x_0 < 2/a$. This last relation sets the interval of convergence for the starting value x_0.

The sequence of errors is given by $E_{n+1} = aE_n^2$, so E_n converges to zero quadratically. The *relative error* satisfies $E_n/(1/a) = (E_n/(1/a))^2$, so that each iteration doubles the number of significant figures in the result. It is not often such a complete analysis can be given of an iterative process!

For completeness we write the above algorithm as a function which accepts the number for which the reciprocal is being sought, and the starting value, and returns the reciprocal – unless the interval of convergence is exceeded, in which case it prints an error message and terminates:

```
double reciprocal(double a, double x)
{      if (x <= 0 || x >= 2/a)
            error("...");
       double r = 1 - a*x;
       while(fabs(r) > eps)
            x *= (r = 1 - a*x, 1 + r);
       return x;
}
```

Notice the use of the comma expression in the loop.

3.10 Speed of computation

The speed of computation, or, inversely, the time taken for a given computation, obviously depends upon the rate of convergence if an iterative process is involved. It also depends, however, on the number of operations performed in each iteration, especially on the number of floating point operations, and the precision—float or double—with which they are carried out. Actual speeds depend upon the machine used and the precise implementation of the various operations. Computers are getting more efficient all the time! On some machines, floating point multiplications and divisions will take about the same time to perform, and each will take considerably longer than addition or subtraction. Comparisons in conditional statements may take a little longer than additions. Purely integer operations are usually much faster than floating point operations. Estimates of the speed of computation of a numerical routine or complete program, therefore, are often based solely upon the number of floating point multiplications and divisions. On other machines, data movement itself may be the more significant factor in determining speed. It is not always easy, therefore, to make machine independent statements about computing speeds, although comparisons may still be possible.

Consider the above routine for the computation of the reciprocal of a floating point number. The first line consists of a single comparison (unless the condition fails, in which case there is little interest in the result anyway). The second line is a definition involving 1 multiplication and 1 subtraction. There follows a `while` loop, where the condition is `fabs(r) > eps`. The function `fabs` itself involves 1 comparison, so the condition involves 2 comparisons in total for each iteration performed. The comma expression in the loop is shorthand for two assignments, each involving 1 multiplication and 1 addition or subtraction. The final `return` statement may be ignored. Thus, if there are p iterations, there will be altogether $p + 1$ comparisons, $2p + 1$ additions or subtractions, and $2p + 1$ multiplications. For large p, the time of computation will be proportional to $p \approx \log_2(\log \varepsilon / \log r_0)$.

This is a very simple case; most practical computations will involve processes varying as the square or cube of some integer parameter n, such as the size of a matrix or the length of a list, or perhaps as $n \log n$. The estimation of the dependence of the time of computation on some suitable parameter is an important part of any computational method.

3.11 Speed and recursion

A function call also takes time; and if it is a very simple function it may take more time to call it than to execute it, for the reasons we gave in 2.5.1. That is why in such cases we may choose to define a function inline so that it is expanded in code rather than called. But a recursive function may call itself many times, and for that reason may be slow to execute.

Consider, for example, the *Fibonacci sequence* defined by:

$$F_0 = 0, \quad F_1 = 1, \quad F_n = F_{n-1} + F_{n-2}, \quad n > 1.$$

The first few numbers are 0, 1, 1, 2, 3, 5, 8, 13, 21, . . ., and it is extremely tempting to write down the obvious recursive function:

```
long   fib(int n)
{       if (n == 0) return 0L;
        if (n == 1) return 1L;
        return fib(n-1) + fib(n-2);
}
```

(We return a long integer because the terms rapidly increase in size.)

This routine certainly works, and it reproduces the sequence indicated; but with increasing n it becomes exceedingly slow. The reason is the *double* recursion in the return statement. For $n = 5$, it requires 15 function calls to produce the correct answer $F_5 = 5$. It takes 177 calls for $F_{10} = 55$, and no fewer than 2692537 to reach $F_{30} = 832040$. The number of function calls itself resembles a Fibonacci sequence:

$$c_0 = 1, \ c_1 = 1, \ c_n = c_{n-1} + c_{n-2} + 1.$$

Let us therefore abandon recursion for a moment and write an iterative function instead. We need to carry *two* variables, the sum of which is the number sought:

```
long   iterfib(int n)
{      long r = 0, s = 1;
       for (int m = 0; m < n; ++m)
       {      long t = r;
              r = s;
              s += t;
       }
       return r;
}
```

This works too, with the important difference that it requires only one function call and n additions, and is extremely fast.

The interesting question then arises whether we can devise a function that is recursive but also fast. We can, and again we must carry two variables. It is convenient to treat them on the same footing in the recursion, so we define the auxiliary function:

```
void   twofib(long &r, long &s, int n)
{      if (n == 0)
       { r = 0; s = 1; }
       else
       {      twofib(s, r, n-1);
              s += r;
       }
}
```

Unfortunately it is not tail-recursive, so there will be some stack requirement; but if it is invoked for given n the number of function calls is just $n + 1$, a very substantial improvement in speed over the original double recursive function `fib`. In order to have a function which returns the number we are seeking, it is convenient finally to define the function:

```
long    fastfib(int n)
{       long r, s;
        twofib(r, s, n);
        return r;
}
```

A further illustration is provided by a recursive function which computes a positive integer power k of a floating point number x. It is tempting to write

```
double pow(double x, int k)          // O(k)
{       if (k == 1) return x;
        return x*pow(k - 1);
}
```

which requires k function calls to evaluate the expression $x \times x \times \ldots \times x$ to k terms. It is better to take advantage of the fact that certain multiplications have already been performed. For instance, we would prefer, symbolically:

$$
\begin{aligned}
\text{pow}(x, 6) &\Rightarrow \text{pow}(x^2 \times x, 3) \\
&\Rightarrow x^2 \times \text{pow}(x^4 = x^2 \times x^2, 1) \\
&\Rightarrow x^2 \times x^4 \\
&\Rightarrow x^6
\end{aligned}
$$

Thus, distinguishing between even and odd powers of x by testing the value of $(k \bmod 2)$, we can obtain a function which is $O(\log k)$ instead:

```
double pow(double x, int k)          // O(log k)
{       if (k == 1) return x;
        if (k % 2) return x*pow(x*x, k/2);
        return pow(x*x, k/2);
}
```

This may be readily generalized to allow k to be negative or even zero by means of the non-recurrent encapsulating function:

```
double intpow(double x, int k)
{       if (k == 0) return 1.0;
        if (k < 0)
        {       k = -k;
                x = 1.0/x;
        }
        return pow(x, k);
}
```

The reader may care to ponder why we include the $k = 0$ case in `intpow` rather than in `pow` itself.

Recursive functions are aesthetically attractive, but they are not always straightforward.

3.12 Mistakes

Another source of error is ubiquitous—human error, or mistakes. These must be eliminated as far as possible by carefully testing routines, especially with extreme values of parameters, against special cases to which the correct solution is known analytically, or otherwise.

There are, however, sources of human error other than sheer carelessness, which arise from failing to understand fully the nature of the problem for which an approximate solution is being sought by computation. For example, the function

$$F_{\text{path}}(x) = x + 0.001 \times \log(x - 2)^2$$

is very nearly linear in x except in the immediate vicinity of $x = 2$ where there is a weak singularity accompanied by two very close roots. Almost any attempt to find values of F_{path}, for example by interpolation from a sequence of regularly spaced values of x, is sure to produce extremely erroneous values near $x = 2$.

It is a dangerous error to suppose that computation is any substitute for understanding the essential nature of a mathematical problem.

4 Roots of non-linear equations

In this chapter we shall build on the foundations so far laid, and illustrate their power, by developing some routines for the extraction of roots of non-linear equations in one variable with real coefficients, of the form $f(x) = 0$, where $f(x)$ is a continuous and differentiable function. Firstly, we shall consider methods which assume no knowledge of the derivative of the function. Next, we shall consider fixed point methods, where the equation $f(x) = 0$ is transformed into $x = g(x)$. Lastly, we return to Newton's method, which uses the derivative explicitly, and shall in particular show how to calculate the roots of a polynomial.

In each method, we begin by assuming an interval (a, b) containing a single root of $f(x) = 0$. If $f(a)$ and $f(b)$ are of opposite sign, then by the intermediate value theorem there must be a number $c \in (a,b)$, such that $f(c) = 0$. It is this number that we seek. We shall illustrate each method by applying it to the function $F(x) = x - e^{-x}$. Since $F(0) = -1$, and $F(1) \approx 0.63$, there is a root (the only real root) in between; it is in fact at $x = \alpha \approx 0.567143$. We shall accept a result which lies within the small interval $(\alpha - \varepsilon, \alpha + \varepsilon)$, where we shall set $\varepsilon = 5.10^{-7}$ in order to obtain a result correct to six decimal digits. For later use, we note at this point that $F'(\alpha) = 1 + \alpha \approx 1.567143$, and $F''(\alpha) = -\alpha \approx -0.567143$.

A related problem is to find the maxima or minima of a function, and we present a simple method for this. Finally, we introduce some constants and functions useful for later work.

4.1 Bisection method

Let us then suppose that $f(a)$ and $f(b)$ have different signs. In the bisection method, we first approximate the root by the midpoint of the interval (a, b), namely $x = (a + b)/2$, and enquire whether $f(x)$ has the same sign as $f(a)$, or as $f(b)$. If it has the same sign as $f(a)$, the root lies in the interval (x, b). If not, it lies in the interval (a, x). Either way, we have halved the interval containing the root, and therefore doubled the precision with which we can state its value. The process may be repeated until the length of the interval is smaller than the fixed interval 2ε. This will require $n = \log_2(|b - a|/\varepsilon) \approx 20$ steps, if $a = 0$, $b = 1$, since $\varepsilon = 5.10^{-7}$.

4.1.1 Simple recursive procedure

The above algorithm may be directly translated into a simple recursive function. Given a suitable definition of the type func, the recursive function may be written

```
void  bisect0(func f, double a, double b, double &x)
{     x = (a + b)/2;
      f(x) * f(a) > 0 ? a = x : b = x;
      if (fabs(a - b) > eps) bisect0(f, a, b, x);
}
```

where we recall that the library function `fabs(x)` returns the absolute value $|x|$. Successive approximations to the root appear in the variable x, which has therefore been passed by reference. Note that for brevity we have employed the ternary statement

```
f(x) * f(a) > 0 ? a = x : b = x;
```

We recall from 2.4.1 that this is merely shorthand for

```
if (f(x) * f(a) > 0)
        a = x;
else
        b = x;
```

and readers may decide for themselves which they prefer.

Provided the interval (a, b) contains the required root, and is not so wide that the number of recursive calls exceeds available memory, this simple function will certainly compute the root to within an absolute error `eps`. (It may be modified to compute the root to within a *relative* error instead (*cf* 3.4); this may be important if the root is expected to be very large or very small.)

4.1.2 Refined recursive procedure

However, the function `bisect0` is not completely satisfactory. It does not allow for the possibility that the process will not converge sufficiently fast, so that the number of recursive calls exceeds a fixed number `maxiter` defined globally. The following function is chosen to return the value of a variable of the enumeration type `state` defined in 2.4.7:

```
state bisect1(func f, double a, double b, double &x)
{     static int iter = 0;
      if (fabs(a - b) < eps) return SUCCESS;
      if (++iter == maxiter) return WONTSTOP;
      x = (a + b)/2;
      f(x) * f(a) > 0 ? a = x : b = x;
      return bisect1(f, a, b, x);
}
```

This certainly terminates if convergence is too slow. But if the function `bisect1` is called several times in the same program, the static variable `iter` will continue to accumulate (*cf* 2.7) and at some stage the program may suffer premature termination. There are two courses open to us. We may define `iter` to be a global variable, or we may encapsulate the recursive function within a non-recursive function, the main purpose of which is to provide the variable `iter`. We shall follow the latter course, because it gives us the additional possibility of checking whether the interval (a, b) does indeed enclose a root.

We define the recursive function, called `interval`, first. The value of `iter` must be passed to it by reference. Then we show the encapsulating function, which we call `bisect`. But for illustration we present the work in the form of a complete program. It is written for an arbitrary function $f(x)$ of one variable, but it is here applied as a test case to the evaluation of the root of the function $F(x) = x - e^{-x}$ introduced above.

```
// #include directives here

typedef double (*func)(double);
state interval(func, double, double, double&, int&);
state bisect(func, double, double, double&);

const double eps  = 5.0E-7;
const int maxiter = 100;

state interval(func f, double a, double b, double &x,
      int &iter)
{     if (fabs(a - b) < eps) return SUCCESS;
      if (++iter == maxiter) return WONTSTOP;
      x = (a + b)/2;
      // put optional internal output here
      f(x) * f(a) > 0 ? a = x : b = x;
      return interval(f, a, b, x, iter);
}

state bisect(func f, double a, double b, double &x)
{     if (f(a) * f(b) > 0) return SAMESIGN;
      int iter = 0;
      return interval(f, a, b, x, iter);
}

inline double F(double x) { return x - exp(-x); }

main()
{     clear;
      double a, b, root;
      cout << "lower estimate = ";
      cin  >> a;
      cout << "upper estimate = ";
      cin  >> b;
```

```
state s = bisect(F, a, b, root);
switch(s) {
        case SUCCESS:     cout << "root = " << root; break;
        case WONTSTOP:    cout << "not convergent!"; break;
        case SAMESIGN:    cout << "unsuitable interval!"; break;
        default:          cout << "that's very peculiar!";
}
}
```

The program begins by including the necessary header files to declare the input and output conventions, the function `fabs`, and the enumeration type `state`. There then follows the definition of the type `func`, which enables us to pass an arbitrary function of that type to the root-finding routine, and declarations (*ie*, prototypes) of the functions we are going to use. (The last are optional, but it is good practice to include them.) These are followed by definitions of the global constants of the program, `eps` and `maxiter`. After these come the two functions `interval` and `bisect`. Since we have already declared them, the order in which these functions are defined is immaterial; otherwise they would have to be in the order given here, so that `bisect` knows what sort of function `interval` is.

The function `main`, which causes the program to execute, is typical of a program devised to test a particular routine. It first clears the screen by calling an appropriate function. This we have indicated by the identifier `clear`, which is in fact a macro. Most implementations will have such a function, but its name is not yet standard. In TurboC++, it is declared to be `void clrscr(void);`, and in this case the macro may be defined by

```
#define clear clrscr()          // NON-STANDARD!
```

If using another implementation, we merely change this definition, or use directly whatever function is appropriate in place of `clear`. But in any case, every executable program should begin by clearing the screen, not least when it is being tested.

Then `main` declares the variables which it will use, in this case the lower and upper limits of the interval within which the root is being sought, and the value of that root, and seeks input values for the former. Finally, it calls the routine `bisect`, and by means of a `switch` statement arranges output according to the state in which the routine terminates.

If run successfully, the program will output to screen something like the following:

```
lower estimate = 0
upper estimate = 1
root = 0.567143
```

If it is desired that there should in addition be a running output giving the current estimate of the root as the recursion proceeds, an internal output statement may be included in the definition of the function `interval` in the place indicated, such as

```
cout << iter << "\t" << a << "\t" << b << "\t" << x << "\n";
```

where '\t' is the tab character. With this output line included, the above program will show each step of the computation. We print out selected lines below:

iter	a	b	x
1	0	1	0.5
2	0.5	1	0.75
3	0.5	0.75	0.625
4	0.5	0.625	0.5625
. . .	. . . .	. . . .	. . . .
10	0.566406	0.568359	0.567383
15	0.567139	0.567200	0.567169
20	0.567142	0.567144	0.567143
21	0.567142	0.567143	0.567143

4.1.3 Non-recursive procedure

Sometimes it is preferable to avoid the repeated calls to a recursive function. The following procedure, which is an alternative definition of bisect, uses a while loop instead:

```
state bisect(func f, double a, double b, double &x)
{       state s = ITERATING;
        int iter = 0;
        double xmid, fmid, dx;
        double fa = f(a), fb = f(b);
        if (fa * fb > 0) s = SAMESIGN;
        x = fa < 0 ? (dx = b - a, a) : (dx = a - b, b);
        while (s == ITERATING)
        {       fmid = f(xmid = x + (dx *= 0.5));
                if (fmid < 0) x = xmid;
                if (fabs(dx) < eps)     s = SUCCESS;
                if (++iter == maxiter) s = WONTSTOP;
        }
        return s;
}
```

The variables fa, fb and fmid, where fmid = f(xmid) and xmid is the midpoint of the current interval, are defined in order to avoid repeated calls to the function f; dx represents the current length of the interval. The integer variable iter counts the number of iterations of the while loop. We have used a condensed notation twice. The ternary statement

```
x = fa < 0 ? (dx = b - a, a) : (dx = a - b, b);
```

which contains two comma statements, is an abbreviation for the compound conditional statement

```
if (fa < 0)
{       dx = b - a;
        x  = a;
}
else
{       dx = a - b;
        x  = b;
}
```

while the statement

```
fmid = f(xmid = x + (dx *= 0.5));
```

is equivalent to the three statements

```
dx   *= 0.5;
xmid = x + dx;
fmid = f(xmid);
```

It takes a little practice before compact statements of this kind can be used with confidence, but perseverance will save a lot of trees.

The purpose of the state variable s is now apparent. It is to trap error conditions without abrupt termination of the program. There is no need to call upon a function such as error (*cf* 3.3) to terminate it prematurely. No matter what happens to the computation of x, the function bisect always returns a value of s. The while loop repeats as long as the state s remains ITERATING. It terminates successfully (s == SUCCESS) when the interval $|dx| < $ eps, and unsuccessfully (s == WONTSTOP) if the number of iterations iter reaches the preset maximum maxiter.

The bisection method has been constructed to converge linearly with a rate of ½. To see that this is so we merely have to note that the sequence, $\{x_n\}$, of approximations to the root α, satisfies $|x_n - \alpha| \leq (1/2^n)|b - a|$.

4.2 Regula falsi method

Although the bisection method is certain to converge, it may do so very slowly. It uses the *sign* of f, but no other knowledge of the function. The midpoint of the interval is chosen as the next approximation no matter how close the real root might be to one or other extremity. In the regula falsi method (method of *false positions*), further knowledge of the function is introduced, and this will in many cases speed up convergence, although not in all. The first approximation is obtained by drawing a straight line between the points $(a, f(a))$ and $(b, f(b))$. This line intersects the x-axis at the point:

$$x = a - f(a)(b - a)/(f(b) - f(a))$$

which is the required approximation. Apart from that, the method proceeds as in the bisection method, always choosing an interval that includes the root. A simple recursive representation of this algorithm is:

```
void   regfalsi(func f, double a, double b, double &x)
{      double fa = f(a), fb = f(b);
       x = a - fa*(b - a)/(fb - fa);
       f(x) * fa > 0 ? a = x : b = x;
       if (fabs(a - b) > eps) regfalsi(f, a, b, x);
}
```

This simple routine also is inadequate. In addition to the checks which had to be included in the routine bisect, there is the possibility that the slope of the line,

$$(f(b) - f(a))/(b - a)$$

might be small or even zero. If the slope becomes too small the method fails, or the rate of convergence is seriously affected. We must therefore exclude such conditions. More importantly, however, the magnitude of the interval $|a - b|$ often does not tend to zero, so the routine does not terminate appropriately. This is because in the neighbourhood of a root most functions are strictly convex or concave, so that one end of the interval approaches the root while the other remains fixed. If we run this routine with the function F and step by step output, we obtain:

Call	a	b	x
1	0	1	0.6127
2	0	0.6127	0.572181
3	0	0.572181	0.567703
4	0	0.567703	0.567206
5	0	0.567206	0.567150
6	0	0.567150	0.567144
7	0	0.567144	0.567143
8	0	0.567143	0.567143

. . . .

The convergence is much faster than bisect, although still only linear. (It might be much worse with some other function.) But in any case the routine as written does not terminate because the interval does not tend to zero, the lower end of the interval remaining fixed. This is because the function F is strictly concave in the immediate vicinity of the root. The same behaviour would result if it were strictly convex: only if $F''(\alpha) \approx 0$, which is the case, for example, with the function $\sin x$ in the vicinity of $x = \pi$, do both ends change their values.

4.3 Secant method

In the secant method we overcome this difficulty, but introduce another one. We first present the simple recursive version. We use the same approximation to the root given the ends of the interval, namely,

$$x = a - f(a) \times (b-a)/(f(b) - f(a));$$

but we do not insist that at each step this estimate of the root shall remain within the interval. This may mean that we have to be careful that the starting interval is not too wide lest the successive approximations chase off to infinity. Instead, we enquire whether the increment in x, namely,

$$-f(a) \times (b-a)/(f(b) - f(a)),$$

exceeds the tolerance ε. If it does, we repeat the step; otherwise we have the result to the required precision. In repeating the step we have to choose afresh the ends of the interval. This may be done in more than one way, but we shall choose to replace a by b and b by x, as an approximation (as we shall see) to Newton's method:

```
void   secant(func f, double a, double b, double &x)
{       double fa = f(a), fb = f(b), dx;
        x = a + (dx = -fa*(b - a)/(fb - fa));
        a = b; b = x;
        if (fabs(dx) > eps) secant(f, a, b, x);
}
```

The routine now terminates properly. For the function F defined before, again setting $\varepsilon = 5.10^{-7}$, and with starting values $a = 0$, $b = 1$, we obtain:

Call	a	b	x
1	0	1	0.6127
2	1	0.6127	0.563838
3	0.6127	0.563838	0.567170
4	0.563838	0.567170	0.567143
5	0.567170	0.567143	0.567143
6	0.567143	0.567143	0.567143

The following is a non-recursive version of `secant`. It takes avoiding action if the slope
(`fb` - `fa`)/(`b` - `a`) should become less than a small constant number `del`:

```
state secant(func f, double a, double b, double &x)
{       state s = ITERATING;
        int iter = 0;
        double fa = f(a), fb = f(b), dx, fx;
        if (fa * fb > 0) s = SAMESIGN;
        if (fabs(fb) < fabs(fa))        // smaller function value
        {       swap(a, b);
                swap(fa, fb);
        }
        x = a; fx = fa;
        while (s == ITERATING)
        {       double t = (fb - fx)/(b - x);
                if (fabs(t) < del) s = NEARZERO;        // f too flat
                else
                {       x += (dx = -(b-x)*fx/t);
                        fx = f(x);
                        if (fabs(dx) < eps)     s = SUCCESS;
                        if (++iter == maxiter) s = WONTSTOP;
                }
        }
        return s;
}
```

The function `swap` was introduced in 2.5.4.

4.4 Convergence

The bisection method has been constructed to be linear. Nevertheless, let us establish
it formally. We begin knowing that there is a root $\alpha \in (a, b)$. At each step we have an
approximation to the root given by $x_n = (a_n + b_n)/2$, where a_n and b_n have been chosen so
that $\alpha \in (a_n, b_n)$. Therefore, $|\alpha - x_n| \leq \frac{1}{2}|b_n - a_n| = (1/2)^n|b - a|$. Defining the error by
$E_n = (\alpha - x_n)$, then $|E_n| \leq \frac{1}{2}|b_n - a_n| = (1/2)^n|b - a|$. The sequence $\{x_n\}$ converges linearly
to α with a rate of 0.5.

Now let us consider the rate of convergence of the regula falsi and secant methods. We
follow the treatment presented in the book by K. E. Atkinson. Given an interval (a, b), they
both define an approximation to the root given by

$$c = a - f(a)(b - a)/(f(b) - f(a))\}.$$

If the actual root is again denoted by α, then

$$\alpha - c = \alpha - a + f(a)(b - a)/(f(b) - f(a)).$$

With some rather tedious algebraic manipulation, this may be written in terms of the divided differences introduced in 3.7. We obtain:

$$\alpha - c = -(\alpha - a)(\alpha - b)f[a, b, \alpha]/f[a, b]$$

$$= -(\alpha - a)(\alpha - b)f''(\xi_2)/2f'(\xi_1).$$

where $\xi_1 \in (a, b)$, and $\xi_2 \in I(a, b, \alpha)$ (*cf.* 3.7).

In the regula falsi method one of the endpoints, a, say, remains fixed, the other being the previous approximation to the root. That being so, the error formula can be written:

$$E_{n+1} \leq KE_n, \text{ where } K = |\alpha - a| . \text{Max}(f''(\xi_2))/2\text{Min}(f'(\xi_1)).$$

Here, K should be evaluated in the close vicinity of the root α provided that $f(x)$ does not vary rapidly around the root. Thus $K \approx |\alpha - a| |f''(\alpha)|/2|f'(\alpha)|$. The method is linear with rate less than K. For the function F defined above, $|f''(\alpha)|/2|f'(\alpha)| \approx 0.18$. If $K \approx 0$, because $f''(\alpha) \approx 0$, the convergence is said to be *superlinear*. This is the case with a root of the sine function, for example, where $f''(\alpha) = 0$.

For the secant method, on the other hand, each endpoint reflects the previous approximation alternately. The error formula becomes

$$E_{n+1} \leq KE_n E_{n-1}$$

which is clearly faster than linear. To take the analysis a little further, consider a small interval about the root, $I \equiv (\alpha - \mu, \alpha - \mu)$ where μ is a small number, which contains the first two approximations to the root, $x_0, x_1 \in I$. Or, if it does not, let us start applying the analysis at a step where it becomes true; if it is never true, the sequence of approximations fails to converge. Note that $|E_0| \equiv |\alpha - x_0| \leq \mu$, and $|E_1| \equiv |\alpha - x_1| \leq \mu$.

Multiply both sides of the error formula by K to obtain

$$K|E_{n+1}| \leq K|E_n| . K|E_{n-1}|,$$

and define $$\lambda \equiv \text{Max}(K|E_0|, K|E_1|).$$

We assume the starting values x_0, x_1 are chosen so that $\lambda < 1$. Then $K|E_2| \leq \lambda^2 < \lambda < 1$, and $|E_2| < \lambda/K$, so that $x_2 \in I$. Similarly, $x_n \in I$ for all $n \geq 0$. The interval I contains all approximations to the root. Furthermore,

$$K|E_3| \leq K|E_2| . K|E_1| \leq \lambda^2 . \lambda = \lambda^3,$$

and in general, if $$K|E_n| \leq \lambda^u, K|E_{n+1}| \leq \lambda^v,$$

then $$K|E_{n+2}| \leq \lambda^w, \text{ where } w = u + v.$$

Setting $u = p_n$, $v = p_{n+1}$, $w = p_{n+2}$, we have

$$p_{n+2} = p_{n+1} + p_n, \qquad p_0 = p_1 = 1,$$

which is a Fibonacci sequence. The solution may be written

$$p_n = (r^{n+1} - s^{n+1})/\sqrt{5},$$

where $r = (1 + \sqrt{5})/2 \approx 1.618$ and $s = (1 - \sqrt{5})/2 \approx -0.618$ are the solutions of $x^2 - x - 1 = 0$. Thus, for large n,

$$p_n \approx (1.618)^{n+1}/\sqrt{5}.$$

In fact, n does not need to be very large: if $n = 8$, then $p_8 \approx 8.025$. We therefore have a bound on the error, $|E_n| \le \lambda^m/K$, where $m = p_n$, and $n \ge 0$.

It may be shown that

$$|E_{n+1}|/|E_n|^r \to |f''(\alpha)/2f'(\alpha)|^{-s}, \quad \text{as } n \to \infty.$$

Thus,

$$|E_{n+1}| \propto |E_n|^{1.618},$$

so that the secant sequence converges much faster than linearly.

4.5 Fixed point method

In the fixed point method we consider how to compute the solution of an equation of the form $x = g(x)$. Such a solution is called a *fixed point* of the function g. In passing, we note that, if $g(x) \equiv x - cf(x)$, where c is a constant, and if α is a fixed point of g so that $g(\alpha) = \alpha$, then $f(\alpha) = 0$, and α is a root of f. There are, however, many other ways of relating g and f, as we shall see. We have first to study the circumstances in which a function can have a fixed point. This will serve as a model for the study of iterative methods in general, because if we define the sequence $\{x_n\}$ by $x_{n+1} = g(x_n)$, and if $x_n \to \alpha$ as $n \to \infty$, then α is a fixed point of g.

Fixed point theorem

Let $g(x)$ be continuous and differentiable in $[a, b]$, where $a < b$, and suppose $g(x) \in [a, b]$ for all $x \in [a, b]$. We write $g([a, b]) \in [a, b]$. Then g has a fixed point $\alpha \in [a, b]$. If $|g'(x)| \le K < 1$, for all $x \in (a, b)$, where K is a positive constant, then α is unique. Further, if the sequence $\{x_n\}$ is defined by $x_n = g(x_{n-1})$, $n \ge 1$, where $x_0 \in (a, b)$, then $x_n \to \alpha = g(\alpha)$.

We now prove this theorem. If $g(a) = a$ or $g(b) = b$ there is a fixed point at a or b, respectively. Otherwise, $g(a) > a$ and $g(b) < b$. Let $h(x) = g(x) - x$. Then h is continuous on $[a, b]$, and $h(a) = g(a) - a > 0$, $h(b) = g(b) - b < 0$; so by the intermediate value theorem there must be a number α such that $h(\alpha) = 0$. Therefore $\alpha = g(\alpha)$ and α is a fixed point of g. To show that it is unique, suppose on the contrary that α and β are both fixed points of g in the interval $[a, b]$, and that $\alpha \ne \beta$. Then by the mean value theorem there exists a number $\xi \in [a, b]$ such that:

$$|\alpha - \beta| = |g(\alpha) - g(\beta)| = |g'(\xi)|.|\alpha - \beta| \le K|\alpha - \beta| < |\alpha - \beta|,$$

which is a contradiction. Therefore $\alpha = \beta$, and the fixed point is unique.

Since $x_0 \in [a, b]$, $x_1 = g(x_0) \in [a, b]$, and by induction $x_n = g(x_{n-1}) \in [a, b]$, $n \geq 1$.

Therefore, $\alpha - x_n = g(\alpha) - g(x_{n-1}) = g'(\xi)(\alpha - x_{n-1})$, $\alpha \leq \xi \leq x_{n-1}$,

and since $|\alpha - x_n| \leq K|\alpha - x_{n-1}| \leq K^n|\alpha - x_0|$,

the sequence converges to α because $K < 1$.

It may further be shown that $|\alpha - x_n| \leq (K^n/(1 - K)).|x_1 - x_0|$. The rate of convergence is determined by the parameter K; it will be very slow if K is close to 1.

Now let us consider how to transform the problem of finding a root of $f(x) = 0$ into a problem of finding a fixed point of some function $g(x)$. For purposes of illustration we shall continue to take for f the function $F(x) = x - e^{-x}$ which we have already studied, and which has a root at $x = \alpha \approx 0.567143$.

The simplest scheme is to set $g(x) \equiv x - cF(x)$, where c is an arbitrary constant. Consider the interval $[0, 1]$, and find the conditions under which $g([a, b]) \in [a, b]$. We have $g(0) = c$, so we must require $0 \leq c \leq 1$. Also $g(1) = 1 - c(1 - 1/e)$, so we require $0 \leq c(1 - 1/e) \leq 1$, which is satisfied if $c \in [0, 1]$. Between these limits g might have a maximum or minimum, which might fall outside the interval $[0, 1]$. We have $g'(x) = (1 - c) - ce^{-x}$, so there is a single extremum at $x_e = \log(c/(1 - c))$, which falls in the interval $[0, 1]$ if $c > \frac{1}{2}$. Moreover, $g''(x_e) = c\exp(-x_e) = 1 - c \geq 0$, so the extremum is a minimum. But $g(x_e) = x_e(1 - c) + c\exp(-x_e) \geq 0$ if $c \in [0, 1]$. Hence, $g([0, 1]) \in [0, 1]$. The conditions of the fixed point theorem are therefore met by $g(x)$ if $c \in [0, 1]$, and it follows that the sequence $\{x_n\}$, where $x_{n+1} = g(x_n)$, converges to $\alpha = g(\alpha)$.

We now examine the rate of convergence. We have:

$$K \equiv |g'(\alpha)| = |1 - cF'(\alpha)| = |1 - c(1 + \alpha)|.$$

The condition $K < 1$ is satisfied if $c \in (0, 1)$, but the fastest convergence will be obtained when $K \approx 0$, that is when $c \approx 1/(1 + \alpha) \approx 0.64$ in this case.

The following routine gives effect to the fixed point algorithm:

```
typedef double (*func)(double, double);

const double eps = 5.0E-7;
const int maxiter = 100;

void  fixedpoint(func g, double x, double c)
{     static int iter = 0;
      cout << iter << "\t" << x << "\n";
      double y = g(x, c);
      if (fabs(y - x) > eps && ++iter < maxiter)
            fixedpoint(g, y, c);

}
```

Taking $x = 0.5$ as the starting approximation in each case, the following iterates are obtained for $c = 0.4, 0.6, 0.8$:

iter	$c = 0.4$	$c = 0.6$	$c = 0.8$
0	0.5	0.5	0.5
1	0.542612	0.563918	0.585225
2	0.558059	0.566952	0.562630
3	0.563763	0.567132	0.568293
4	0.565883	0.567143	0.566852
5	0.566673		0.567217
6	0.566968		0.567125
7	0.567078		0.567148
8	0.567119		0.567142
9	0.567134		0.567144
10	0.567140		0.567143
11	0.567142		
12	0.567143		

Note how much more rapid convergence is with $c = 0.6$, for which K is very small. If we set $c = 0.64$, corresponding to $K \approx 0$, only three iterations are required.

To see how this comes about, suppose that $g'(x) = 0$ at the root $x = \alpha$, and expand $g(x)$ about α as a Taylor series,

$$g(x) = g(\alpha) + g'(\alpha) \cdot (x - \alpha) + (1/2!)g''(\xi) \cdot (x - \alpha)^2,$$

where $g(\alpha) = \alpha$, $g'(\alpha) = 0$, and ξ is a number close to α. Then,

$$x_{n+1} = g(x_n) = \alpha + \tfrac{1}{2}g''(\xi_n) \cdot (x_n - \alpha)^2,$$

where ξ_n is between x_n and α. Since $|g'(x)| < 1$ in the neighbourhood of $x = \alpha$, the sequence $\{x_n\}$ converges to α by the fixed point theorem, as does the sequence $\{\xi_n\}$ also.

Therefore, $|x_{n+1} - \alpha| / |x_n - \alpha| = |E_{n+1}| / |E_n|^2 \rightarrow \tfrac{1}{2}|g''(\alpha)|, \quad n \rightarrow \infty,$

and the sequence converges *quadratically*. In principle, even higher degrees of convergence are possible, corresponding to the simultaneous vanishing of higher derivatives, but such fortune does not often smile in practice.

4.6 Choice of fixed point function

The choice of function $g(x)$ corresponding to the equation $f(x) = 0$ is a matter of judgement. Suppose we have:

$$f(x) = x^4 + 2x^2 - x - 3.$$

The equation $f(x) = 0$ has a root at $x = \alpha \approx 1.124123$. In contrast to functions g of the form $g(x) = x - cf(x)$, we may at first sight consider

$$g_1(x) = (3 + x - 2x^2)^{1/4},$$
$$g_2(x) = [(x + 3 - x^4)/2]^{1/2},$$
$$g_3(x) = [(x + 3)/(x^2 + 2)]^{1/2},$$
$$g_4(x) = (3x^4 + 2x^2 + 3)/(4x^3 + 4x - 1),$$

because if $g_i(x)$ has a fixed point at $x = \alpha$, then $f(\alpha) = 0$. However, not all these functions give convergent sequences, as may be seen by examining whether they satisfy the conditions of the fixed point theorem.

4.7 Newton's method

A further fixed point method for finding the root α of $f(x) = 0$ may be derived by setting $g(x) = x - h(x)f(x)$, where it is supposed that $h(\alpha) \neq 0$. Then, if $g(\alpha) = \alpha$, $f(\alpha) = 0$. Let us choose $h(x)$ so that the iteration $x_{n+1} = g(x_n)$ is quadratically convergent. It is sufficient that $g'(\alpha) = 0$. But

$$g'(x) = 1 - h'(x)f(x) - h(x)f'(x),$$

so that $g'(\alpha) = 1 - h(\alpha)f'(\alpha)$. If $g'(\alpha) = 0$, then we must choose $h(\alpha) = 1/f'(\alpha)$. However, it must not be assumed that at this stage α, and therefore $f'(\alpha)$, are known. It is sufficient that we set $h(x) = 1/f'(x)$. The resulting iteration,

$$x_{n+1} = x_n - f(x_n)/f'(x_n),$$

will be quadratically convergent. This is the basis of Newton's method, without doubt the most celebrated quadratically convergent process. It does, however, require the explicit form $f'(x)$ of the derivative of $f(x)$. We offer an iterative version:

```
state newton(func f, func f1, double &x)
//     f1(x) = f'(x)
{      int iter = 0;
       state s = ITERATING;
       while (s == ITERATING)
       {      double fx = f(x), f1x = f1(x), dx;
              if (++iter == maxiter)
                     s = WONTSTOP;
```

```
            else if (fabs(f1x) < toosmall)
                    s = NEARZERO;
            else
            {      x += (dx = -fx/f1x);
                   if (fabs(dx) < tolerance)
                         s = SUCCESS;
            }
      }
      return s;
}
```

where we have again taken avoiding action in case $f'(x) \approx 0$, or in case the convergence is too slow. Note that if we set $f(x) = x^2 - A$, we regain the original algorithm of chapter 2 for the square root of A, except that the condition for success is now more natural and efficient. It is a condition on the value of the increment `dx` rather than the function `f(x)`, which is much more satisfactory, especially if the derivative `f1(x)` is small near the root.

Although Newton's method is quadratic, convergence may not be rapid, at least initially, unless the initial value of x is chosen sufficiently close to the root being sought. For this reason, it may be desirable in practice to precede the application of Newton's method with a few iterations of the linear bisection method, because that is less sensitive to initial values and may be guaranteed to provide an adequate starting approximation for Newton's method. Of course, if $f(x)$ has several roots the starting approximation, even for the bisection method, must be accurate enough at least to distinguish between them.

4.8 Multiple roots

If $f(x) = (x - \alpha)^p.h(x)$, where $h(\alpha) \neq 0$, we say that $f(x)$ has a root $x = \alpha$ of *multiplicity p*. If $p = 1$, a is a *simple root*; if $p = 2$ it is a *double root*. (There is no need to confine p to integers.) If we apply Newton's method to a multiple root, the rate of convergence reverts to linear. Moreover, the value of $f'(x)$ near the root α is zero or small.

To see why this should be so, consider Newton's method applied to the multiple root as a fixed point method, with

$$g(x) = x - f(x)/f'(x) = x - (x - \alpha)h(x)/[ph(x) + (x - \alpha)h'(x)].$$

Then,

$$g'(x) = 1 - [1 + (x - \alpha)d/dx]\{h(x)/[ph(x) + (x - \alpha)h'(x)]\},$$

so that $g'(\alpha) = 1 - 1/p$. If $p = 1$, $g'(\alpha) = 0$, and the convergence is quadratic as we have shown above. But if $p > 1$, $g'(\alpha) \neq 0$, and convergence is linear. However, if we change the definition of the fixed point function to $g(x) = x - pf(x)/f'(x)$, then $g'(\alpha) = 0$, and the process converges quadratically again.

To illustrate this behaviour, consider the function $F(x) = 1 - x - e^{-x} \approx -\frac{1}{2}x^2$ near $x = 0$, which is therefore a double root. We have $F'(x) = -1 + e^{-x}$, so that $F'(0) = 0$. Choosing a starting value $x = 0.5$, we obtain the following results with $p = 1$ and $p = 2$, respectively:

iter	$p = 1$	$p = 2$
0	0.50.5	
1	0.229253	−0.041494
2	0.110251	−0.000287
3	0.054113	−1.4E−8
5	0.013346	
10	0.000415	
15	1.3E−5	
20	4.1E−7	

4.9 Aitken's extrapolation

Suppose that $\{x_n\}$ is a linearly convergent sequence, and that $x_n \to \alpha$, so that

$$(\alpha - x_{n+1})/(\alpha - x_n) \approx \lambda, \text{ where } |\lambda| < 1.$$

Then

$$(\alpha - x_{n+2})/(\alpha - x_{n+1}) \approx (\alpha - x_{n+1})/(\alpha - x_n),$$

and

$$(\alpha - x_{n+1})^2 \approx (\alpha - x_n)(\alpha - x_{n+2});$$

so that

$$\alpha_{n+2} \equiv x_{n+2} - (x_{n+2} - x_{n+1})^2/[(x_n - x_{n+1}) - (x_{n+1} - x_{n+2})]$$

is a better approximation to α than x_{n+1}. This is the Aitken extrapolation process, which may be realized by the routine:

```
double aitken(func g, int iter, double t[])
{      switch(iter % 3) {
       case 0: t[0] -= (t[1]-t[0])*(t[1]-t[0])/(t[0]-2*t[1]+t[2]);
               return t[0];
       case 1: t[1] = g(t[0]);
               return t[1];
```

```
        case 3: t[2] = g(t[1]);
                return t[2];
        }
}
```

For illustration we shall apply this routine to the fixed point evaluation of the root of $F(x) = x - e^{-x}$, using $g(x) = x - cF(x)$ with $c = 0.4$, comparing the results to those obtained by the normal linear fixed point process:

iter	linear	Aitken
0	0.5	0.5
1	0.542612	0.542612
2	0.558059	0.558059
3	0.563763	0.566841
4	0.565883	0.567031
5	0.566673	0.567101
6	0.566968	0.567143
7	0.567078	
8	0.567119	
9	0.567134	
10	0.567140	
11	0.567142	
12	0.567143	

The marked improvements in iterations 3 and 6 may be noted. The Aitken process is particularly useful when the derivative $f'(x)$ is not known and Newton's method therefore cannot be used.

4.10 Roots of polynomials

Polynomials are used for many purposes in mathematics and in computation. It is therefore important to be able to handle them efficiently. We shall consider how to find a real root of a polynomial with real coefficients.

A polynomial of degree n may be written

$$P_n(x) = a_n x^n + a_{n-1} x^{n-1} + \ldots + a_1 x + a_0,$$

where it is assumed that $a_n \neq 0$. Its derivative is

$$P'(x) = n a_n x^{n-1} + (n-1)a_{n-1} x^{n-2} + \ldots + a_1.$$

Since we know the derivative explicitly it is natural to apply Newton's algorithm to the problem, and this requires us to calculate values of the polynomial and its derivative at each iteration. Both involve the calculation of powers of x repeatedly. We first need an efficient way of doing this.

4.10.1 Horner's algorithm

The polynomial may alternatively be expressed:

$$P(x) = (\ldots (a_n x + a_{n-1})x + a_{n-2})x + \ldots + a_1)x + a_0,$$

and similarly with $P'(x)$.

We may now write a procedure which evaluates the polynomial and its derivative efficiently for any given x. The coefficients a_i are stored in an array of size $n + 1$. This is Horner's algorithm:

```
void   horner(double a[], int n, double x,
                          double &p, double &q)
//     on output, p = P(x), q = P'(x)
{      if (n == 0)
       { p = a[0]; q = 0; return; }
       p = q = a[n];
       for (int i = n-1; i > 0; --i)
       {      p = x*p + a[i];
              q = x*q + p;
       }
       p = x*p + a[0];
}
```

In terms of Horner's algorithm, the following function computes a root **x**, given a starting value, by Newton's method:

```
state polyroot(double a[], int n, double &x)
{      int iter = 0;
       state s = ITERATING;
       while (s == ITERATING)
       {      double p, q, dx;
              horner(a, n, x, p, q);
              if (++iter == maxiter)
                     s = WONTSTOP;
              else if (fabs(q) < toosmall)
                     s = NEARZERO;
```

```
                    else
                    {       x += (dx = -p/q);
                            if (fabs(dx) < tolerance)
                                    s = SUCCESS;
                    }
            }
            return s;
}
```

Of course, rough values of the roots must first be found before the function `polyroot` may be used to refine them. These methods may be extended to the case of complex roots, but complex arithmetic must then be used. We shall see in 5.1.5 how to compute the complex roots of a quadratic equation with real coefficients.

4.10.2 Stability

The computation of roots of a polynomial are notoriously prone to instability, as we remarked in 3.5, especially if the coefficients of the polynomial are large. A very small change in the coefficients may perturb the computed roots greatly, or even make them complex.

Let us see how this may happen. Again, we follow the treatment of K. E. Atkinson. Let $P(x)$ be a function (not necessarily a polynomial) with a root at $x = \alpha$, so that $P(\alpha) = 0$. Let $Q(x)$ be some other function, and suppose that the perturbed function

$$P(x; \varepsilon) \equiv P(x) + \varepsilon Q(x)$$

has a root at $x = \alpha(\varepsilon)$. We shall assume that α is a continuous function of ε; this may be shown to be the case for polynomials, at least in the complex plane. Then we may by Taylor's theorem (which also applies in the complex plane) expand $\alpha(\varepsilon)$ in powers of ε:

$$\alpha(\varepsilon) = \alpha + \alpha'(0)\varepsilon + O(\varepsilon^2).$$

Now $P(\alpha(\varepsilon); \varepsilon) \equiv 0$, because $\alpha(\varepsilon)$ is a root. Therefore,

$$P(\alpha(\varepsilon)) + \varepsilon Q(\alpha(\varepsilon)) \equiv 0, \text{ all } \varepsilon.$$

Differentiation of this relation with respect to ε gives

$$P'(\alpha(\varepsilon))\alpha'(\varepsilon) + Q(\alpha(\varepsilon)) + \varepsilon Q'(\alpha(\varepsilon))\alpha'(\varepsilon) = 0,$$

so that
$$\alpha'(\varepsilon) = \alpha - Q(\alpha(\varepsilon))/[P'(\alpha(\varepsilon)) + \varepsilon Q'(\alpha(\varepsilon))].$$

Therefore,
$$\alpha(\varepsilon) = \alpha - \varepsilon Q(\alpha)/P'(\alpha) + O(\varepsilon^2).$$

If we take $P(x) = (x - 1)(x - 2) \ldots (x - n)$, and $Q(x) = x^m$, with $m < n$, so that we perturb just the coefficient of x^m in the expansion of $P(x)$, then

$$P'(i) = \prod_{j \neq i} (i - j) \text{ and } Q(i) = i^m,$$

so that the perturbed roots are given by $\alpha_i(\varepsilon) = i + \varepsilon \mu_i + O(\varepsilon^2)$,

where $\mu_i = (-1)^{i-1} i^m / (i - 1)!(n - i)!$.

It is clear that the largest roots may be quite seriously perturbed. Even with small values of ε, the perturbations may be so large that the linear approximation no longer suffices. For example, if we take $n = 7$, $m = 6$ and $\varepsilon = 0.002$, we obtain, using the above routines:

i	$\varepsilon \mu_i$	$\alpha_i(\varepsilon)$
1	0.000003	1.000003
2	−0.0001	1.998938
3	0.003	3.033125
4	−0.23	3.819569
5	0.65	complex
6	−0.78	complex
7	0.33	7.233013

The polynomial is clearly ill-conditioned.

4.11 Maxima and minima of functions

If the first derivative of a function is known explicitly the maxima and minima may be found as the zeros of the derivative. But it is possible to find a maximum or minimum without this knowledge. We present here a method which is a generalization of the bisection method for roots.

In the bisection method we sought to bracket a root of the function $f(x)$ within an interval (a, b), where $f(a)$ and $f(b)$ are of opposite sign. To be definite let us seek a minimum of f: we can always find a maximum as the minimum of $-f$. To bracket a minimum, we need three numbers instead of two. For if $f(x)$ has a minimum somewhere in the interval (a, c), then there must exist values of b such that $f(b) < f(a)$ and $f(b) < f(c)$. We seek to shrink the width of the interval until it confines the minimum within sufficiently narrow limits.

We shall seek the minimum of the function $f(x) = \exp(-x) \sin(2\pi x)$ which lies in the interval $(1, 2)$. It is at $x = (1/2\pi)\tan^{-1}(2\pi) + (3/2) \approx 1.72488$. It is tempting to propose the obvious generalization of the recursive function `bisect0` defined in 4.1.1. However, we require it to return the value of x which minimizes `f(x)`:

```
double zerosect(func f, double a, double b, double c)
{       static double x   = 0.0;
        if (fabs(c-a) < tol*fabs(b)) return b;
        x = (b + c)/2;
        if (f(x) < f(b)) return zerosect(f, b, x, c);
        else return zerosect(f, x, b, a);
}
```

where the usual remark should be made about the use of a static variable if the function is called more than once in the same program (*cf* 4.1.2). If the width of the current interval (a, c) is less than some small multiple of b (which is assumed not to be zero), determined by the constant tol, which may be taken to be, say, 10^{-7}, then we have found the minimum to the precision required and it is b, the value of which is returned. Otherwise, the interval must be divided, a decision made about which sub-interval contains the minimum, and the test repeated. We assume that we already have starting values of the parameters which bracket a single minimum. Divide the higher interval (b, c) in halves by setting $x = (b + c)/2$. We already know that $f(x) < f(c)$, but we ask whether $f(x) < f(b)$ also. If so, the minimum must be bracketed by the numbers b, x, c. If not, it is bracketed by x, b, a instead. Taking $a = 1, b = 1.5, c = 2$, and tol $= 10^{-7}$, we obtain the correct result with 90 evaluations of the function f. A large number of evaluations is to be expected because $f(x)$ varies slowly in the neighbourhood of its minimum.

However, the function zerosect requires twice as many evaluations of $f(x)$ as are really necessary: two per recursive call to zerosect. If we carry the current value of $f(b)$ as a parameter, we may divide the number of calls by two:

```
double halfsect(func f, double a, double b, double c, double fb)
{       static double x   = 0.0;
        static double fx = 0.0;
        if (fabs(c-a) < tol*fabs(b)) return b;
        x = (b + c)/2;
        fx = f(x);
        if (fx < fb) return halfsect(f, b, x, c, fx);
        else return halfsect(f, x, b, a, fb);
}
```

The revised function halfsect requires only one function evaluation per recursive call, plus the initial evaluation of $f(b)$. It therefore obtains the same result with only 46 function evaluations.

This routine is still not optimum, however. The reason is that by splitting the higher sub-interval (b, c) into two equal parts separated by $x = (b + c)/2$ the original interval has been reduced to one-half (assuming that b was chosen in the same fashion) if the minimum lies in the higher sub-interval, or three-quarters if it is in the complementary sub-interval (a, x). This lack of symmetry is inefficient. We now seek to redress it.

We wish to choose x so that (a, b) and (x, c) are of equal length. Let:

$$u = (x - b)/(c - a), \quad v = (b - a)/(c - a), \quad 1 - v = (c - b)/(c - a)$$

Then clearly we want to set $u = 1 - 2v$, which is positive if $v < 0.5$ and the higher sub-interval is the longer. We then have $|b - a| = |x - c|$. To determine v we note that we are seeking a rule which applies on every recursive call. If it applies to the current one, it applied to the previous one also. Therefore $v = u/(1 - v)$, or $v^2 - 3v + 1 = 0$. The solution is $v = (3 - \sqrt{5})/2 \approx 0.38197 \ldots = 1 - \phi$, where $\phi \approx 0.61803 \ldots$ is the "golden mean" introduced in 4.4. Setting sect $= 1 - \phi$, the algorithm that results is known as the *golden section*:

```
double goldsect(func f, double a, double b, double c, double fb)
{       static double x  = 0.0;
        static double fx = 0.0;
        if (fabs(c-a) < tol*fabs(b)) return b;
        if (fabs(c-b) > fabs(b-a))
        {       x = b + sect*(c-b);
                fx = f(x);
                if (fx < fb) return goldsect(f, b, x, c, fx);
                else return goldsect(f, x, b, a, fb);
        }       else
        {       x = b - sect*(b-a);
                fx = f(x);
                if (f(x) < f(b)) return goldsect(f, b, x, a, fx);
                else return goldsect(f, x, b, c, fb);
        }
}
```

Notice that we have first to determine which sub-interval is the larger. The function goldsect obtains the minimum with only 36 function evaluations.

Our main interest here has been to generalize the routine bisect0 in an interesting fashion. Other methods exist which take explicit account of the quadratic nature of a function in the neighbourhood of a minimum or maximum (assuming its second derivative there is not too small). The method which follows is a simple version of that proposed by Brent (see Press *et al.*), and it, too, resembles the bisection method in the nature of its recursion.

We assume again that we have already found three points $a < b < c$ such that $f(b) < f(a)$ and $f(b) < f(c)$, so that there exists a minimum in the interval (a, c). We fit a parabola to these three points, and seek its minimum, x, say. If $x = b$, that is the end of the matter, and we return the value of x. If not, we consider the sub-interval $a < b < x$, or $b < x < c$, as the case may be, depending on the relative value of $f(x)$, and continue recursively. This time we shall need to carry the values of both $f(a)$ and $f(c)$ as parameters:

```
double parabola(func f, double a, double b, double c,
                double fa, double fc)
//      fa = f(a), fc = f(c)
{       double fb = f(b);
        double p = (b-a)*(fb-fc);// fit parabola
        double q = (b-c)*(fb-fa);
```

```
        double r = (b-a)*p - (b-c)*q;
        double s =  p-q;
        if (fabs(s) < toosmall) s = sign(s)*toosmall;
        r /= (2.0*s);
        double x = b - r;              // minimum of parabola
        if (fabs(r) < tol) return x;
        if (r > 0) return parabola(f, a, x, b, fa, fb);
        return parabola(f, b, x, c, fb, fc);
}
```

Note that this recursive function finds a maximum just as well as a minimum. We then encapsulate it so as to initialize parameters and to perform obvious checks on the values provided:

```
double brent(func f, double a, double b, double c)
{       if (a > c) swap(a, c);         // ensure a < c
        if (b < a || b > c) error("interval not inclusive");
        double fa = f(a), fc = f(c);   // initialize parabola
        return parabola(f, a, b, c, fa, fc);
}
```

Working with the same function used above, `brent` requires only 9 function calls, a distinct improvement over the methods derived from `bisect0`. The success of this routine depends, however—rather like the Newton method for roots—on there being a distinct maximum or minimum sufficiently close to the starting point.

4.12 Some standard things

In this and earlier chapters we have begun to introduce a number of conventions and functions which we have found generally useful. More will occur in later chapters. We have found it convenient to collect declarations of functions and macros and definitions of constants, of a standard nature, in a header file `standard.h`, with the accompanying source code compiled into a file `standard.obj` (we use the DOS convention) which may be linked into the compilation of any source file that uses it.

The header file `standard.h` is presented in Appendix A. It uses TurboC++ conventions, because that is the particular implementation we have used in the preparation of this book. If any other implementation is used, however, it should be relatively easy to translate the routines declared in Appendix A into that implementation. It begins with some header files required for almost any numerical computation; they include the input and output conventions which in TurboC++ are given in `iostream.h`, and the standard mathematical functions such as `fabs`, the exponential and logarithmic functions, and the trigonometric functions, which in TurboC++ are given in `math.h`. There follow definitions of the constant characters `tab` and `bell`, and the constant character string `sep` (= `", "`). It is, of course, only a matter of taste, but in a simple tabulation it seems more pleasing, and is certainly easier to read, if we write

```
cout << tab << x << sep << y << endl;
```

rather than

```
cout << "\t" << x << ", " << y << "\n";
```

Here, `endl` is defined in TurboC++, where it is equivalent to the newline character `'\n'` for most purposes.

There follow two useful macros. The first is `newline`, which conveys what is required even more clearly than `cout << endl`. The second is `clear`, which, as we explained in 4.1.2, it is better to write in order to clear the screen than the non-standard `clrscr()` of TurboC++.

Then there are two functions useful especially for "high level" output in the function `main`. The first is `statement`, which simply outputs a string and is equivalent to `cout << str;`. The second is defined as the function template `output`; it allows one in particular to output in the form `cout << "value = " << x;` the value of an object `x` of any type for which the operator `<<` has been defined. There is a corresponding function template `input`, which allows one similarly to write `input ("enter value: ", x);` for an object of any type for which the operator `>>` has been defined. (The reader may care to include a `bell` statement in order to issue a "beep" when input is awaited).

There are next some commonly occurring mathematical constants to 15 decimals, followed by the definitions of the identifiers `byte`, `word` and `longword`, and the enumerations `bool`, introduced in 1.4.2, and `state`, introduced in 2.4.7 and used to some effect in this chapter.

There follow some standard functions, such as `error` and `quadroots`, both defined in 3.3, and the simple function templates `sign`, `max` and `min`, defined in 2.5.7, and another template, `swap`, which interchanges two variables of given type and was first introduced as an ordinary function in 2.5.4. There is a function `mod` which returns the value of $(x \bmod y)$ for floating point variables. Other functions, mostly concerned with output, are also included, but they will be defined when we first have need of them. One very useful function, however, is `getyes`. Its definition is:

```
bool  getyes(const char *str)
{      cputs(str);
       for (;;)
       {      int c = getch();
```

```
        if (c == 'Y' || c == 'y')
             return TRUE;
        if (c == 'N' || c == 'n')
             return FALSE;
    }
}
```

The TurboC++ function `cputs` is a low level function that prints the given string `str` to screen; it could readily be replaced by the expression `cout << str`. The function `getch` is a standard C library function that pauses until a character is entered from the keyboard; it does not echo the entered character to screen. The `for` loop runs until the character entered is `'Y'` or `'y'`, or `'N'` or `'n'`, returning TRUE or FALSE as appropriate.

The function `getyes` may be used, probably in the function `main`, in the manner illustrated by the following fragment:

```
........
if (getyes("do this? (Y/N): ")
      call_this_function;
else  call_that_function;
........
```

It enables the user to decide in person which branch to take in a program.

Having defined the header file `standard.h` in Appendix A, we are free to include it in any program in place of the possibly long list of individual `#include` and `#define` directives, constant definitions, and the declaration of favourite homemade functions, that we shall use widely. Of course, Appendix A has been written according to TurboC++ conventions. It is strongly recommended that any serious user write a version of Appendix A for personal use in the chosen implementation of C++. It will save a lot of tedious repetition, and therefore the possibility of error.

If the function `error`, `getyes`, `mod` and `quadroots` (cf 5.1.5) are to be used in programs their definitions, together with those of other standard functions desired by the user, must first be written to a file `standard.cpp`, say, which must be separately compiled (cf 5.5).

5 Classes

The standard types we have dealt with so far all have the property that the usual rules of combination are defined: the language allows us, for example, to add or divide integers and floating point numbers, and to convert between them automatically or at will. We saw, in 1.4.3, how to define *structures*, and how thereby to define new types; but, as we left them, these types were not fully satisfactory because we did not consider how to combine two or more objects defined in terms of them. In C++, a *class* is a user-defined type, a generalization of a structure, which includes appropriate rules for construction, initialization and destruction of objects, for combining them into expressions, and for input and output; and which, if it is properly defined, can be used like any standard type. Classes are sometimes referred to as *abstract types* because their data members may be hidden, but they are no less real than any other type! In this chapter, we shall illustrate the concept of class by constructing class representations of the complex numbers and of strings. We shall also consider arrays of class objects. In this way we shall introduce many of the features and problems of representing data structures by classes, which will be applied in later chapters.

5.1 Complex numbers

The simplest way to represent complex numbers is to declare a structure with two floating point members representing real and imaginary parts:

```
struct complex {
     double re;
     double im;
};
```

much as we declared the cartesian type `point` in 1.4.3, except that here we prefer to use `double` instead of `float`, and we do not choose to associate names with numbers. We may then declare complex numbers in terms of this structure:

```
complex z, z1, z2;
```

accessing the real and imaginary components through the dot operator. That is what we would do in C. Thus, if we want z to represent the sum of z1 and z2, we set

```
z.re = z1.re + z2.re;
z.im = z1.im + z2.im;
```

As for the square root of −1 itself, it may be defined as const complex i = {0, 1}; where we have made assignments to the real and imaginary parts by listing their values, as we are permitted to do (*cf* 1.5.4), rather than by using the dot operator.

It would, however, be much more natural if we could simply write

```
z = z1 + z2;
```

without explicit reference to real and imaginary parts; and similarly for all other expressions which occur in complex arithmetic. This means that the type declaration must itself include appropriate definitions of the arithmetic operators. In addition to the *data members*, representing real and imaginary parts, we shall need *function members* representing the usual arithmetic operators and any other functions in common use. We shall also require suitable rules for initialization and for assignment. The data members may be hidden, but it must nevertheless be possible to access them when required. All of this the C++ concept of class allows us to do.

5.1.1 Declaration

We declare the type complex as a class in the following way:

```
class complex {
private:
        double re;
        double im;
public:
        complex();
        complex(double, double=0);
        double real();
        double imag();
        complex operator+();
        complex operator+=(complex&);
        friend complex operator+(complex&, complex&);
        friend ostream& operator<<(ostream&, complex&);
        friend istream& operator>>(istream&, complex&);
        ........
};
```

Note that a class declaration *always* begins with the keyword `class`, and *always* ends with a semi-colon. (Recall the similar convention for structures (*cf* 1.4.3).) Furthermore, we should observe that this is a greatly simplified declaration of the class `complex`, omitting most of the arithmetic operators and all the mathematical functions, such as exponential and logarithm, and conversion to polar variables, commonly applied to complex numbers. The omissions are indicated by the row of dots.

Most implementations of C++ will provide a full definition of the class `complex` as a standard package, and there will be a corresponding header file called `complex.h` or something similar. The present simplified definition will suffice to illustrate the class concept at work, but it must be replaced by a full definition when complex numbers are required in an actual application. When we include the header file `complex.h` in later work, it is the full implementation that we refer to.

Note next the appearance of the important new keywords `private`, `public`, `friend` and `operator`. The data members are declared `private` in order that they shall be directly accessible only within the class definition. The keyword `private` is not provided in all implementations, but with care it may be taken as read, because class members are private by default unless declared otherwise. Thus, dropping the keyword `private` in the above declaration of `complex` would have no effect.

We shall see presently that private members are accessible also to a function, or a whole class, declared within the class declaration to be a `friend` of that class. Friend functions and friend classes are not members of the class: they merely have access to its private members. The members declared `public`, on the other hand, are accessible outside the class declaration by using the dot operator (or in the case of a pointer, the `->` operator), and provide a public interface to the type so declared. The keyword `operator` denotes a special type of function which enables us to define operations, arithmetic or otherwise, upon objects of the class.

At first sight, it might appear that the class representation is extremely inefficient, because each time an object is created the member functions seem to be created along with it. That is not so. Only the data members are created; the function members belong to the whole class, not to particular objects of the class. The compiler provides only one copy of the function members, and each object of that class accesses the function members by means of the dot or pointer operators. Note that the size of an object is the size of its data members only: `sizeof(complex)` is the number of bytes in two doubles. (Recall, however, the remark about "padding" in 1.4.3.)

5.1.2 Implementation

To take matters further, we must provide suitable *definitions* of the member functions which occur in the class declaration. Let us consider in turn what we may reasonably require of an interface. First, we must be able to *construct* an object of the type defined; that is, we must be able to declare and initialize it. We should expect to be able to do that by writing statements such as:

```
complex z;   double x, y;   z = complex(x,y);   complex z1 = z;
```

which are the analogues for complex numbers of the similar statements which can be made for the standard types, such as:

```
float f;    int i;    f = float(i);    float f1 = f;
```

Here, the variable i must be initialized before it can be used in the following statement, and so must the variables x and y above. What we wish to emphasize, however, is that in both cases the first statement declares an object of given type, the second declares (and initializes) auxiliary variables of another type, the third converts from one type to the other, and the fourth assigns to a new object the value of an old one of the same type.

For the type complex declared above, all these requirements are met with the aid of the first two public function members, which are declared complex() and complex (double, double), respectively. They are known as *constructors*. Note that *the name of a constructor must be identical with the name of the class*. When appropriately defined, the first constructor enables us to make the declaration complex z; while the second allows us to declare z and then to assign to it the complex number (x,y): namely, complex z(x,y);. The statement complex z1 = z;, on the other hand, is taken care of by default. Offerred no alternative definition of the assignment operator, the compiler simply performs a bit copy of z into z1. There are dangers associated with bit copying, however, as we shall see in the section on strings, and one tries to avoid it altogether.

Let us now *define* these constructors. They are functions with no return type, not even void, and they serve to create an object of type complex. The first constructor we shall initialize by convention so that both real and imaginary parts are zero:

```
inline complex::complex()
{      re = 0; im = 0; }
```

We have defined it to be an inline function because it is so simple: it merely reserves in memory an object consisting of two doubles, assigning to them both the value zero. Notice that we have used the scope operator :: to indicate that complex() is a member function of the class complex. Had we not done so, we should not have been able to access the private data members re and im; indeed, we should not even have been able to refer to the constructor outside the class declaration. The declaration complex z; then *defines* the complex number z as an object having zero real and imaginary parts.

(In passing, we note than the comparable declarations float f;, int i;, and so on, may or may not be initialized to zero. In some implementations they are, but other compilers may behave differently, and it is therefore best always to initialize objects of standard type explicitly.)

The second constructor requires us to assign our own values to the data members:

```
inline complex::complex(double x, double y)
{      re = x; im = y; }
```

The declaration `complex z(x,y);` then defines a complex number z having real part x and imaginary part y. This may alternatively be written `complex z = complex(x,y);` which must be interpreted in terms of a temporary variable: `complex t(x,y); complex z = t;`, and where, in addition, we have taken advantage of the default interpretation of the assignment operator. (In the absence of an explicit definition of assignment in a class, as we have already observed, the compiler simply performs a bit copy of one object into the other.)

Evidently, there is a certain redundancy. Instead of `complex z;` we could have used `complex z(0,0);`. Both would have defined z to have zero real and imaginary parts. Indeed, the first constructor may be disposed of altogether, while still maintaining the convenience of being able to call simply `complex z;`, by declaring a single constructor having default values. In the class declaration we could eliminate the first contructor `complex()`, replacing the second constructor by:

```
complex::complex(double=0, double=0);
```

Maintaining the same *definition* of the second constructor, the statement `complex z;` then calls `complex(0,0)` by default, which is exactly what the first constructor would have given. In a call to a function in the absence of the values expected to be passed to the function, the default values, if any, are automatically assigned. If there are no default values, it is an error. However, it is more usual to maintain the explicit default constructor. It then becomes possible, without ambiguity, to declare the second constructor as:

```
complex::complex(double, double=0);
```

whereupon the statement `complex(a);` converts the (real) constant a to the corresponding complex number `(a,0)` with zero imaginary part. This is the form in which we chose to declare the second constructor. Note that the constructors *overload* the class name `complex`; the same name is used for functions having the same purpose but different signatures.

The data members of the class `complex` are hidden: they may not be directly accessed outside the class definition. If a class is well designed there should be little need to access them. However, they can be accessed indirectly by the use of *access functions*. In the declaration of class `complex` we have included the member functions `real()` and `imag()`. The real and imaginary parts of the complex number z are then returned by the function calls `z.real()` and `z.imag()`. The implementation of these functions is extremely simple:

```
inline double complex::real() { return re; }
inline double complex::imag() { return im; }
```

Inline definitions of member functions are often provided within the class declaration itself, without the use of the keyword `inline`. Thus, we could equivalently have declared the class `complex` in the following way:

```
class complex {
private:
      double re, im;
public:
      complex()                          { re=0; im=0; }
      complex(double x, double y = 0)    { re=x; im=y; }
      double real()                      { return re;  }
      double imag()                      { return im;  }
      ........
};
```

Only very simple functions can be written in this way. Apart from the size of code generated by the repeated expansion of large `inline` functions, the declaration of the class becomes difficult to read and therefore liable to error, at least of interpretation. Larger member functions should always be defined outside the class declaration, to be declared `inline` only if they are still not too large: the balance is a matter of judgement in the unremitting tension between code length and function call overheads. As we remarked in 2.5.1, however, the keyword `inline` is only a recommendation to the compiler.

In passing, we note that there is still another way to make an inline definition of a constructor; it is to *list* the parameters:

```
class complex {
privete:
      double re, im;
public:
      complex(double x, double y)  :  re(x), im(y)  {}
      .....................
};
```

This method is especially useful when we come to consider derived classes in the next chapter. The conventions may also be mixed:

```
class complex {
private:
      double re, im;
public:
      complex(double x, double y)  :  re(x)  { im=y; }
      .....................
};
```

5.1.3 Friend functions

The access functions `real` and `imag` can be written in another way. They can be declared *friend functions* of the class. Although no longer class members, and therefore no longer accessible through the dot and pointer operators, they still have access to the private members of the class. The advantage is that we may now use notation resembling ordinary mathematics. In mathematics, we would write the real and imaginary parts of the complex number z as `real(z)` and `imag(z)`, respectively, rather than `z.real()` and `z.imag()`. To do that, we have to declare these functions as friends within the class declaration, and then define them externally just as if they were ordinary functions, excepting only that they may use the private members. Thus:

```
class complex {
private:
      double re, im;
public:
      ........
      friend double real(complex&);
      friend double imag(complex&);
      ........
};
```

following which the definitions become simply

```
inline double real(complex &z) { return z.re; }
```

and

```
inline double imag(complex &z) { return z.im; }
```

The functions are still small enough to be inline. Notice, however, that the scope operator `::` is not required for a friend function: to include it would be an error, because they are not member functions. Note also that we have chosen to pass the variable z by reference to save copying overheads during execution (although for complex numbers it is at most a marginal advantage). We could also have declared it a reference to a constant, but in this case it is sufficiently obvious that the value of z is not changed during execution.

In this case, the choice between friend and member is largely aesthetic, motivated by a desire to write expressions which resemble mathematics rather than computing science. Similar considerations suggest the use of friend functions for `log(z)` and `exp(z)`, and all the other functions most commonly applied to complex variables. Strict C++ programmers might not approve! In the next paragraph, however, we shall introduce friend functions for more substantial reasons.

5.1.4 Operator functions

The most distinctive feature of classes is that they can contain within themselves all the common operations for combining objects together into expressions. In this simplified exposition of the class `complex`, we shall confine ourselves to the operators for addition; but similar considerations apply to all the other arithmetic operators, and we first note the complete list of operators that may be redefined (overloaded) in the context of an arbitrary class. It would be difficult or perverse to find appropriate meanings for all of them in all classes. They are:

```
+       -       *       /       %       ^       &       |       ~       !
=       <       >       +=      -=      *=      /=      %=      ^=      &=
|=      <<      >>      <<=     >>=     ==      !=      <=      >=      &&
||      ++      --      []      ()      new     delete
```

of which the last four are index, function call, free memory allocation and de-allocation. The precedence of these operators is set by their usual meanings and cannot be changed. Some of them are unary operators, some binary, some both; but their syntax in expressions is also set by their usual meanings and cannot be changed.

Operator functions are denoted by the keyword `operator`. The first operator function to occur in the declaration of the class `complex` is the *unary plus*. It is declared to be a member function: `complex::operator+();`. Although its use can always be avoided, it is helpful to be able to reflect the mathematical convenience of being able to write:

```
complex a = +z;   complex b = -z;
```

The unary plus operator *could* be defined by:

```
complex complex::operator+()              // DON'T!
{ return complex(re, im); }
```

but this would require the creation of a temporary object. The problem can be avoided with the aid of the special pointer `this`, which is part of the C++ language, a keyword in fact. It is a hidden variable which is *implicitly* declared for all member functions (except static members, *cf* 5.3) of all classes, and it points to whatever object of the class is being invoked. In the case of the class `complex` it is implicitly declared as `complex *const this;`. For the definition of unary plus we evidently wish to return the *value* of the complex number itself, that is, the value pointed to by `this`, namely `*this`. Thus a better definition, which does not involve the creation of a temporary object, is

```
inline complex complex::operator+() { return *this; }
```

which we have additionally made inline.

The same trick is not available for unary minus, however, and we have to accept a temporary object. The definition of unary minus as a member function is

```
inline complex complex::operator-() { return complex(-re, -im); }
```

We can use operators defined in this manner in *prefix* or *postfix* form: for a unary operator @, and an object o, @o means the same as o@. Notice that unary operators take no arguments if they are members, and one argument if they are friends. The standard increment and decrement operators of the C language, however, have *different* meanings for their prefix and postfix forms; if redefined they lose this property.

Similarly, binary operators can be defined as member functions taking one argument or as friend functions taking two. Since for any binary operator @, a@b can be interpreted as a.operator@b or as operator@(a, b), which in general might have different meanings, it is an error to define both within the same class. No operator may be defined having more than two arguments. (The ternary operator ?: may not be redefined.)

Let us now consider the sum of two complex numbers. We wish to be able to write, for example, complex z = a + b;. Naturally, we first try a member function, defining it as:

```
complex complex::operator+(complex &z)      // WRONG!
{ return complex(re+z.re, im+z.im); }
```

The call, z1.operator+(z2);, which by convention may also be written in the shorthand *infix* form z1 + z2;, changes the value of z1. What we have done is to add the two numbers and assign their sum to z1. We have in fact just defined, not the arithmetic + operator, but the "plus and assign" operator +=. Let us therefore define this latter operator properly, already declared as a member function in 5.1.1:

```
inline complex complex::operator+=(complex &z)
{ return complex(re+z.re, im+z.im); }
```

What of our original quest? We must define a *friend* function instead, as already declared:

```
complex operator+(complex &z1, complex &z2)
{ return complex(z1.re+z2.re, z1.im+z2.im); }
```

where the references to both z1 and z2 could be declared references to const if there were any doubt about the interpretation. We may now write complex z = z1 + z2;, as desired.

Finally, input and output operators must be defined if the class is to be of practical use. The appropriate operators are << and >>, which occur in the well-known expressions cout << x and cin >> x. Postponing to a later chapter a fuller discussion of input and output in C++, we content ourselves for the time being with offering suitable definitions for the class complex. Both data members of a class object must be input or output, with whatever formatting arrangements we desire. We choose them to be:

```
ostream& operator<<(ostream &s, complex &z)
{     s << "(" << z.re << ", " << z.im << ")"; return s; }
```

and

```
istream& operator>>(istream &s, complex &z)
{      s >> z.re >> z.im; return s; }
```

Both are declared in 5.1.1 as friend functions of the class `complex`. The first outputs a complex number in the form (`1.234, 2.345`), for example; the second requires input of real and imaginary parts to be separated by white space. In terms of them we can write the following little program which tests out the class insofar as we have developed it:

```
#include <iostream.h>
#include <complex.h>
main()
{      complex x, y;
       cout << "\nEnter x: "; cin   >> x;
       cout << "\nEnter y: "; cin   >> y;
       cout << "\nx + y = " << x + y;
       complex z = -x;
       cout << "\n-x    = " << z;
       z += x;
       cout << "\nx - x = " << z;
}
```

On the screen we shall see something like:

```
Enter x: 1 2
Enter y: 3 4
x + y = (4, 6)
-x     = (-1, -2)
x - x = (0, 0)
```

5.1.5 Complex roots of quadratic equations

To illustrate that it is possible to write useful programs using the class `complex`, we show how to extend the function `quadroots` defined in 3.3 to compute the roots of a quadratic equation with real coefficients when these roots may be a complex conjugate pair. Corresponding to the equation

$$ax^2 + bx + c = 0,$$

there is a discriminant $d = b^2 - 4ac$. In our earlier treatment, if $d < 0$ there was an error situation which immediately terminated the program. If complex roots are allowed, there is no error; we merely have to calculate the complex roots. The following function does so: notice that if $d \geq 0$ the new function expresses the real roots as complex numbers with zero imaginary part.

```
void   quadroots(double a, double b, double c, complex &u,
       complex &v)
{      double d = b*b - 4*a*c;
       if (d < 0)
       {     u.re = v.re = -b/(2*a);
             u.im = sqrt(-d)/(2*a);
             v.im = -u.im;
       } else
       {     u.im = v.im = 0;
             double s = -(b + sqrt(d)*sign(b))/2;
             u.re = s/a;   v.re = c/s;
       }
}
```

We leave it the reader to generalize quadroots still further, in order to cover the case where the coefficients a, b, c may be complex.

5.2 Class string

Let us now consider how to represent character strings as a class. This will introduce several new features of the class concept that will be useful in later work.

5.2.1 Declaration

It is sufficient for our immediate purposes to declare the class string as follows:

```
class string {
private:
       int  size;
       char *str;
public:
       string();                    // string x;
       string(int)                  // string x(10);
       string(char*);               // string x("...");
       string(string&);             // string x = y;
       ~string();                   // destructor
       string& operator=(char*);    // assignment x = "...";
       string& operator=(string&);  // assignment x = y;
       char&   operator[](int);     // indexing
       string& operator+(char*);    // concatenation x + "..."
       string& operator+(string&);  // concatenation x + y
       friend ostream& operator<<(ostream&, string&);
       friend istream& operator>>(istream&, string&);
};
```

where, except for input and output, we have indicated in a comment the purpose of each function.

First, however, we must consider the hidden data members declared `private`. We provide the size of the string and a pointer to its first character. The size is intended to be one more than the length of the string to allow for the terminal character ' \0 '. Thus, in any particular case, `size` will be given by `strlen(str) + 1`, where we shall assume that the function `strlen` is declared in a header file called `string.h` containing all the standard library functions for manipulating strings in C. We shall use these functions freely in what follows. The functions `strlen` and `strcpy` we have met already, the latter copying one string into another. We offered simple definitions of them both in 2.8.

In the C representation of strings we must explicitly define arrays to hold any desired input string; schematically:

```
char buf[64];
cin >> buf;
```

where it is assumed that the input operator has been defined so that the terminal character ' \0 ' is provided automatically on entering the string from the keyboard. The length of the string, on the other hand, is implicit, always available from the function `strlen`. In the class representation of strings, by contrast, we choose to include the size of the string explicitly, but privately. This increases the memory requirement by a few, namely, `sizeof(int)`, bytes per string, but it has the advantage that we never have to refer explicitly to the size of a string, provided only that it is less than the preset buffer size. The buffer is built into the definition of the input operator, as we shall see.

One word of clarification may be helpful. The data member `size` is the length of the string pointed to by `str`, plus one. The size of an *object* of the class `string`, however, is a constant, namely, `sizeof(string)`, which equals `sizeof(int) + sizeof(char*)`, ignoring any "padding". The *real* data are the character array pointed to by `str` in the C representation. The class `string` is merely a device, albeit a powerful one, for handling character strings as defined in C, and it carries a small overhead.

5.2.2 Implementation

We have provided four constructors, corresponding to four different ways in which we might wish to declare a string. Let us consider first the "conversion constructor" which converts a C string into an object of class `string`. We might try to define it as follows:

```
string::string(char *s)
{       size = strlen(s) + 1;
        str  = new char[size];
        strcpy(str, s);
}
```

The first statement initializes `size` to the length of the string, plus one; it needs no further comment. The second statement, if it can, makes `str` point to the first character in a reserved block of memory exactly `size` characters long (*cf* 1.5.4). The third statement *initializes* `str` by copying `s` into it.

Our explanation of the role of the second statement was conditional. The block of memory to be pointed to by `str` has to be available in the free store (the "heap"). If it is not, because it has been used up already by other operations of the program, the operator `new` returns the value 0 (it returns a NULL pointer) instead of the expected address of the newly reserved block of memory. A potential error situation must therefore be allowed for. Since failure to allocate memory as the program demands is a fatal error, we shall do this by terminating execution of the program after printing out an appropriate error message. After the second statement we shall insert a conditional statement:

```
if (!str) error("...");
```

where the error function was defined in 3.3. A suitable error message would be:

```
"Memory allocation failure in string::string(char*)"
```

Allowing for this error condition the conversion constructor may now be written:

```
string::string(char *s)
{       str = new char[size=strlen(s)+1];
        if (!str) error("...");
        strcpy(str, s);
}
```

where we have also combined the originally first and second statements into one by using an assignment expression in place of the index.

It is now a simple matter to write down the three other constructors. They are:

```
string::string()                // default constructor
{       str = new char[size=1];
        if (!str) error("...");
        *str = '\0';
}
```

```
string::string(int len)         // fixed length constructor
{       str = new char[size=len+1];
        if (!str) error("...");
        for (int i=0; i<size; ++i)
                str[i] = '\0';
}
```

```
string::string(string &x)        // copying constructor
{       str = new char[size=x.size];
        if (!str) error("...");
        strcpy(str, x.str);
}
```

In the first two constructors we have here explicity initialized the character arrays `str[]` because the `new` operator only provides uninitialized memory. It is not necessary to do so in the other two because `str` is there initialized by copying. Note that we do not choose to make these functions inline since they are rather too long for that. The default constructor has size one corresponding to the null string. It could have been omitted had we preferred to give the second constructor the default value 0 in the class declaration, namely `string::string(int=0);`, but our preference was for an explicit default constructor.

Any object created by means of a constructor which uses the operator `new`, unlike a normal automatic variable, will continue in scope until the end of the program unless other measures are taken. It can be made to behave like an automatic variable if we provide a *destructor* which is called *implicitly* when the object is required to go out of scope by the normal rules. Essentially, a destructor frees the memory reserved for an object by destroying the pointer which points to it. The destructor name *must* be the class name preceded by a tilde. In this case its definition is

```
inline string::~string() { delete str; }
```

which uses the special operator `delete` (a keyword) to perform the necessary manipulation to free the memory pointed to and to put the pointer out of scope. Given a properly defined destructor, the scope rules for a class are the same as those for a standard type.

An explicit destructor was not necessary for the class `complex` because the operator `new` was not used, and the scope of a complex number as we defined it is the normal scope of its data members, which are doubles. We *could* have defined the class differently, for instance by using a pointer to a pair of doubles; in that case an explicit destructor would have been necessary.

Many of the complexities of the class `string,` and of any other classes in which memory is allocated by means of the operator `new`, arise from the implicit use of the operator `delete` as an object goes out of scope.

5.2.3 Assignment

In the case of the class `complex`, we did not define the assignment operator, choosing to rely upon bit copying in default of assignment. It would be dangerous to do that for strings, however, which are of varying length. More importantly, in the absence of explicit assignment operators for the class `string`, the destructor can cause great damage.

Consider the following programme fragment:

```
string x(20);
string y(10);
y = x;
........
```

The initializations are taken care of by the constructor `string::string(int)`, and the effect is to create two blocks of memory, of size 20 and 10 bytes, respectively, the first pointed to by `x.str`, the second by `y.str`. In the absence of an explicit assignment operator, the effect of the assignment `y = x;` is not only to attempt to bit-copy a longer string into a shorter one, but also to destroy one of the pointers and to duplicate the other. But at the end of the intended scope of `x` and `y` the destructor will be called *twice*; it will attempt to free the same block of memory twice, with potentially disastrous consequences to the continuing execution of the program. Explicit assignment operators are required that will remove this difficulty. The two operators offered are as follows:

```
string& string::operator=(char *s)      // y = "...";
{     delete str;                        // redefine str
      str = new char[size=strlen(s)+1];
      if (!str) error("...");
      strcpy(str, s);                    // copy s into str
      return *this;
}
```

and

```
string& string::operator=(string &x)    // y = x
{     if (this == &x) return *this;      // don't self-copy!
      delete str;                        // redefine str
      str = new char[size=x.size];
      if (!str) error("...");
      strcpy(str, x.str);                // copy x.str into str
      return *this;
}
```

These ensure that on assignment one of the pointers is destroyed immediately; only one remains to be destroyed on leaving scope.

Suppose instead that we had written

```
string x = "abdce";
string y = x;
........
```

A similar situation might have occurred. However, in this case we have already defined the initialization `string::string(string&);` and this will have dealt with the matter adequately. The exercise illustrates the fact that initialization and assignment are quite different activities.

It is important to note that we have chosen to *return a reference* (*cf* 2.5.4) in the definitions of assignment. Just as the passing of a variable to a function by reference avoids unnecessary copying, so the return by reference also avoids copying. When working with objects where copying represents a considerable overhead it is better to use that convention.

5.2.4 Indexing

The next operator declared in the class **string** is the indexing operator [ ]. This enables us to determine a particular character along the string. Regarding a string as an array, **char array[];,** we are seeking the *i*th character given by **array[i]**. The operator may be written:

```
char& string::operator[](int i)
{       return str[i]; }
```

Notice that we have returned a reference to the required character rather than the character itself. We have not done this to avoid copying: the copying of a character is a trivial matter. It is so that without further ceremony we can make assignments to individual characters in the string:

```
string x = "abcde";     // x.str == "abcde"
char ch = x[2];         // ch == 'c'
x[2] = 'h';             // x.str == "abhde"
```

If we had returned the character itself we should have had to define a specific function allowing us to assign to individual elements of the string, such as:

```
void string::setelem(int i, char ch)        // NO NEED!
{       str[i] = ch; }
```

Using a reference as return makes that superfluous.

In passing, we note that we *could* have dispensed with the operator [ ] altogether, because the data can always be accessed by an ordinary member function:

```
char string::getelem(int i)                 // NO NEED!
{       return str[i]; }
```

Proper definition of the operator [ ] makes the writing of programs much less cumbersome, and enables us to use notation somewhat closer to ordinary mathematics.

There remains one glaring inadequacy in the definition of the indexing operator. The index i must lie in the range 0...(size-1) or there is an error. In defining the operator we therefore have to perform a check on the value of i and terminate execution of the program with a suitable error message if the range is exceeded. It being a matter entirely internal to the class, we may declare a private member function range, say, which returns the value of i if it is within range, but causes termination otherwise:

```
class string {
private:
        int   size;
        char *str;
        int range(int);            // check range
public:
        ........
        char& operator[](int);
        ........
};
```

We may define range as follows:

```
inline int string::range(int i)
{ return (i<0 || i>=size) ? (error("..."), -1) : i; }
```

where we have again used the error function defined in 3.3. The boolean condition tests whether the index lies outside the range. If it does, the function returns the comma expression; if it does not (*ie*, the index is within range), it simply returns the value of the index. The comma expression is evaluated from left to right, so error is called, and the program terminated, before the impossible value of −1 can be assigned and returned: it is included only so that *formally* there is an integer return value whatever the value of the boolean condition. Not to do so might lead the compiler to protest. The error message could be something like "Index out of range in string::operator[]()".

We can now refine the definition of the indexing operator:

```
char& string::operator[](int i)
{      return str[range(i)];    }
```

and we no longer have to concern ourselves with range checks in objects of the class string.

5.2.5 Concatenation

The most complicated operation we shall consider is the *concatenation* of strings, represented by suitable definitions of the + operator. We emphasize that the choice of operator is largely arbitrary. What we are about to define might more properly be denoted by +=, but + is the natural and customary choice, and it accords with the C convention to which it gives effect.

The operator functions + are defined to be:

```
string& string::operator+(char *s)          // x + "..."
{       char *tmp = new char[size+=strlen(s)]; // create new array
        if (!tmp) error("...");
        strcpy(tmp, str);                    // copy in str
        delete str;
        strcat(tmp, s);                      // concatenate
                                             // strings
        str = tmp;                           // redefine str
        tmp = 0;                             // disconnect tmp
        return *this;
}

string& string::operator+(string &x)         // x + y
{       char *tmp = new char[size+=x.size];  // create new array
        if (!tmp) error("...");
        strcpy(tmp, str);                    // copy in str
        delete str;                          // destroy str
        strcat(tmp, x.str);                  // concatenate
                                             //     strings
        str = tmp;                           // redefine str
        tmp = 0;                             // disconnect tmp
        return *this;
}
```

Again, we have chosen to return by reference to avoid copying. In the second function, notice that the new `size` is one less than the sum of the sizes of the separate strings. This is because only one terminal character is now required, that at the end of the concatenated string. The first of these functions is not essential, only desirable. If it is omitted, however, the concatenation x + "..." will require a call to the constructor `string(char*)` to create a temporary variable before the remaining concatenation function can be brought into play; this also requires unnecessary copying. Note that in each case we must create a new buffer `tmp` just long enough to hold the concatenated strings. We then copy the original string `str` into `tmp`, after which we may delete `str` to avoid the build-up of memory demand. The two strings are then concatenated in the array `tmp`. We are now free to redefine the class member `str` to point to `tmp`, after which we set `tmp` equal to the null pointer. This latter step is important, for otherwise we should have *two* pointers pointing to the array, and this might cause confusion. Finally, we return *this as usual.

In addition to the functions `strlen` and `strcpy`, we have here used the C concatenation function `strcat`. In the concise style so beloved of C programmers it might be defined as follows:

```
char   *strcat(char *s, char *t)
{      char *p = s;                 // hold pointer value
       while (*s++) ;               // find end of s
       while ((*s++=*t++)!='\0') ;  // copy t onto end
       return p;                    // return original pointer
}
```

where it is assumed that the character array pointed to by s is long enough to hold t as well.

5.2.6 Input and output

The remaining operators in the class declaration are for input and output. They may be defined quite simply as friend functions in terms of the corresponding operators for C strings:

```
ostream& operator<<(ostream &s, stream &x)
{ s << x.str; return s; }
```

and

```
istream& operator>>(istream &s, string &x)
{ char buf[64]; s >> buf; x = buf; return s; }
```

For input we need a buffer array of characters, which we have chosen to be of size 64, sufficient to input strings of less than that number of characters. (We ignore the possibility of overflow.) Note that this buffer, being local to the input function, is recreated for each input operation, and goes out of existence when the operation is complete.

5.2.7 Example

The example we shall give has no intrinsic merit except to test in simple use the class we have so painstakingly constructed. It is essential that every fragment of every program should be tested with an appropriately designed, and preferably simple, program. In this case we offer as one test:

```
main()
{      string x, y;
       cout << "Enter name and surname\n";
       cin >> x >> y;
       string z = x + " " + y;
       cout << "Your name is " << z << "\n";
}
```

We have seen this program before in 1.1.1, where it interacted with the redoubtable Mr Peter Bloggs. It is interesting to compare the two versions. Note in particular that there is no overt reference to any input buffer; as we have already observed, that is now taken care of in the input operation itself. Note also that the concatenation operator chains properly in spite of the difference of definition of the strings involved in the declaration of z.

5.3 Static members

As we have seen, the function members of a class belong to the class, the objects of which have access to them. In contrast, each object of the class has its own set of data members created by the constructors. Sometimes, however, it may be desirable to make certain data available to every object, and it would be both wasteful of memory and liable to error if such common information had to be provided separately in each object. A global variable would satisfy the need, but it would be available to every other part of the program also. As far as possible global variables should be avoided, especially if there might be confusion about two or more of them performing similar roles in different parts of a program (in two different class definitions, say), but having different values.

The solution is to provide a *static* data member, public or private as appropriate, which will have class scope. In the class `string`, for example, the input buffer size could be provided as a private static member. We could redeclare the class:

```
class string {
private:
        int  size;
        char *str;
        static int bufsize;
public:
        ........
};
```

The input function would then be redefined:

```
istream& operator>>(istream& s, string &x)
{       char *buf = new char[string::bufsize];
        if (!buf) error("...");
        s >> buf;
        x = buf;
        delete buf;
        return s;
}
```

where the constant buffer size would have been given its chosen value globally:

```
const int string::bufsize = 64;
```

Note that the scope operator : : must be used when referring to a static class member. Note also that the buffer is deleted as soon as it has been assigned to x; having been created by the new operator, it would otherwise remain in scope until the end of any user program.

To take a more interesting example, suppose that for diagnostic reasons we wish to know how many objects of a class are in action at any point in a program. We must increase the value of an integer variable count by one each time a constructor is called, and decrease it by one for every (implicit) call to a destructor. If we furthermore declare count to be private, we shall need to define a function giving us access to the value of count. A function that uses only the static data members of a class can be declared static also. Thus, we could declare the class string:

```
class string {
private:
        int   size;
        char *str;
        static int count;
public:
        string(...);
        ~string();
        static void printcount();
        ........
};
```

where the definition of each constructor now contains an additional statement ++count; and the destructor contains the statement --count;. We require the static function printcount only to print out the current value of count. Its definition can be:

```
void string::printcount()
{ cout << "\ncount = " << count; }
```

and it should be accessed by means of the class scope operator, although in principle it can be accessed by any class object using the dot operator (which would somewhat beg the question in the present context).

Since we are concerned with a diagnostic application we might as well make the implicit calls to the destructor visible by means of an output statement. We can turn this visibility on or off at will: compiler directives may be used inside constructors and destructors, although not inside ordinary functions (except the function main). We redefine the destructor:

```
string::~string()
{       --count;
        #ifdef SHOWIT
                cout << "\ndeleted " << str;
        #endif
        delete str;
}
```

If we now recast the example of the previous section to display the number of string objects in scope at various stages of the program, we shall have something like the following:

```
void   showit()
{      string x, y;
       string::printcount();
       cout << "\nEnter name and surname\n";
       cin >> x >> y;
       cout << "Your name is " << x + " " + y;
}
```

When we call `showit` from the function `main`,

```
#define SHOWIT 1
main()
{      string::printcount():
       showit();
       string::printcount();
}
```

the screen will show:

```
count = 0
count = 2
Enter name and surname
Peter Bloggs
Your name is Peter Bloggs
deleted Bloggs
deleted Peter Bloggs
count = 0
```

It is interesting to see what happens to the output if the assignment and concatenation operator functions are changed so that they no longer receive or return variables by reference. The additional copying becomes starkly obvious. For this reason it is strongly recommended that static data members and access functions should be employed diagnostically as part of systematic class design. They can be deleted when the class has been sufficiently tested. Output controlled by compiler directives can be switched by defining or undefining a constant such as SHOWIT at the head of the program.

5.4 Class arrays

Types can be derived from classes much as they can from standard types. In particular, arrays of objects of class *type* can be declared as *type* `array[arraysize];` and accessed using the index operator in the usual way: *type* `ti = array[i];`. Furthermore, we can define array types just as before:

```
typedef type array[arraysize];
array a, b;
........
```

where the arrays `a, b` both have `arraysize` elements.

It would be convenient, however, if we could use pointers instead, allocating free memory accordingly. That is to say, we would like to be able to write

```
type *array = new type[arraysize];
```

in order to declare a pointer to a block of memory of sufficient size to hold `arraysize` objects of the class `type`. (As already noted, if sufficient free memory is not available, a NULL pointer is returned.) Clearly, we have to consider how the operator `new` acts in these circumstances, and what subsequent meaning should be given to the operator `delete`. We shall consider arrays of complex numbers first, then arrays of strings.

5.4.1 Array of complex numbers

Suppose we wish to store three complex numbers in a fixed array. We can declare an array of constant length 3:

```
const int n = 3;
complex c[n];
```

This will have been initialized by default, being equivalent to

```
complex c[] = { complex(), complex(), complex() };
```

where the size of the array is taken to be 3 by default, and where we have used the default constructor. The individual elements of the array can then be changed by assignment; for example,

```
c[0] = complex(1,2);   c[1] = complex(3,4);   c[2] = complex(5,6);
```

Alternatively, the array can be initialized from the beginning:

```
complex c[] = { complex(1,2), complex(3,4), complex(5,6) };
```

It would often be convenient, however, if we could instead declare a pointer to an array. In the case of an array of `complex`, we would write `complex *cp;` but we have to initialize this so that it points to a block of memory sufficient to hold the array object pointed to. If m is an integer, the operation new [m] allocates just m bytes of free memory by default. If we wish to create an array of n complex numbers, we might therefore try to write

```
complex *cp = new char [n * sizeof(complex)];        // WRONG!
```

However, this is not correct; cp, so defined, points to a single object containing n * sizeof(complex) bytes. To indicate that it is intended to point to an array of n complex numbers we must instead use the cast notation to write:

```
complex *cp = (complex*) new char[n*sizeof(complex)]; // CORRECT!
```

and this is correct. Alternatively, we can take advantage of the fact that new is defined so that, applied to a type identifier type, it allocates memory in blocks of sizeof(type) bytes. Therefore, we can also write:

```
complex *cp = new complex[n];             // ALSO CORRECT!
```

The last two statements are equivalent, but the meaning of the second is clearer.

The scope of an object pointed to by a pointer defined by new is until the end of the program unless specific action is taken to limit it. In the present case we can try to write delete cp; but it is not clear whether we have deleted thereby a pointer to a single complex number or to an array of complex numbers. In the case of an array of n elements we must in fact write:

```
delete [n] cp;
```

to free the memory allocated to each element of the array.

5.4.2 Array of strings

Consider now the case of an array of strings. We may declare the pointer strarray as follows:

```
string *strarray = new string[arraysize];
```

which points to a block of arraysize * sizeof(string) bytes of free memory. It may be initialized element by element by assignment from the left; for example:

```
strarray = {"abcde", "fgh", ... };
```

but any remaining elements will be initialized using the default constructor string:: string(). In the absence of a default constructor there would be an error; it is therefore always best to have one. We are here speaking of the storage space required for the data members of each object of class string contained in the array strarray. Each of these objects, `strarray[i]`, will require buffer storage, `strarray[i].str`, in which it stores the array of characters contained in the string. This is provided automatically by each constructor called by the definition of the array of strings. This is made plainer if we define the above array by the more explicit statement:

```
string strarray[] = { string("abcde"), string("fgh"), ... };
```

In the case of the class `string` we also have a destructor, which will implicitly delete each string as the array goes out of scope. It will still be necessary in addition to delete the pointer to the array, however:

```
delete [arraysize] strarray;
```

5.4.3   Array template

It is sometimes helpful to be able to declare an array before it has been decided how many elements it is to have, or even what their type should be. This may be done using the keywords `template` and `class`, and the angle brackets < and >, as we did in 2.5.7, where we defined function templates, but this time we shall declare a class template instead:

```
template <class T, int size> class array;
```

This declares `array` to be a "parameterized type"—there being two parameters in this case, namely the type of element, `T`, and the number of elements, `size`.

We next have to define `array`, and we do it just as we defined `string` in 5.2, only more simply:

```
template <class T, int size> class array {
private:
        T *data;
public:
        array()  { data = new T [size]; }
        ~array() { delete data; }
        T& operator[](int i) { return data[i]; }
};
```

Now one may call for specific arrays by the declarations

```
array<int, 6> ia;
array<float, 128> fa;
```

and so on. Thanks to the operator [ ] they may be indexed in the usual way. The reader may care to include some further functionality to make `array` less prone to accident. We shall not use parameterized arrays in this book, but the concept of class templates will be further developed in chapter 20.

5.5 Classes and header files

In any self-respecting implementation of C++, the class `complex` is a standard library item, declared in full in some header file, and pre-compiled into an *object file*, which can be linked together with other files which call for complex numbers. How this is done in practice is a matter for the compiler writer to explain; but schematically it goes as follows. Let the class `complex` be fully declared in the header file `complex.h`, which includes all its inline function definitions, whether internal or external to the class declaration. The remaining function definitions are gathered into a file, `complex.cpp`, say, which does not contain a function `main`, and which therefore is not executable. It can nevertheless be compiled to an object file, `complex.obj`, say, and maintained in good condition so that it may be used whenever complex numbers are called for. Now suppose `program.cpp` is an executable program file which calls for complex numbers. We suppose it has a directive to include the header file `complex.h`, so that any references to complex numbers will be understood. Then the compiler can be asked to compile `program.cpp` into an object file `program.obj`, and to link it to `complex.obj`. Schematically, we may write:

```
compile program.cpp + complex.obj
```

and the result will be an executable file, `program.exe`, say, which may be run in DOS by calling PROGRAM.

It is not our purpose to explain the compilation process, nor the eccentricities of operating systems; it is merely to point out the usefulness of *modules*, namely self-contained collections of functions which have been compiled to object files and which may be used as standard items in the construction of complicated programs. Class definitions are particularly important collections of functions which will often be pre-compiled into modules for general application. For this reason, if no other, they should be prepared with great care, bearing in mind as far as possible all the different applications to which they are likely to be put.

Likewise, we may record the declaration and inline definitions of the class `string` in a header file, `mystring.h`, say (it cannot be called `string.h` because we have assumed the latter to contain all the C string functions), and compile to `mystring.obj` all the functions defining the class `string` in the source code `mystring.cpp`, say. Actually, we reconsider the class `string` in the next chapter; the results of that reconsideration are to be found in the class `restring` declared in Appendix B. Other class declarations are to be found in later appendices. Since many classes are standard weapons in the armoury, we are guaranteed consistency of class definition provided we always include the appropriate header files in the source code of any program we are writing. Indeed, the module `standard` (*cf* 4.12) is handled in this way.

6 Derived classes and streams

The C++ concept of derivation allows the programmer to develop a number of classes which, although distinct, share some common features. The common features may be incorporated in one or more *base classes* from which the other classes are *derived*. In this way much repetitive labour is saved, clarity and consistency are improved, and the possibility of error is diminished. We begin with some simple examples, and then reconsider the class `string`, this time as a derived class. We are led to a simplified discussion of the concept of input and output streams as derived classes. In particular, we show how to format streams, for purposes of tabulation, for example. Finally, we describe how to deal with input and output operations on external files.

6.1 The base class `duple`

Suppose we wish to define a class of objects representing points in the euclidean plane, and another class representing complex numbers. Each of these classes has a pair of real numbers as a common feature: the x and y coordinates in the first case, the real and imaginary parts in the second. Both classes may be derived from a single base class, the data members of which are a pair of real numbers. In both cases we shall wish to access the respective data members, and to deal with their input and output. It is natural, therefore, to include all these features in the base class.

We shall define the base class `duple` in the following way:

```
class duple {
friend ostream& operator<<(ostream&, duple&);
friend istream& operator>>(istream&, duple&);
protected:
      double dx, dy;
public:
      duple(double x=0, double y=0)     { dx = x; dy = y; }
      double getx()                     { return dx; }
      double gety()                     { return dy; }
      void putx(double x)               { dx = x; }
      void puty(double y)               { dy = y; }
};
```

where for simplicity we have provided only one constructor (although allowing default para-meter values), and where the input and output operators are defined without sophisticated for-matting:

```
ostream& operator<<(ostream &s, duple &d)
{   s << d.dx << tab << d.dy; return s;   }

istream& operator>>(istream &s, duple &d)
{       s >> d.dx >> d.dy; return s;       }
```

The keyword `protected` means the same thing as `private`, except that `protected` members, unlike `private` members, are accessible to the members of a derived class. The derivation may itself be `public` or `private`, the latter being the default. If derivation is `public`, then `public` and `protected` members of the base class are `public` and `protected` members, respectively, of the derived class. If the base class has `private` members, however, they are inaccessible to the derived class. On the other hand, if the deriva-tion is `private`, both `public` and `protected` members of the base class become `private` members of the derived class, and the `private` members, if any, of the base class remain inaccessible to the derived class. Access to the members of a base class can be made *more* restrictive under derivation, but never less. The position is summarized in the following table.

Base class access	Derivation	Inherited access
public	public	public
private	public	inaccessible
protected	public	protected
public	private	private
private	private	inaccessible
protected	private	private

Clearly, we wish the data members of the base class `duple` to be accessible to the derived classes, and the `public` member functions of `duple` to be `public` members of the derived classes. That is why the data members of `duple` have been declared `protected` rather than `private`. It is also why derivation must be `public` in this case.

The keyword `protected` has not been provided in some implementations. Assuming that one wishes the base class members concerned to be available to derived classes, they must then be declared `public` instead, even though this gives them greater accessibility than is needed.

6.1.1 The derived class point

With these preliminaries we now define the class `point`. We shall need members reflecting some of the usual operations carried out on points in a plane: for example, we may wish to dilate or contract the coordinates, and we may wish to add and subtract points as if they were vectors emanating from the origin. Such functions are also common to the class `complex` and may be used when we define it. Thus, we shall take for the class `point`:

```
class point : public duple {
public:
      point(double x=0, double y=0) : duple(x, y) {}
      point& operator*=(double a) {dx*=a;dy*=a;return *this; }
      point& operator/=(double a) {dx/=a;dy/=a;return *this; }
      point& operator+=(point &p) {dx+=p.dx;dy+=p.dx;return *this; }
      point& operator-=(point &p) {dx-=p.dx;dy-=p.dy;return *this; }
};
```

Of course, other operations could be added, such as rotation about the origin, or reflection in the coordinate axes, but we are not interested here in completeness. Note that the constructor merely assigns the data members of the base class, there being nothing else for it to do in this case. We could equally have written

```
point::point(double x=0, double y=0) { dx = x;   dy = y; }
```

since the data members of `duple` are accessible to `point`—indeed, they *are* the data members of `point`. The two conventions are equivalent, and the choice is largely a matter of taste. The convention we have adopted is preferable only insofar as it emphasizes the fact that `point` is derived from `duple`.

Because the data members of `point` are precisely the data members of `duple`, we need no further input or output functions for `point`; those of `duple` will suffice. Nor is it necessary to declare input and output functions for `point` as friends of `point`: friendship survives derivation. Thus, the fragment

```
point p = ...;
cout << p;
```

outputs the data members of p derived from `duple`.

A word of warning is perhaps necessary, however. If the derived class were to possess additional data members *not* derived from the base class, we might wish to redefine the input and output operators in order to include the additional members. Consider the class `town`, which in addition to having map coordinates x and y, also has a name, which for present purposes we shall represent by an integer. Defined as a class derived from `duple`, we would write:

```
class  town : public duple {
friend ostream& operator<<(ostream&, town&);
friend istream& operator>>(istream&, town&);
private:
       int name;
public:
       town(double x, double y, int nm) : duple(x, y), name(nm) {}
       .........................................................
};
```

where, as before, the constructor may equivalently be written

```
town::town(double x, double y, int nm)
{     dx = x;   dy = y;   name = nm;     }
```

The new input and output operators must now be declared to be friends of the derived class. They may, for example, be defined as:

```
inline ostream& operator<<(ostream &s, town &t)
{ s << t.nm << tab << *(duple*)&t; return s;  }

inline istream& operator>>(istream &s, town &t)
{     s >> t.nm >> *(duple*)&t; return s;         }
```

Some further explanation is called for. The objective is to write simply cout << t; or cin >> t;. The name of the town is easily accessed using the dot operator, and is so dealt with above. The problem arises with the data members derived from the base class. For that we require the base class data of the object t. But t is a town, not a duple. Moreover, town has a size greater than that of duple; neglecting "padding",

```
sizeof(town) = sizeof(duple) + sizeof(int).
```

We therefore need to extract the duple part of town explicitly. Since t is an object of type town, its address, &t, may be interpreted as a pointer to town. This may be reinterpreted as a pointer to duple instead using the cast notation (duple*)&t. The object pointed to by (duple*)&t is *(duple*)&t, which is the duple part of the object t of type town, and both input and output have already been defined for objects of the base class duple. This expression is therefore included above in the definition of the input and output operators for the class town. This kind of type conversion should only be used with derived classes.

Although unnecessary for the class point, as we have seen, it is good practice to make the pointer conversion nevertheless. On some future occasion it might be desirable to add data members to `point`; if pointer conversion has been included there is then no need to redefine input and output for the base class data. We therefore redefine:

```
inline ostream& operator<<(ostream &s, point &p)
{          s << *(duple*)&p; return s;          }

inline istream& operator>>(istream &s, point &p)
{          s >> *(duple*)&p; return s;          }
```

6.1.2 The derived class `complex`

Consider now the class of complex numbers as a class derived from the base class `duple`. There are two courses open to us. First, the class `complex` may be derived directly from the class `duple`; we would then write

```
class complex : public duple { ........ };
```

and proceed just as we did for the class `point`.

But this would mean that we had to redefine the operations already defined for the class `point`, as well as defining other operations peculiar to the class `complex`. That would be a waste of programming effort. (In more substantial examples, it might amount to a great deal of waste!) The second course is to derive the class `complex` from `point` as a base class, for derivation is transitive. In this case, we may use the functions already defined for `point`, adding only the functions peculiar to `complex`. This is the course we shall follow:

```
class complex : public point {
public:
        complex(double x=0, double y=0) : point(x, y) {}
        complex& operator*=(double);
        complex& operator+=(complex&);
        complex& operator*=(complex&);
        .............................
};
```

where the operations derived by pointer conversion from `point` are, for example, first the multiply assignment, where the argument is a `double`:

```
inline complex& complex::operator*=(double d)
{       *(point*)this *= d; return *this;     }
```

and second the plus assignment, where the argument is `complex`:

```
inline complex& complex::operator+=(complex &z)
{ *(point*)this += *(point*)&z; return *this; }
```

The operators peculiar to the class `complex` which may not be derived from the base class by pointer conversion are multiply and assign, where the argument is `complex`:

```
complex& complex::operator*=(complex &z)
{       double re = px*z.px - py*z.py;
        double im = py*z.px + px*z.py;
        px = re; py = im;
        return *this;
}
```

and similarly for divide and assign with `complex` argument.

Input and output may be derived from the base class `point`:

```
inline ostream& operator<<(ostream &s, complex &z)
{           s << *(point*)&z; return s;               }

inline istream& operator>>(istream &s, complex &z)
{           s >> *(point*)&z; return s;               }
```

or directly from `duple` as the "ultimate" base class. Perhaps the former is the better practice, although the latter avoids one set of pointer conversions.

We emphasize that the treatment of the class `complex` offered here is only by way of illustration of the concept of derivation. It is *not* the way the class is defined in standard implementations. Note also that we could have defined `point` and `complex` as structures rather than classes: they have no private members other than those of `duple`, and the default access category of a structure is `public`.

6.2 Derivation and access

It is important that the rules of access to class members should be understood in the context of derivation from a base class. We shall assume that all members are public to simplify the discussion; the extension to protected and private members is straightforward. Let us define a simple base class

```
class base {
public:
        int x, y;
        base(int a, int b)        { x = a; y = b; }
        int f()                   { return 5 * y; }
};
```

and an equally simple derived class

```
class derive : public base {
public:
      int y, z;
      derive(int a, int b, int c, int d) \
            : base(a,b)          { y = c; z = d; }
      int f()                    { return 6 * y; }
};
```

where the "backslash" character '\' merely allows us to spread an inline definition over more than one line.

Now define an object of the derived class, and a pointer to that object, say:

```
derive d(1,2,3,4);
derive* dp = &d;
```

The rules of member access under derivation then allow us unambiguously to deduce that the value of d.x is 1, and of d.z is 4. The value of d.y is unambiguously 3. That is because the member identifier y in the class declaration derive "hides" the member identifier y in base. If we wish to find the value of that member of the object d, we have to access it using the class scope operator. The value of d.base::y is unambiguously 2. Similarly, the value returned by the function d.f() is 18, while that returned by the function d.base::f() is 10.

The rules for pointer access are exactly the same: for example, the value of dp->y is 3, and the value of dp->base::y is 2.

We may perform a standard pointer conversion from the derived to the base class by writing:

```
base *bp = &d;
```

If we then ask the value of bp->y it is 2, because it is not possible to refer by means of a base class pointer to a member only defined in the derived class. To try to find the value of bp->z is an error. Alternatively, pointer conversion may be forced by means of a cast: the value of ((base*)dp)->f() is 10. Using casts, the pointer conversion may even be implicit: the value of (*(base*)&d).y is 2. As we have seen, such expressions are useful to us. In what follows, we do not hesitate to use a cast, even if it is strictly unnecessary, if by doing so we make the meaning clearer.

For completeness we remark that a class may be derived simultaneously from more than one base class. Schematically, if we have base classes onebase and twobase, say:

```
class onebase {
protected:
      int x;
public:
      onebase(int i)    { x = i; }
};


class twobase {
protected:
      int y;
public:
      twobase(int i)    { y = i; }
};
```

then we may define a class derived from both of them, for example:

```
class derive : public onebase, public twobase {
private:
      int a, b;
public:
      derive(int i, int j) : onebase(i+1), twobase(j+2), a(x+3), b(y+4);
};
```

If we now call `derive(1,2)` there results a call to `onebase(2)` and `twobase(4)`, so that x has the value 2, y the value 4, a the value 5 and b the value 8. Note also that when an object of a derived class is created, base class constructors are called in the order of their listing before the derived class constructor itself. The destructors operate in the reverse order.

6.3 Class `string` revisited

In the last chapter we introduced a treatment of strings in C++ with the aid of the class `string`, within which the normal C string operations were embedded. While satisfactory as far as it went, there was a somewhat inelegant mingling of concepts. On the one hand, there were the creation and destruction of arrays of characters which constituted the data members; on the other, there was the use of the C string functions to copy and concatenate strings—and in principle, to perform any other string operation. One of the advantages of C++ is that it allows (and encourages) the separation of distinct concepts. In the case of strings it seems natural, and it is certainly instructive, first to define a base class which creates and manages arrays of characters, and then to derive from it a class containing the usual string operations. There is, of course, a certain flexibility in deciding whether a given operation should belong to the base class or to the derived class, but the desired separation of concepts will be our guide. We begin, therefore, by constructing the base class `strbuf` which provides, and in principle manages, the "buffer" space required by a string:

```
class strbuf {
private:
      int range(int);                         // index in range
protected:
      char *buf;                              // char array
      int size;                               // string length + 1
public:
      strbuf(int sz);
      ~strbuf()                     { delete buf; }
      char& operator[](int i)       { return buf[range(i)]; }
      void swap(int, int);                    // swap elements
};
```

The function `range`, as defined in 5.2.4, is declared `private` because it is required only for operations internal to the class `strbuf`. The data members, however, must not be `private` because they will be used by the derived class; they are declared to be `protected`, and as before they are a pointer to a character array `buf`, and the `size` of the array.

The `public` members are, first, a single constructor and a destructor. The constructor constructs a zero initialized buffer of given length:

```
strbuf::strbuf(int sz)
{     buf = new char[size = sz];
      if (!buf) error("...");
      for (int i = 0; i < size; ++i)
            buf[i] = '\0';
}
```

The destructor has already been defined inline in the class declaration. There follow two functions typical of those required for manipulating character arrays. The first is the indexing operator which allows us to write to or read from a given element of the buffer provided that it is within range; it may be regarded as a primitive input and output operation, and we have defined it inline. The second by way of illustration is a function which interchanges two elements of the buffer, having ascertained that the elements concerned are both within range:

```
void  strbuf::swap(int i, int j)
{     char tmp = buf[range(i)];
      buf[i]  = buf[range(j)];
      buf[j]  = tmp;
}
```

Of course, many other functions could be added, such as functions to flush a predetermined portion of a buffer, to perform some more complicated permutation of elements, or to declare whether a buffer is full or empty. What we have already is more than sufficient for the purpose in hand, however.

We now re-define the class `string` as a class derived from the base class `strbuf`:

```
class string : public strbuf {
friend   ostream& operator<<(ostream&, string&);
friend   istream& operator>>(istream&, string&);
private:
        static int inbufsize;                       // input buffer
public:
        string()            : strbuf(1)        {}   // string x;
        string(int n)       : strbuf(n+1)      {}   // string x(n);
        string(char*);                              // string
                                                    // x("...");
        string(string&);                            // string x = y;
        ~string ()                             {}
        string& operator=(char*);                   // assignment
        string& operator=(string&);                 // assignment
        string& operator+(char*);                   // concatenation
        string& operator+(string&);                 // concatenation
};
```

There are the same four constructors and the destructor that we had before, but this time they are derived inline from the two already provided for the base class `strbuf`.

As before, there are two assignment operators and two concatenation operators. All involve the creation of local string objects, the reassignment of data members, and the rendering harmless of pointers left with nothing to do. The first assignment operator is

```
string& string::operator=(char *str)
{       delete buf;            // delete original string
        string tmp(str);       // copy str into tmp
        size = tmp.size;       // reassign size
        buf  = tmp.buf;        // reassign buf
        tmp.buf = 0;           // disconnect tmp
        return *this;
}
```

and the second is similar. We leave the concatenation operator functions to the reader. Note that having destroyed `buf` to dispose of any string initially present (*ie*, in `*this`, the object which invokes the function), we provide just enough space for the new string by creating a temporary variable of type `string`. That variable has then to be assigned to `*this`, so that buf now points to the same object as `tmp.buf` does. The presence of a destructor will destroy that object when `tmp` goes out of scope, and that is why we "disconnect" `tmp.buf` by setting it equal to zero—the destructor has no effect on a null pointer.

Finally, we must define input and output. We *could* define them in terms of the primitive indexing operator, but it is more satisfactory to use the standard C++ string operations << and >>. The output operator is then very simple:

```
ostream& operator<<(ostream &s, string &x)
{      s << x.buf;   return s;        }
```

Input is a little more involved because we first have to create an input buffer. This requires us to initialize the length of the buffer already declared in the class declaration. This time, we arbitrarily choose it to be just long enough to contain a single line of text in a standard PC text editor:

```
const int string::inbufsize = 81;
```

An input operator may then be written down:

```
istream& operator>>(istream &s, string &x)
{      string instr(string::inbufsize);
       s >> instr.buf;
       x = instr;
       instr.buf = 0;                  // disconnect
       return s;
}
```

Input and output *could* have been defined instead in the base class, there being no additional data members in the derived class. We have chosen to put it in `string` because of the appearance of the statement `x = instr;`, and the assignment operator is to be found in `string` not in `strbuf`. Had we defined input in `strbuf` instead, we should either have had to transfer the assignment operators also, which did not seem appropriate, or rely upon bit copying to provide assignment by default. We therefore chose to put it in `string`.

Objects of class `string` may use the `public` members of `strbuf` as well as of `string`: for example, if we want to interchange the elements 2 and 4 of a given string and output the result we have only to write:

```
string x = "...";   x.swap(2,4);   cout << x;
```

This treatment of strings is far from complete; for example, there is still no provision for overflow and underflow. (The reader might care to see what happens if the user makes a null entry, and try to compensate for it.) But it is sufficient for our purpose: it provides a concrete model for the following discussion of streams, and it deals adequately with the input and output requirements of this book, which so far as text is concerned are very simple. The declarations of the classes `strbuf` and `string` are repeated in the header file `restring.h` in Appendix B.

6.4 Streams

We have already remarked that in C++ input and output are not part of the language, and are therefore, to a greater or lesser extent, implementation dependent. For that reason, if for no other, the manual must be consulted for any particular implementation. The purpose of this section is, firstly, to give some understanding of how input and output work in C++; and, secondly, to show how formatted input and output may be provided whenever occasion demands. The subject flows on naturally from what we have just done, because in C++ input and output streams are defined as classes, and derived classes in particular. In the class `string`, we were adapting standard C methods of handling strings to the language of a class. Here, we shall be adapting the basic concepts of input and output in C to the stream classes of C++. Our starting point is the treatment of streams proposed in the book by Stroustrup.

6.4.1 The stream buffer

In order to manage input and output in an orderly way it is first necessary to define a buffer, namely a dynamic and properly managed array of (signed) characters. It is in the spirit of C++ that the data defining the buffer, and the functions allowing access to it, should constitute a class. Very schematically, and on no account to be compiled, we define the class `streambuf` as follows:

```
class streambuf {
private:
        char *base;             // start of buffer
        char *eptr;             // end of buffer
        char *pptr;             // next put character
        char *gptr;             // next get character
public:
        streambuf();            // initialized buffer
        ~streambuf();           // destructor
        int sputc(int);         // put character
        int sgetc();            // get character
};
```

The private data members are pointers which keep track of the beginning (`base`) and end (`eptr`) of the buffer, and the locations of the first available space where a new character may be put (`pptr`), and of the next character to get (`gptr`). There will in practice be other members, not least some functions to check whether the buffer is full or empty. For example, the value of the boolean expression `gptr==base` tells us whether or not the buffer is empty.

There is a constructor which creates a buffer area by means of the `new` operator, and a destructor. The former might be written

```
streambuf::streambuf()
{      base = new char [bufsize];
       eptr = base + bufsize;
       pptr = gptr = base;
}
```

where `bufsize` is a given constant integer. There are then the working functions for putting and getting characters:

```
inline void streambuf::sputc(int c) { *pptr++ = c; }
inline char streambuf::sgetc()       { return *gptr; }
```

Notice that we have been extremely cavalier in offering definitions for these functions; in particular, we have taken no account of overflow (attempting to insert a character into a full buffer) and underflow (attempting to extract a character from an empty buffer).

6.4.2 Output streams

Let us now consider the class `ostream`, which is essentially a pointer to `streambuf`:

```
class ostream {
private:
       streambuf *buf;
       . . . . . . . . . . . . . . .
public:
       ostream(streambuf*);
       ~ostream();
       void put(char);
       ostream& operator<<(char*);
       . . . . . . . . . . . . . . .
};
```

The constructor creates a buffer and takes care of all the data members we have omitted; we shall not define it further. We require output operations for all the standard data types.

Suppose in particular that we wish to output a character string. We may define

```
void  ostream::put(char ch)
{ buf->sputc(ch); }
```

and then

```
ostream& ostream::operator<<(char *s)
{ while (*s) put(*s++);   return *this; }
```

The output operator is similarly defined for the other basic types, `int`, `float`, and so on, reading in the appropriate number of bytes. However, a stream is a pointer to a succession of signed characters. A fragment such as `int i = 12345; cout << i;` converts the (possibly two-byte, but in any case of fixed size) integer `12345` to the six-byte character string `"12345"` before displaying the result on screen. It may seem an obvious point, but it has important ramifications when we come to study operations involving external files.

6.4.3 Input streams

Likewise, we may define the class `istream`, and consider the input of a character string. We have:

```
class istream {
private;
        streambuf *buf;
        . . . . . . . . . . . . . .
public:
        istream(streambuf*);
        ~istream();
        void get(char&);
        istream& operator>>(char*);
        . . . . . . . . . . . . . .
};
```

where
```
        void  istream::get(char &ch)
        {     ch = buf->sgetc();    }
```

Then,
```
        istream& istream::operator>>(char *s)
        { while (*s) get(s++); return *this; }
```

where, in this last function, it is assumed that the return character ending an input string is first converted into a zero to provide the terminator `'\0'`.

Any particular implementation of C++ will provide full library definitions of the classes `istream` and `ostream`, declared in the header file we have called `iostream.h`.

6.4.4 Formatted input and output

The base class `streambuf` allows us to deal with an unformatted stream. Often we shall wish to do differently: for example, we may be reading formatted data from an experiment, or wishing to present results of computations in tabular form. For the latter we have so far used the tab character for this purpose, but it is far from adequate as readers may by now have discovered for themselves. In order to deal with formatted data in a consistent manner we need to introduce yet another class, which we call the class `ios`, a full declaration of which might also be found in the header file `iostream.h`. It is from `ios` that the input and output streams will in future both be derived: it may be regarded as a prototype of `ostream` and `istream`, but one which allows data to be formatted at pleasure. We have already seen how input and output operations may be defined for user-defined types; these may now be formatted by the methods presented here.

Note, however, that the class `ios`, and some of the details which follow, are implementation dependent, and are therefore presented only for illustrative purposes. Some of the functions we shall declare are those defined in the TurboC++ implementation, but they may have their equivalents in other implementations. The final result is a prescription for formatting data that we can use immediately only in TurboC++, but it should not be too difficult to translate it into other implementations which provide `ios`'s primical formatting facilities.

Let us for present purposes define the class `ios` as follows:

```
class ios {
protected:
      streambuf *buf;
      long x_flags;
      int  x_width;
      int  x_precision;
      int  state;                              // status of stream
      ........
public:
      ios(streambuf*);
      ~ios();
      enum      {
                left       = 0x0002,     // left-adjust output
                right      = 0x0004,     // right-adjust output
                showpoint  = 0x0100,     // force decimal point
                showpos    = 0x0400,     // prefix '+' to
                                         // positive numbers
                fixed      = 0x1000,     // require 'fixed
                                         // point' notation
      };
      long      flags(long);
      int       width(int);
      int       precision(int);
};
```

where `streambuf` is the stream buffer introduced in 6.4.1; it is here declared `protected` because it will be used by the classes `ostream` and `istream` when derived from `ios`. The protected data member `state` will be used for file operations in 6.5 (it is not to be confused with the enumeration introduced in 2.4.7 and Appendix A). The other protected members will be used by `istream` and `ostream` when redefined as derived classes.

First note the anonymous enumeration included in the public part of the class. It is presented in hexadecimal form, and the values given, when converted to binary, switch on or off successive bits of a long integer. These bits are used as flags to determine such features of the format as whether the number concerned it to be flush with the left-hand or right-hand side of the field, whether the decimal point is to be shown, whether the plus sign is to be shown with positive numbers and whether fixed point notation is to be used. (There are other values, not noted here, which could be used to switch on the remaining bits.) If one wishes to present a tabulation in right-adjusted fixed point notation, one defines:

```
long L = ios::right + ios::fixed;
```

If the function `flags` is defined by

```
inline void ios::flags(long n) { x_flags = n; }
```

then the call `ios::flags(L)` will set the flags accordingly. Similarly, if it is desired that the field should be w characters wide, and that a floating point number should be presented to a precision of p digits after the decimal point, the functions

```
inline void ios::width(int n)          { x_width = n; }
inline void ios::precision(int n)      { x_precision = n; }
```

will achieve that when `ios::width(w)` and `ios::precision(p)` are called.

The output and input streams are now redefined so as to be derived from the class `ios`, and all but the private members of the class `ios` will therefore be accessible to them:

```
class ostream : public ios { ... };
class istream : public ios { ... };
```

The derivation has to be public so that stream functions may refer directly to the non-private members of the class `ios` as if they were members of the derived classes, and therefore to the formatting functions that it contains.

It is convenient to introduce a class to take care of all these matters once the parameters have been decided. In this book, we shall call it the class `format`:

```
class format {
friend ostream& operator<<(ostream&, format&);
private:
      long lb;
      int pr, wd;
      ios& iosfunc(ios&, long, int, int);
public:
      format(long, int, int);
};
```

where the private function iosfunc is given by

```
ios& format::iosfunc(ios& s, long L, int p, int w)
{  s.flags(L);  s.precision(p);  s.width(w);  return s; }
```

and the constructor by

```
inline format::format(long L, int p, int w)
{       lb = L;    pr = p;    wd = d;              }
```

  Finally, we can define the output operator for an object of the class format,

```
inline ostream& operator<<(ostream &s, format &f)
{   f.iosfunc(s, f.lb, f.pr, f.wd);  return s;  }
```

and similarly for input,

```
inline istream& operator>>(istream &s, format &f)
{   f.iosfunc(s, f.lb, f.pr, f.wd);  return s;  }
```

  These operators allow us to declare *manipulators*, which enforce the desired format on any quantity that is to be input or output. For example, suppose that we wish to output successively two numbers, x and y, which might be an int and a float, respectively. If we declare the following objects of the class format:

```
format setx(Lx, px, wx), sety(Ly, py, wy), reset(0, 0, 0);
```

where L, p, w have been chosen in each case to suit the type concerned, then, since we have defined the output operation for these objects, we may write

```
cout << setx << x << sety << y << reset;
```

and the numbers x and y will be displayed on screen with the individual formats defined by setx and sety. Finally, the manipulator reset returns the formatting parameters to their default values.

6.4.5 Tabulation

Although it requires no further development of the concept of streams it is convenient at this point to indicate how tabulations may most readily be effected in C++.

The first step is to define a structure to hold the various quantities to be displayed on a line of the intended tabulation. For example, we might wish to display the results of a routine which computes an approximate solution ycalc to a differential equation in the variable x, comparing it with the exact solution ytrue and also displaying the error ytrue - ycalc. A suitable structure in this case would be:

```
struct table { double x, y, f, e; };
```

Then a single line of the tabulation may be defined by

```
table t = { x, y, fx, fx - y };
```

where fx = F(x) is the exact solution.

An output operator may now be defined for a variable of type table. If the value of x is to be shown to one place of decimals, the other values to six with plus signs included, then a suitable operator might be:

```
ostream& operator<<(ostream &s, table &t) // formatted line
{       long Lx = ios::left + ios::fixed;
        long Ly = ios::left + ios::fixed + ios::showpos;
        format xpw(Lx, 1, 8), ypw(Ly, 6, 12), rst(0, 0, 0);
        s << xpw << t.x << ypw << t.y
                << ypw << t.f << ypw << t.e << rst;
        return s;
}
```

A single line of the formatted table may now be obtained by writing cout << t << endl;. This, however, is too simple for most purposes, because it fails to make allowances for tabulations which exceed the number of lines in a single screen, normally 25. The trick is to count the number of lines already tabulated and to pause with a message when the screen is full, proceeding to a fresh screenful ("page") when the message is obeyed. Thus we write instead:

```
ostream& operator<<(ostream &s, table &t)    // formatted
                                             // tabulation
{      const int pgln = 24;                  // page length
       static int line = 0;
       int row = line % pgln;                // position on
                                             // screen
       long Lx = ios::left + ios::fixed;
       long Ly = ios::left + ios::fixed + ios::showpos;
       format xpw(L, 1, 8), ypw(L, 6, 12), rst(0, 0, 0);
       if (line == 0)
       {      clear;                          // clear screen
              s << xpw << "x" << ypw << "ycalc" << ypw << "ytrue"
                     << ypw << "error" << rst << "\n\n";
              line++; line++;
       }      else
       if (row == 0)
       {      pressanykey();                  // defined below
              clear;                          // clear new page
       }
       s << xpw << t.x << ypw << t.y << ypw << t.f << ypw << t.e << rst;
       line++;
       return s;
}
```

Notice that we have also made provision for the tabulation to be appropriately headed. The function `pressanykey()` causes the whole program to pause until a key is pressed, after which the screen is cleared for the next page. This function is declared in the header file `standard.h` and may be defined as follows:

```
void   pressanykey()
{      int x = wherex();         // store position of cursor
       int y = wherey();
       gotoxy(1, 25);            // cursor to bottom left
       insline();               // insert blank line
       textcolor(WHITE);        // bold for message
       cputs("Press ANY KEY to continue: ");
       normvideo();
       putch(bell);             // beep
       getch();                 // await key press
       delline();               // delete message
       gotoxy(x, y);            // restore cursor
}
```

The ten routines called by this function are library functions and will not be further explained here. The tabulations shown in this book have mostly been produced using this method. Some implementations, however, do not provide such functions, nor the primitive formatting functions used by the base class `ios`, preferring to rely upon the facilities provided by a commercial user interface to provide formatting and tabulation as required.

6.5 File input and output

In this section we consider how to write to external files and how to read from them. The standard C++ streams `cout` and `cin` are objects of the stream types `ostream` and `istream`, respectively, which, together with the class `streambuf`, have been designed with standard operations from keyboard and to screen in mind. However, these stream types already contain many of the features we require for input and output more generally, of which the most important are the definitions of the operators `<<` and `>>`. They also allow us to use formatted input and output, as we have just seen. We would obviously like to retain all these properties, so we define two new classes, `ofstream` and `ifstream`, which are derived from `ostream` and `istream`, respectively, but which allow writing and reading operations on external files. We continue with our schematic presentation, assuming that full library definitions have been provided, with declarations given in a header file that we shall call `fstream.h`. It is necessary, however, to repeat the warning given at the end of the last section, namely that some implementations may prefer to rely upon user interfaces for handling the operations described here.

The class `ios` contains, we suppose, two further public enumerations of constants. In the Turbo C++ implementation, the first describes the status of the stream:

```
enum    ios::io_state
{       goodbit  = 0x00;   // all is ok
        eofbit   = 0x01;   // reached end of file
        failbit  = 0x02;   // last I/O operation failed
        badbit   = 0x04;   // tried invalid operation
        hardfail = 0x08;   // hopeless error!
};
```

together with access functions such as

```
int    ios::eof() {  return state & eofbit: }
```

where `state` is the protected data member of `ios` which records the current status of the stream; its bits may be set in a similar way.

The second enumeration describes how a file is to be prepared:

```
enum   ios::openmode
{      in       = 0x01;   // open for reading
       out      = 0x02;   // open for writing
       app      = 0x08;   // open and append
       binary   = 0x80;   // binary file
       ..............
};
```

It is to be used in conjunction with member functions defined to open files in the reading mode indicated, and later to close them. If our new classes are derived from the `iostream` classes—themselves, we have supposed, derived from `ios`—all but the private members of `ios` will be available to them, and in particular enumerations of constants contained therein. To write to a file we therefore define:

```
class ofstream : public ostream {
public:
       ofstream();
       ofstream(const char*, int=ios::out);
       ~ofstream();
       void open(const char*, int=ios::out);
       void close();
       ..............
};
```

The first constructor allows us to declare an object of type `ofstream` without specifying which file it is attached to. We might write

```
ofstream outfile;
```

and later in the program, when the file is opened for writing, we would perhaps write

```
outfile.open("OUTFILE.FIL");
```

We shall assume that the DOS file OUTFILE.FIL is then "open for writing" by default, because `outfile` is an object of type `ofstream` for which the open mode is preset to open. If a file of that name already exists its contents should be deleted. Alternatively, we might use the second constructor to write

```
ofstream outfile("OUTFILE.FIL");
```

thus achieving the objective in one rather than two steps. The choice is largely one of clarity.

We might instead wish to append what we have to write to the end of an existing file. In that case we would use the binary OR operator to overwrite the default:

```
outfile.open("OUTFILE.FIL", ios::in | ios::app);
```

To read from a file we similarly define:

```
class ifstream : public istream {
public:
      ifstream();
      ifstream(const char*, int=ios::in);
      ~ifstream();
      void open(const char*, int);
      void close();
      . . . . . . . . . . . . . .
};
```

in which the default mode is open for reading. Accordingly, we could write

```
ifstream infile;
```

and later

```
infile.open("INFILE.FIL");
```

or more simply

```
ifstream infile("INFILE.FIL");
```

In what follows we shall consider input and output operations, firstly on text files, and then on binary (non-text) files.

6.5.1 Text files

First, suppose that we wish to write a character string to a file, and subsequently to read from it, confirming that we have been successful by echoing the result to screen. The following program achieves this objective:

```
#include <restring.h>
#include <fstream.h>
ofstream ofile; ifstream ifile;

char   *filename = "TEXTFILE.TXT";
string msg = "Peter's pointless message!";

main()
{     clear;
      cout << msg << endl;
      ofile.open(filename);
      if (!ofile) error("cannot open file for output!");
      ofile << msg;
      ofile.close();
      ifile.open(filename);
      if (!ifile) error("cannot open file for input!");
      string s;
      ifile >> s;
      while (!ifile.eof())
      {     string x;
            ifile >> x;
            s + " " + x;
      }
      ifile.close();
      cout << s << endl;
}
```

We first echo the message to screen; then open the file for writing, write to it the predefined string msg with the statement ofile << msg;, and close it. Next, we re-open the file as a file opened for reading, read its contents, and close it. Finally, we write the retrieved message to screen in the usual way. But reading is a little more complicated than writing! In this case, we know the length of the string we have written to file, but in general we do not. Moreover, the input operator >> is so defined (as we know from chapter 1) that it stops reading a character string when it encounters "white space", which it also skips. Only the first word of the string msg is therefore read by the statements string x; ifile >> x;. If we wish to read the whole string, which in general may contain many words, we have to repeat this pair of statements, building up the complete string—which we have denoted by s—as we go, using the concatenation operator, and not forgetting to re-insert the white spaces. The boolean function ios::eof(), not defined here, tells us whether the end of file has been reached, whereupon the while loop terminates, and the complete string is written to screen. Alternatively, we could have written the string to screen word by word as the words are read, inserting a space after each word.

6.5.2 Binary files

We have already remarked that an expression like cout << x writes to screen the character string representation of the quantity x: for example, if we have declared x to be an integer and assigned to it the value 12345, it will cause five characters to appear on the screen. Similarly, if we were to write the number given by int x = 12345; to file by means of the statement ofile << x; we should be writing five characters to the file, and in general as many characters as the number has digits. That is all very well for writing, but when it comes to reading we would not know beforehand how long the character representation of the number is, and therefore if there were a succession of numbers we could not read them except by the wasteful process of putting white space in between each pair of numbers when writing them. That is to represent numbers when written to file by character strings, and the problem becomes much worse if we consider floating point numbers.

But we normally represent integers in computing by a fixed number of bytes (*eg*, two or four), and we know there are similar conventions for floating point numbers. We have to read and write, not the character representation of numbers, but the *binary* representation—sizeof(int) bytes for an integer, and so on. That means that for file input and output we have to abandon the character based operations >> and <<, respectively, replacing them by operations which copy a given number of bytes to or from the stream. These we take to be the functions

```
istream& istream::read(char*, int);
ostream& ostream::write(char*, int);
```

which we shall also expect to find declared in the header file iostream.h. Since the fstream classes are derived from the iostream classes, these also serve as file input and output functions.

The following program writes three floating point numbers to file, and subsequently reads them from file, echoing the outcome to screen:

```
#include <standard.h>
#include <fstream.h>

char   *filename = "DATAFILE.DTA";

ofstream ofile;
ifstream ifile;

main()
{       clear;
        int floatsize = sizeof(float);
        float f1 = 1.2345, f2 = 31.415, f3 = 0.27182;
        ofile.open(filename, ios::out | ios::binary);
        if (!ofile) error("cannot open file on writing!");
```

```
ofile.write((char*)&f1, floatsize);
ofile.write((char*)&f2, floatsize);
ofile.write((char*)&f3, floatsize);
ofile.close();
cout << "writing = " << f1 << sep << f2 << sep << f3 << endl;
ifile.open(filename, ios::in | ios::binary);
if (!ifile) error("cannot open file on reading!");
float ff1, ff2, ff3;
ifile.read((char*)&ff1, floatsize);
ifile.read((char*)&ff2, floatsize);
ifile.read((char*)&ff3, floatsize);
ifile.close();
cout << "reading = " << ff1 << sep << ff2 << sep << ff3  << endl;
}
```

Note first that we can no longer accept the default constructors `istream::istream`
`(char*)` and `ofstream::ofstream(char*)` as we did for text files; we have to
declare output and input files using the second parameter of the constructors, ORing the
default value with `ios::binary` as indicated. Moreover, in writing or reading an object (in
this case a `float`) we must interpret the address of it as a pointer to `char`, following that
conversion with the integer number of bytes which have to be written or read.

Finally, from a file containing a long sequence of numbers and/or characters, we may wish
to extract a particular one, or alternatively we might wish to overwrite one at a particular
place. For these purposes, an additional enumeration should be included in the definition of
the class `ios` which determines the starting point and direction of an offset to the position
required:

```
enum ios::seek_dir { beg = 0, cur = 1, end = 2 };
```

and additional member functions in the stream classes:

```
ostream& ostream::seekg(long, seek_dir);
istream& istream::seekg(long, seek_dir);
```

Using these functions in conjunction with the `read` and `write` functions it is possible to
read or write a block of bytes of given length, starting at a particular point in the file, as
measured from the beginning, the current position, or the end of the file by means of a
long integer.

We emphasize, however, that in the whole of 6.5, as in much of 6.4, we have used
TurboC++ conventions; but we hope that by so doing we have at least clarified the processes
involved in file operations using C++.

7 Integer arithmetic

In this chapter we briefly study three branches of mathematics in which arithmetic is conducted mostly in terms of integers: prime number factorization, rational arithmetic, and the generation of sequences of numbers which for practical purposes, and in a sense to be investigated, may be regarded as random. In the next chapter we study the extent to which such sequences come up to expectations. We begin this chapter, however, by recalling some of the well-known properties of the prime numbers, the foundations of any theory of arithmetic.

7.1 Prime numbers

If an integer a divides the integer b with zero remainder we say that a is a *divisor* of b and we write $a \mid b$; otherwise we write $a \nmid b$. A prime number, or prime, is then defined as a positive integer of which the only divisors are unity and the number itself. Since unity is a divisor of all numbers it is not itself regarded as a prime number. Thus, the first few primes are:

$$2, 3, 5, 7, 11, 13, 17, 19, 23, 29, 31, \ldots.$$

The *fundamental theorem of arithmetic*, the earliest proof of which is attributed to Euclid, asserts that any integer may be represented as the product of a sequence of primes, and that apart from the order in which they occur the product is unique. A number which is not itself prime is said to be *composite*. Thus, for example, $31\,050 = 2.3^3.5^2.23$, and in general $n = p_1^{k_1} p_2^{k_2} \ldots p_m^{k_m}$. We shall adopt the usual convention that in a product of this kind the sequence of primes is such that $p_1 < p_2 < \ldots < p_m$, where $2 \leq p_1$. Note that, if $ab = n$, a and b cannot both exceed $\sqrt{n}$. It follows that any composite number n has a prime factor which does not exceed $\sqrt{n}$. This is a very useful guide to obtaining the prime factors of composite numbers.

It is easy to see that the number of primes is infinite. For suppose, following Euclid, that the largest prime is p_M: then the larger number $N = p_1 p_2 \ldots p_M + 1$ is not divisible by any of the primes p_i. Therefore N is prime, which is a contradiction; or, alternatively, N is divisible by a number larger than p_M, itself not divisible by any prime, which must therefore be prime, again a contradiction.

The sequence of primes is irregular. For example, there are sequences of consecutive integers all of which are composite: for if $N = p_1 p_2 \ldots p_M$, then each of the numbers in the sequence $N+2, N+3, N+4, \ldots, N+p_M$ is composite; and since we can take M to be as large as we like, such sequences of numbers each of which is composite can be as long as we like. On the other hand, there appear to be an infinite number of pairs of primes of the form $(p, p+2)$, such as (3, 5), (11, 13), (17, 19), (29, 31). There are over 8000 such pairs among the first 1 000 000 integers, but they occur irregularly.

In spite of such irregularities, the "average" incidence of primes is remarkably regular. If we define, as is usual, $\pi(x)$ to be the number of primes which do not exceed x, then, for example, $\pi(1) = 0$, $\pi(2) = 1$, $\pi(20) = 8$, $\pi(1\ 000) = 168$, $\pi(1\ 000\ 000) = 78\ 498$. It can be shown that in the limit as $x \to \infty$ the function $\pi(x)$ behaves asymptotically as $x/\log x$. Indeed, if the Riemann hypothesis concerning the zeros of the zeta function is true, then

$$\pi(x) = L(x) + O(x^{1/2}\log x),$$

where

$$L(x) \equiv \int_2^x dt/\log t = x/\log x + x/(\log x)^2 + 2!x/(\log x)^3 + \ldots$$

The closeness of approximation is illustrated in the following table:

x	$\pi(x)$	$x/\log x$	$L(x)$
10^3	168	144.8	176.6
10^6	78498	72382.4	78626.5
10^9	50847534	48254942.4	50849233.9

Note that the ratio $x/\log x$ is large: many numbers are prime! The average number of primes in an interval Δx at x can be obtained by differentiating $L(x)$, and is about $\Delta x/\log x$. In a very cavalier fashion, we may say that the probability that a very large number x taken at random is a prime, is about $1/\log x$.

We may also estimate a rough value of the nth prime if n is large. If p_n is the nth prime, then $\pi(p_n) = n$, or $p_n = \pi^{-1}(n)$. But if $y = x/\log x$, then $\log y = \log x - \log\log x$. If x is very large, loglog x may be neglected in comparison with $\log x$. Hence, in the limit of very large x, $x \approx y\log y$: the function inverse to $x/\log x$ is asymptotic to $x\log x$. Thus $p_n \approx n\log n$ for very large n.

7.1.1 A table of primes

We present a simple routine which prepares a table of the first 6499 primes, the largest of which should be about $6500 \log 6500 \approx 57\ 000$ (actually it is 65 053). We shall use unsigned integers, which we have called "words". If many more primes are needed, they may be represented by unsigned long integers, which we have called "longwords", or by methods where the length is limited only by available memory (*cf* 20.4). The primes are stored in ascending order in an array called `primes`, and in order that `primes[n]` should be the nth prime we have chosen to make `primes[0]` = 1:

```
const int numprimes = 6500;
word   primes[numprimes];

void   genprimes()
{      primes[0] = 1; primes[1] = 2;
       word n; int i;
       for (n = 3, i = 2; i < numprimes; n+=2)    // odd only
       {     int j;
             for (j = 1; j < i; ++j)
                   if (!(n % primes[j])) break;   // not prime
             if (j == i) primes[i++] = n;         // new prime
       }
}
```

Notice that after `primes[0]` and `primes[1]` have been set, only odd integers are tested for primality, and they are tested against all the primes already found. The condition `!(n % primes[j])`, where the operator `%` stands for modulo (*cf* 2.1.1), succeeds if any of the primes already discovered divides n; the first to do so causes the `for` loop to break. Much faster methods are available for finding large primes, but they form a subject in their own right and we shall not discuss them here.

The table of primes having been constructed, it is a simple matter to write it to file (*cf* 6.5.2) so that it can be used whenever needed without having to compute it again:

```
char   *filename = "PRIMES.DTA";
const in t wordsize = sizeof(word);
ofstream ofile;

void   writeprimes()
{      ofile.open(filename, ios::out | ios::binary);
       for (int i = 0; i < numprimes; ++i)
       {     word w = primes[i];
             ofile.write((char*)&w, wordsize)
       }
       ofile.close();
}
```

Whenever the table is needed it may be read:

```
ifstream ifile;

void   readprimes()
{      ifile.open(filename, ios::in | ios::binary);
       for (int i = 0; i < numprimes; ++i)
       {     word w;
```

```
            ifile.read((char*)&w, wordsize);
            primes[i] = w;
        }
}
```

7.1.2 Prime factors

Using the table of primes we can find the prime factors of any number less than the square of the largest prime in the table, and of many others still larger provided they are not prime. We shall therefore accept the number to be tested in the form of a longword. The following routine tests whether the number is prime, and if not finds the number of times successive primes are divisors. We shall need to store these powers:

```
int     powers[numprimes];        // stores powers

bool  factors(longword &n)      // is n prime?
{       int numfac = 0;
        for (int i = 1; i < numprimes; ++i)
        {     word psi = primes[i];
              while (!(n % psi))
              {     ++numfac;
                    ++powers[i];
                    n /= psi;
              }
        }
        return (numfac < 2) ? TRUE : FALSE;
}
```

Note that the number is passed to the routine by reference, so its value is reduced by the routine if it is not a prime, the final value being either 1, in which case all prime factors have been found, or is a prime number greater than the largest tabulated but less than the square of the largest tabulated, or is simply too large to handle and therefore may not be a prime after all. These various possibilities must be taken care of in the output.

We shall need to tabulate the prime factors and the number of times each occurs:

```
void  tabulate()
{     cout << endl << tab << "prime"
            << tab << "power" << endl << endl;
      for (int i = 1; i < numprimes; ++i)
      {       if (powers[i])
            cout << tab << primes[i]
                  << tab << powers[i] << endl;
      }
      cout << endl;
}
```

The output function may then be written:

```
void  putout(longword n, bool prime)      // n is final quotient
{     word biggest = primes[numprimes - 1];
      longword limit = (long) biggest * biggest;
      if (n <= limit)
      {     if (prime) statement("is prime\n");
            else
            {     tabulate();
                  if (n == 1) statement("list complete\n");
                  else
                  {     output("final quotient = ", n);
                        statement("\nis prime\n");
                  }
            }
      }     else
      {     tabulate();
            output("final quotient = ", n);
            statement("\nis too large!\n");
      }     newline;
}
```

where the optional macros, `statement`, `output` and `input`, were defined in 4.12, and are to be found in the header file `standard.h` reproduced in Appendix A.

Finally, we offer a user program to find the prime factors of a succession of numbers chosen in turn by the user. It is necessary each time to initialize the array `powers`. This may be done with the simple routine:

```
void  clearpowers()
{     for (int i = 0; i < numprimes; ++i)
            powers[i] = 0;
}
```

The user program might then contain:

```
main()
{     clear;
      readprimes();
      do
      {     clear;
            clearpowers();
            longword n;
            input("number = ", n);
            bool prime = factors(n);
```

```
                putout(n, prime);
        }       while (getyes("another number? - (Y/N): "));
}
```

where the interactive boolean function `getyes` also was defined in 4.12 and declared in the header file `standard.h` of Appendix A. On entering a number for factorization, the screen will typically display:

```
number = 185562
      prime  power
      2      1
      3      2
      13     2
      61     1

list complete

another number? - (Y/N):
```

There are much faster methods available for factoring large integers, including some which involve addition and subtraction only, but we shall not describe them here.

7.1.3 The greatest common divisor

We can easily obtain closed formulae for the *greatest common divisor* and the *least common multiple* of two or more integers in terms of their prime factors. For if $m = \prod_p p^\alpha$, in which the product is over all prime divisors of m and we have suppressed the suffix on α_p, and similarly if $n = \prod_p p^\beta$, then:

$$\gcd(m,n) = \prod_p p^{\min(\alpha,\beta)}, \quad \mathrm{lcm}(m,n) = \prod_p p^{\max(\alpha,\beta)},$$

so that $$\mathrm{lcm}(m, n) = mn/\gcd(m, n).$$

We *could* therefore obtain these arithmetic functions by first factoring the pair of numbers into primes; for example, $84 = 2^2.3.7$, $90 = 2.3^2.5$, so $\gcd(84,90) = 6$ and $\mathrm{lcm}(84,90) = 1260$. It is easy to write a routine to perform these computations for us. It is, however, much faster to compute the greatest common divisor directly using Euclid's recursive algorithm which was briefly introduced in 2.7.

Euclid's algorithm is based upon the following simple facts. Firstly, if $u \neq 0$, the largest integer that divides u without remainder is $|u|$; u and v are both divisible by 1; therefore there exists a greatest common divisor between 1 and the lesser of $|u|$ and $|v|$. Second, if $\gcd(u, v)$ is the greatest common divisor of the integers u and v, and q is any integer, then $\gcd(v, u - v) = \gcd(u, v)$. For if there is an integer d such that $d|u$ and $d|v$, then $d|(u - qv)$

also; and conversely, if $d \mid v$ and $d \mid (u - qv)$, then $d \mid u$ also. In particular, $\gcd(u, v) = \gcd(v, u)$. It is customary to define $\gcd(u, 0) = |u|$.

In terms of long integers, and with local output suppressed, Euclid's algorithm becomes:

```
long gcd(long u, long v)
{       if (v == 0) return labs(u);
        return gcd(v, u % v);
}
```

where `long labs(long)` is a function returning the absolute value of a long integer. This routine may also be used to compute the least common multiple.

Two integers u, v having no common divisor other than 1, so that $\gcd(u, v) = 1$, are said to be prime to each other, or relatively prime, or *coprime*. Note that to be coprime neither number need be prime; for example, 9 and 10 are coprime. A set of numbers, $(u, v, \ldots, w)$ is said to be coprime if and only if every pair in the set is coprime.

It is very likely that a pair of numbers will turn out to be coprime. To obtain some feeling for this, assume (without proof!) that there exists a well-defined probability p that $\gcd(u, v) = 1$ for pairs of integers u,v chosen at random. Now $\gcd(u, v) = d \, (\neq 1)$ if and only if $d \mid u$, $d \mid v$ and $\gcd(u/d, v/d) = 1$. Therefore the probability that $\gcd(u, v) = d$ is p/d^2. If we sum this probability over all possible values of d we must obtain unity for the total probability:

$$1 = (p/1^2) + (p/2^2) + (p/3^2) + (p/4^2) + \ldots = (\pi^2/6)p.$$

Thus, the probability that two integers, chosen at random, are coprime is $6/\pi^2 \approx 0.61$, a result which in less cavalier fashion was first obtained by Dirichlet. There is a 61% probability that two random integers have no common divisor other than unity. In most practical problems (except those which deliberately use random sequences) numbers are unlikely to arise completely at random; nevertheless, we have good reason to suppose that numbers will frequently turn out to be relatively prime.

7.2 Rational numbers

A rational number is defined as a pair of integers (u, v) which are relatively prime. They represent the fraction u/v. Hitherto we have represented fractions by floating point numbers. Since these are subject to rounding errors it is sometimes useful to be able to calculate directly in terms of rational numbers instead, thus obtaining *exact* answers just as we do with integers (provided that overflow is avoided).

The number pair (u, v) represents not only the fraction u/v, but also all fractions ud/vd, where d is any integer. Given a fraction, 15/6, say, it is first necessary to divide numerator and denominator by their greatest common divisor in order that the pair of numbers which represent the fraction as a rational number be relatively prime. The greatest common divisor of 15 and 6 is 3, and the rational number representing 15/6 is therefore (5,2), which of course we write 5/2. This process may be called reduction to canonical form. Thus, given *any* pair of integers, x and y, the rational number representing the fraction x/y is the canonical fraction $(x/g)/(y/g)$, where $g = \gcd(x, y)$.

7.2.1 The class rational

As we have seen, it is very likely that the integers occurring in rational arithmetic will be coprime. That being so, there is a marked tendency for numbers to become very large in the course of arithmetical calculations. Let us therefore define the class which represents rational numbers by pairs (num, den) of long integers, with the convention that den $\geq$ 1:

```
class rational {
private:
        long num, den;
        long gcd(long, long);
public:
        rational()               { num = 0; den = 1; }
        rational(long n)         { num = n; den = 1; }
        rational(long, long);
        rational(rational &r)    { num = r.num; den = r.den; }
        operator double()        { return double(num)/den; }
        . . . . . . . . . . . . . . . . . . . . . . . . . . . . . . . .
};
```

The header file rational.h is shown in Appendix C; it contains the full range of arithmetic operations appropriate to rational arithmetic. We have redefined gcd as a private member function so that it is local to the class. This means that we can define gcd differently for use elsewhere without affecting its use in the class rational.

We provide four constructors, three defined inline. The first is the default constructor which defines the rational zero, with num = 0, den = 1. The second converts an ordinary long integer into a rational number in canonical form, with den = 1. The fourth is the usual copy constructor.

The third constructor defines a rational number by stating a pair of long integers which must be reduced to canonical form. It may be written:

```
rational::rational(long n, long d)
{       long g = gcd(n, d);
        if (d < 0) { n = -n; d = -d; }
        num = (n == 0) ? 0 : n/g;
        den = (n == 0) ? 1 : d/g;
}
```

We also include a new kind of operation—the conversion of a rational number to its floating point approximation. C++ allows the keyword operator to be used with the keyword `double` in the combination `operator double( )` in order to achieve this conversion. We have defined the operator inline. In a user program we may then write

```
rational r = ...;   double d = double(r);
```

No return type is given for the function `double( )`: it is already implicit, and it is an error to provide for it explicitly. Likewise, we could provide for conversions to other fundamental types by defining `operator` *type*`( )` appropriately, but we have no need to do so here.

Constructors such as `rational::rational(int n, int d)` are not needed because C++ provides automatic conversion from `int` to `long`.

In the header file we provide for the usual unary plus and minus. There follow the boolean operators ==, >, <, >=, <=. The equality operator may be written:

```
bool   rational::operator==(rational &r)
{      return (num==r.num && den==r.den) ? TRUE : FALSE;
}
```

The operator > may be written:

```
bool   rational::operator>(rational &r)
{      double s = double(*this);
       double t = double(r);
       return (s > t) ? TRUE : FALSE;
}
```

In a sense, this is a cheat: in order to effect the comparison it converts the rational numbers into floating point numbers using the conversion operator just described. The operator >= may be constructed from the previous two:

```
bool   rational::operator>=(rational &r)
{      bool s = (*this > r);
       bool t = (*this == r);
       return (s || t) ? TRUE : FALSE;
}
```

The operators <, <= are similar. Next comes the assignment operator:

```
rational& rational::operator=(rational &r)
{      num = r.num;
       den = r.den;
       return *this;
}
```

The arithmetic operators with or without assignment present a problem. We are concerned to reduce to canonical form as efficiently as possible expressions like $r = s$ @ t, where r, s, t are rational numbers and @ represents any of the operators $+$, $-$, $*$, $/$. Consider first the case of multiplication. In an obvious notation:

$$\frac{r_n}{r_d} = \frac{s_n}{s_d} \times \frac{t_n}{t_d}.$$

It is tempting to set $r_n = s_n \times t_n$, $r_d = s_d \times t_d$, and then reduce this to canonical form by dividing both r_n and r_d by their greatest common divisor. These products, however, are in general larger than they need be, because s_n and t_d may have a common divisor greater than one, and so may t_n and s_d. It is therefore better to proceed as follows. Let $g_1 = \gcd(s_n, t_d)$ and $g_2 = \gcd(t_n, s_d)$. Then the rational number r given by

$$r_n = (s_n/g_1) \times (t_n/g_2), \quad r_d = (s_d/g_2) \times (t_d/g_1),$$

is in canonical form. Thus, premature overflow is avoided.

The case of division is similar, except that an error message should be generated if $t_n = 0$, and due account must be taken of signs. Addition and subtraction are more complicated. Let $g_1 = \gcd(s_d, t_d)$. Then, if $g_1 = 1$ (and this is very likely!),

$$r_n = s_n t_d \pm s_d t_n \text{ and } r_d = s_d t_d.$$

But if $g_1 > 1$, let $z = s_n(t_d/g_1) \pm t_n(s_d/g_1)$, and $g_2 = \gcd(z, g_1)$. Then the rational number given by

$$r_n = t_n/g_2 \text{ and } r_d = (s_d/g_1) \times (t_d/g_2)$$

is in canonical form.

The arithmetic assignment operators may therefore be defined as follows; for addition:

```
rational& rational::operator+=(rational &r)
{      long g1 = gcd(den, r.den);
       if (g1 == 1)                      // 61% probability!
       {      num = num*r.den + den*r.num;
              den = den*r.den;
       }      else
       {      long t = num*(r.den/g1) + (den/g1)*r.num;
              long g2 = gcd(t, g1);
              num = t/g2;
              den = (den/g1)*(r.den/g2);
       }
       return *this;
}
```

and similarly for subtraction. For multiplication by a rational, we have

```
rational& rational::operator*=(rational &r)
{      long g1 = gcd(num, r.den);
       long g2 = gcd(den, r.num);
       num = (num/g1)*(r.num/g2);
       den = (den/g2)*(r.den/g1);
       return *this;
}
```

and by an integer,

```
rational& rational::operator*=(long i)
{      long g = gcd(den, i);
       num *= (i/g);
       den /= g;
       return *this;
}
```

For division we can write:

```
rational& rational::operator/=(rational &r)
{      if (r == 0)
              error("zero divisor in rational::op/=(...)");
       long g1 = gcd(num, r.num);
       long g2 = gcd(den, r.den);
       num = (num/g1)*(r.den/g2)*sign(r.num);
       den = labs((den/g2)*(r.num/g1));
       return *this;
}
```

```
rational& rational::operator/=(long i)
{      if (i == 0) error("zero divisor in rational::op/=(...)");
       long g = gcd(num, i);
       num = (num/g)*sign(i);
       den *= labs(i/g);
       return *this;
}
```

Finally, there are some friend operators, beginning with the rational equivalent of `fabs`:

```
rational rabs(rational &r)
{      if (r < 0) return -r;
       return r;
}
```

The arithmetical operators may be obtained in the usual way from the arithmetical assignment operators. For example:

```
rational operator*(rational &r1, rational &r2)
{       rational r;
        long g1 = gcd(r1.num, r2.den);
        long g2 = gcd(r1.den, r2.num);
        r.num = (r1.num/g1)*(r2.num/g2);
        r.den = (r1.den/g2)*(r2.den/g1);
        return r;
}
```

Last of all come the output and input operators, of which output at least should be formatted. We need to print the numerator right-justified so that it may be followed immediately by the division sign, and then by the denominator left-justified (*cf* 6.4.4):

```
long ln = ios::right, ld = ios::left;
format fn(ln,0,6), fd(ld,0,6), rs(0,0,0);
```

where we have chosen a fieldwidth of 6. No value need be offered for the precision because there is no decimal point.

The output operator may now be written

```
ostream& operator<<(ostream &s, rational &r)
{       s << fn << r.num << rs << "/"
          << fd << r.den << rs;
        return s;
}
```

The input operator need not be formatted if input is to be taken only from the keyboard:

```
istream& operator>>(istream &s, rational &r)
{       long rn, rd;
        s >> rn >> rd;
        r = rational(rn, rd);
        return s;
}
```

This completes the class `rational`. The reader might wish to consider, however, what advantages might accrue to the treatment of rational numbers from admitting as valid the three special constants ($\pm 1, 0$) and ($0, 0$).

In passing, it may be wondered why, in a language that permits the overloading of function names, we have to use `fabs` for the absolute value of a floating point number, `abs` for an integer, `labs` for a long integer, and now `rabs` for a rational. It is because it is customary to follow the C convention in which overloading was not available. In C++ they could all have been named, say, `abs`; only the signatures and return types would differ.

7.2.2 The harmonic numbers

The harmonic numbers occur in many applications. They are defined by

$$H_n = (1/1) + (1/2) + (1/3) + \ldots + (1/n).$$

It is well-known that the sequence $H_1, H_2, H_3, \ldots$ does not converge, but there is an asymptotic approximation given by

$$H_n = \log n + \gamma + (1/2n) - (1/12n^2) + O(1/n^4),$$

where γ is Euler's constant, $\gamma = 0.5772156649. \ldots$ We shall write a simple routine for the harmonic numbers and use it to estimate γ. This will illustrate the ease with which rational numbers may be used. We choose a simple recursive routine

```
rational harmonic(int n)
{       if (n == 1) return 1;
        return rational(1,n) + harmonic(n-1);
}
```

The following program then prints out each harmonic number, together with an estimate of γ:

```
main()
{       clear;
        for (int n = 1; n < 25; ++n)
        {       rational r = harmonic(n);
                double g = double(r) - log(n);
                g -= (1.0/(2*n))(1.0 - (1.0/(6*n)));
                cout << n << tab << g << tab << r << endl;
        }
}
```

The program is eventually limited by overflow, but the first few approximations are shown below, with the corresponding harmonic numbers:

n	γ_{est}	H_n
1	0.583333	1/1
2	0.577686	3/2
3	0.577314	11/6
4	0.577247	25/12
5	0.577229	137/60
6	0.577222	49/20
7	0.577219	363/140
8	0.577218	761/280
9	0.577217	7129/2520
10	0.577216	7381/2520

7.2.3 Bernoulli's numbers

Another example of the use of rational numbers is the calculation of Bernoulli's numbers B_n, which are of great importance for the construction of asymptotic series. They are defined in terms of a particular series expansion:

$$x/(e^x - 1) = B_0 + B_1 x + (B_2 x^2/2!) + (B_3 x^3/3!) + \ldots = \sum_{k \geq 0} B_k x^k/k!$$

Multiplying both sides by $e^x - 1$ and equating powers of x, we obtain:

$$\sum_{k<n} \binom{n}{k} B_k = 1 \text{ if } n = 1, 0 \text{ otherwise}$$

where $\binom{n}{k}$ is a binomial coefficient. From this equation the $(n-1)$-th Bernoulli number may be obtained from the lower numbers, presumed already known. Evidently, $B_0 = 1$ and $B_1 = -1/2$. Apart from B_1, all the odd Bernoulli numbers are zero.

We shall write a routine to compute these numbers. First, however, we need a routine to compute the binomial coefficients. They satisfy the relationship:

$$\binom{n}{k} = \binom{n-1}{k} + \binom{n-1}{k-1},$$

and it seems obvious that we should use a recursive routine to compute them:

```
long   recbin(int n, int k)
{      if (k == 0 || k == n) return 1;
       if (k == 1 || k == n-1) return n;
       return recbin(n-1, k)
              + recbin(n-1, k-1);

}
```

We have represented the binomial coefficients as long integers because they become very large for large n.

However, this function is "doubly recursive", resembling the function we first wrote down for the Fibonacci sequence in 3.11. Like the function fib, it becomes exceedingly slow for large n because of the excessive number of function calls it generates. It is possible to convert it into a singly recursive function that is not tail-recursive, as we did for the Fibonacci sequence, and this produces a much faster function. In this case, however, it is better simply to write a tail-recursive function based instead on the definition $\binom{n}{k} = n!/k!(n-k)!$. The recursive formula needed is $\binom{n}{k} = [(n-k+1)/k]\binom{n}{k-1}$. Moreover, we shall need to use the binomial coefficients in the form of rationals. In the function that results we include suitable error conditions and take advantage of the symmetry property $\binom{n}{k} = \binom{n}{n-k}$:

```
rational binomial(int n, int k)
{       if (n < 0)
                error("1st index out of range in binomial(...)");
        if (k < 0 || k > n)
                error("2nd index out of range in binomial(...)");
        if (k > n-k) k = n-k;
        if (k == 0) return 1;
        return rational(n-k+1, k)*binomial(n, k-1);

}
```

A function may now be written which returns the Bernoulli numbers. It is recursive, but not tail-recursive. However, overflow may well take place before memory is exhausted. If that were not so, it would again be better to write a non-recursive routine. The recursive function is:

```
rational bernoulli(int n)
{       if (n < 0) error("index out of range in bernoulli(int)");
        if (n == 0) return rational(1,1);
        if (n == 1) return rational(-1,2);
        if (n % 2) return 0;
        rational r = 0;
        for (int k = 0; k < n; ++k)
        {       rational s = binomial(n+1,k);
                s /= rational(n+1, 1);
                r -= s*bernoulli(k);
        }
        return r;
}
```

The lower non-zero Bernoulli numbers are:

$B_0 = 1$, $B_1 = -1/2$, $B_2 = 1/6$, $B_4 = -1/30$, $B_6 = 1/42$,
$B_8 = -1/30$, $B_{10} = 5/66$, $B_{12} = -691/2730$, $B_{14} = 7/6$, $B_{16} = -3617/510$,
$B_{18} = 43867/798$, $B_{20} = -174611/330$, $B_{22} = 854513/138$.

The rapid growth in the size of the numbers beyond about $n = 10$ is characteristic of an asymptotic series, which is strictly speaking divergent. It may nevertheless be used to considerable effect. For example, the sum of a series can often be converted into an integral together with correction terms arising from the successive derivatives of the function at the extremities of the integral. The *Euler–Maclarin summation formula* is:

$$\sum_{0 \leq k < n} f(x) = \int_0^n f(x)\mathrm{d}x + \tfrac{1}{2}[f(n) + f(0)] + B_2[f'(n) - f'(0)]/2!$$

$$+ B_4[f^{(3)}(n) - f^{(3)}(0)]/4! + \ldots$$

This will connect with the work of 16.5 on the asymptotic errors of integration. In particular, setting $f(x) = 1/x$, we find that the harmonic numbers may also be expressed asymptotically in terms of the Bernoulli numbers:

$$H_n = \log n + \gamma + \sum_{1 \leq k \leq m} (-)^{k-1} B_k / k n^k + O(1/n^m).$$

Alternatively, setting $f(x) = \log x$, and after a little manipulation, we obtain Stirling's well-known asymptotic formula for the factorial:

$$n! = (2\pi n)^{1/2}(n/\mathrm{e})^n [1 + (1/12n) + (1/288n^2) + O(1/n^3)].$$

7.3 Congruences and residues

If m, n, q, r are integers, m being positive, such that $n = qm + r$, then $m \mid (n - r)$, which we may write alternatively as $n \equiv r \pmod m$. If, further, $0 \leq r < m$, r (then being the remainder after division of n by m) is called the *residue* of n modulo m, or simply the *remainder*, and it is usual to write $r = n \bmod m$, or in the C language, $r = n \% m$. Clearly, there are m residues, namely the integers $0, 1, \ldots, m - 1$; and any integer whatsoever is congruent $(\bmod \, m)$ to one or other of these integers, which therefore comprise a complete set of residues $(\bmod \, m)$.

If $a \equiv b$ and $x \equiv y \pmod m$, it is easy to see that $a \pm x \equiv b \pm y$ and $ax \equiv by \pmod m$. On the other hand, $x \equiv y \pmod m$ is a necessary consequence of $ax \equiv by$ and $a \equiv b$ if and only if a and m are coprime. Furthermore, if r and s are coprime, $a \equiv b \pmod{rs}$ if and only if $a \equiv b \pmod r$ and $a \equiv b \pmod s$. These are the equivalents for congruence of the ordinary arithmetical operations.

In number theory we are usually dealing with non-negative integers. In numerical computations, however, negative integers may also occur. We shall assume that the expression `n % m` has the same sign as n. The precise conventions regarding signs of the integer operations `/` and `%` unfortunately depend on the implementation, but the ANSI C standard requires that `(a/b)*b + a%b = a` in all cases (provided $b \neq 0$). Conventionally, $x \bmod 0 = x$.

It is sometimes convenient to use the same notation even when the quantities involved are not all integers. For example, we may wish to write $(3/2) \equiv (1/2) \pmod 1$, or $-\pi \equiv \pi \pmod{2\pi}$.

We can even write x mod 1 for the fractional part of a real variable x, namely $\{x\} = x - \text{int}(x)$, although in C++ the expression `int(x)` is implementation dependent if `x` is negative. Mostly, however, we shall be concerned with integers, and indeed, the C operator `%` is defined only for integers.

Congruences are more familiar than is usually supposed. Railway timetables are congruences with moduli 365, 7 and 24 (except in leap year). The congruence notation is intended to resemble that for equations. We may hope, for example, to say something about the solutions of the congruence

$$f(x) \equiv 0 \ (\text{mod } m).$$

However, what we can say may surprise us. Whereas the solutions of the equation $x^2 = 1$ are $x = -1, 1$, the solutions of the congruence $x^2 \equiv 1 \ (\text{mod } 8)$ are $x = 1, 3, 5, 7$, while there is no solution at all to $x^2 \equiv 2 \ (\text{mod } 8)$.

7.3.1 Fermat's theorem

Following the book by Hardy and Wright, we denote by $\phi(m)$ the number of integers n between 0 and m which are prime to m, so that $\gcd(m, n) = 1$. Thus $\phi(9) = 6$, the integers prime to 9 being 1, 2, 4, 5, 7, 8. The function ϕ is called *Euler's function*, and it is customary to set $\phi(1) = 1$. If $m = p$, a prime, then $\phi(p) = p - 1 = p(1 - 1/p)$ (where, however, the division $1/p$ is *not* to be interpreted as integer division). If $m = p^\alpha$, where α is a positive integer, only the multiples hp, with $h = 0, 1, 2, \ldots, p^{\alpha} - 1$, are not prime to p^α. Hence,

$$\phi(p^\alpha) = (p^\alpha - 1) - (p^{\alpha-1} - 1) = p^\alpha(1 - 1/p).$$

Now Euler's function ϕ is *multiplicative* in the sense that, if $\gcd(m, m') = 1$, then $\phi(mm') = \phi(m)\phi(m')$. For let a, a' each run through a complete set of residues (mod m) and (mod m'), respectively: then $a'm + am'$ runs through a complete set of residues (mod mm'). The prime factors of m are coprime: it follows, if $m = \prod_{p|m} p^\alpha$, that

$$\phi(m) = m \prod_{p|m}(1 - 1/p).$$

Thus, in the case of $m = 9$, the only prime to divide 9 is 3, so $\phi(9) = 9 \times 2/3 = 6$, as we have already found. And once the prime factors of m are known we have a ready method for calculating ϕ. Thus, since $10^6 = 2^6.5^6$, it follows that $\phi(10^6) = (2^6 - 2^5)(5^6 - 5^5) = 4.10^5$.

We may now state *Fermat's theorem*. If p is prime, then $a^p \equiv a \ (\text{mod } p)$, or equivalently, $a^{p-1} \equiv 1 \ (\text{mod } p)$ if $p \nmid a$. Both statements are particular cases of the more general theorem usually attributed to Euler:

$$a^{\phi(m)} \equiv 1 \ (\text{mod } m), \quad \gcd(a, m) = 1.$$

For if x runs through a complete set of residues (mod m), then ax does likewise. Taking the product of each set we have $\Pi(ax) \equiv \Pi x$ (mod m), and therefore $a^{\phi(m)}\Pi x \equiv \Pi x$ (mod m). The theorem follows since a and m are coprime.

7.3.2 Periodic sequences

Suppose we have a function $f(k)$ which maps a given range of integers into itself: for example, if $0 \le k < m$, then $0 \le f(k) < m$ also. Apart from that there is no restriction on the function. Obviously, $f(k) = k$ and $f(k) = m - k - 1$ are examples of such functions. Then the sequence $x_0, x_1, x_2, \ldots$ defined by the iteration $x_{n+1} = f(x_n)$ being indefinitely repeated, with x_0 given and $0 \le x_0 < m$, is ultimately periodic. By this we mean that there are integers λ and μ for which the numbers $x_0, x_1, \ldots, x_\mu, \ldots, x_{\mu+\lambda-1}$, are distinct, but $x_{k+\lambda} = x_k$ for $k \ge \mu$. Furthermore, there is an integer n such that $x_n = x_{2n}$.

Since there are only m distinct integers available (mod m) there must come a point when an existing member of the sequence is repeated. Let the first repetition occur at $x_{\lambda+\mu}$, which implies definitions of both λ and μ. Thereafter the sequence repeats itself indefinitely, with "period" λ. Clearly, $0 \le \mu < m$, $0 < \lambda \le m$, and $\lambda + \mu \le m$. Now if $k \ge n$, then $x_k = x_n$ if and only if $\lambda \mid (k - n)$ and $n \ge \mu$. Thus, putting $k = 2n$, we find that $x_{2n} = x_n$ if and only if $\lambda \mid n$ and $n \ge \mu$. Note that the longest possible period is $\lambda_{max} = m$, but whether this is attained in any particular case depends upon the value of x_0 and, of course, on the properties of f.

The residues (mod m) provide us with an example of what we seek. Consider the function $f(x) \equiv x^2 + 13$ (mod 32). If $x_0 = 0$, the first few terms of $x_{n+1} = (x_n^2 + 13)$ mod 32 are:

$i = 0$	1	2	3	4	5	6	7	8
$x_i = 0$	13	22	17	14	17	14	17	14

Thus $\mu = 3$, $\lambda = 2$, $n = 4$, satisfying all the requirements above.

7.4 Random numbers

We wish to generate sequences of numbers which are "random" in the sense that they are uniformly distributed over a certain interval, say [0, 1], and to the degree of accuracy required by any likely application it is not possible to predict the value of the next number from knowledge of those that precede it. We shall in fact be generating non-negative integers n_i; if the maximum integer generated is n_{max}, then the corresponding floating-point number is $x_i = \text{float}(n_i)/n_{max}$. In any case, there are at best only n_{max} different numbers in the sequence, so we shall in practice wish n_{max} to be larger, and preferably much larger, than the number of different numbers required in a user program. The treatment of random sequences given here is based upon Volume 2 of the book by Knuth.

7.4.1 Linear congruence generators

The number of different integers we can produce simply and efficiently with a computer is large but finite, limited by the word length of the machine. A sequence generator is therefore of the kind we have studied in 16.3.2. It *must* be periodic, and therefore not strictly random, although one hopes the period is long enough that any short sub-sequence *appears* to be so; and congruences offer a suitable vehicle. The computation is simplest if the congruence is linear, and although there are other possibilities we therefore consider the linear congruencial sequence

$$x_{n+1} = (ax_n + c) \bmod m,$$

where the *multiplier a*, the *increment c*, and the *modulus m* are constants, and x_0 is the *starting value*.

To get a preliminary feel for the sequences produced by a linear congruence, we write a small program to generate some sequences of chosen modulus m and length n, and then examine the effect of changing the parameters a and c. The parameter s is the *seed* which activates the generator: it is passed by reference to the generator and is automatically updated on every call. The first number generated is x_0. We shall start as we intend to continue, representing all numbers by unsigned long integers, and in the function main we use the macro input introduced in 4.12 and also defined in the header file standard.h reproduced in Appendix A:

```
#include <standard.h>

longword m, a, c, s, n;

inline longword next(longword &s)
{       return s = (a*s + c) % m;  }

main()
{       clear;
        input("m = ", m);                    // modulus
        input("a = ", a);                    // multiplier
        input("c = ", c);                    // increment
        input("s = ", s);                    // seed
        input("n = ", n);                    // length
        for (longword i = 0; i < n; ++i)
                cout << next(s) << tab;
}
```

We intend eventually to use large integers in order to obtain sequences of long period. For the moment, however, let us restrict ourselves to m = 32 and n = 36, varying the remaining parameters. The following are some typical sequences:

(A) $m = 32$, $a = 13$, $c = 1$, $s = 123456$, $n = 36$ (period 32)

1	14	23	12	29	26	19	24	25	6	15	4	21	18	11	16	17	30
7	28	13	10	3	8	9	22	31	20	5	2	27	0	1	14	23	12

(B) $m = 32$, $a = 17$, $c = 1$, $s = 243156$, $n = 36$ (period 32)

21	6	7	24	25	10	11	28	29	14	15	0	1	18	19	4	5	22
23	8	9	26	27	12	13	30	31	16	17	2	3	20	21	6	7	24

(C) $m = 32$, $a = 19$, $c = 1$, $s = 312457$, $n = 36$ (period 16)

12	5	0	1	20	29	8	25	28	21	16	17	4	13	24	9	12	5
0	1	20	29	8	25	28	21	16	17	4	13	24	9	12	5	0	1

(D) $m = 32$, $a = 10$, $c = 4$, $s = 2436578$, $n = 36$ (period 1)

24	20	12	28	28	28	28	28	28	28	28	28	28	28	28	28	28	28
28	28	28	28	28	28	28	28	28	28	28	28	28	28	28	28	28	28

There is no need for statistical tests! Sequence *A* seems fairly random, *B* by no means so, proceeding in ordered pairs; but both *A* and *B* have a period equal to the maximum 32. Sequence *C* looks fairly random, but has a period of only 16, so only half the numbers (mod 32) are present; sequence *D*, after a short settling down sequence, has period 1. Clearly, we need to study how the nature of the sequence depends upon the values of the parameters.

7.4.2 Choice of parameters

If we apply the generator twice we get

$$x_{n+2} = (a(ax_n + c) + c) = (a^2 x_n + [(a_2 - 1)/(a - 1)]c) \bmod m,$$

and if we apply it k times we get

$$x_{n+k} = (a^k x_n + [(a^k - 1)/(a - 1)]c) \bmod m.$$

Let us consider the circumstances in which we get the full period $\lambda = m$. We assume that $0 \leq a, c < m$. It is clear that if $a = 0$ or $a = 1$ we do not get anything approaching random behaviour. We therefore assume $a > 1$. First, if $c = 0$, then $x_{n+1} = ax_n \bmod m$, and if the sequence contains 0, all successive terms are also 0, and the period is 1. Therefore $c > 0$ if $\lambda = m$. Since the period is m we may take $x_0 = 0$, because the sequence will certainly include that value somewhere and starting with it produces the whole sequence. In that case,

$$x_n = [(a^n - 1)/(a - 1)]c \bmod m,$$

which is a closed formula for the entire sequence, very useful for theoretical purposes although not for computation.

Again, if $\lambda = m$, the sequence must somewhere contain the value 1. Denoting the multiplying factor involving a by k, we then have $kc \bmod m = 1$, or $kc = qm + 1$, for some integer q. If $g = \gcd(c, m)$, then $g(kc/g - qm/g) = 1$, an equation of which there is no solution unless $g = 1$. Thus, if the period is to be m, c and m must be coprime. In sequences A and B, which both have period m, c and m are coprime. The converse, however, does not follow. In sequence D, c and m are not coprime, but in sequence C they are. The coprimality of c and m is a necessary but not sufficient condition for the period to be m, the maximum possible.

Next, suppose that $m = p^\alpha$, where p is a prime and α is a positive integer. We shall suppose $p^\alpha > 2$, so that either $p > 2$, or $p = 2$, and $\alpha > 1$. We wish to find the circumstances in which the period $\lambda = p^\alpha$ also. If it is, we can as before set $x_0 = 0$ and use the closed formula for the sequence:

$$x_n = k_n c \bmod p^\alpha, \quad \text{where } k_n = (a^n - 1)/(a - 1).$$

Since c is coprime to m, we are interested in the lowest value of λ for which $k_\lambda \equiv 0 \pmod{p^\alpha}$. We shall prove that $\lambda = p^\alpha$ if and only if $a \equiv 1 \pmod p$ when $p > 2$ and $a \equiv 1 \pmod 4$ when $p = 2$.

Firstly, assume $\lambda = p^\alpha$. Then if $a \not\equiv 1 \pmod p$, $k_n \equiv 0 \pmod{p^\alpha}$ if and only if $a^n - 1 \equiv 0 \pmod{p^\alpha}$. Putting $n = \lambda$, this then implies that $a^\lambda \equiv 1 \pmod{p^\alpha}$, which in turn implies $a^\lambda \equiv 1 \pmod p$. But we know from Fermat's theorem that $a^p \equiv a \pmod p$, and by induction that $a^\lambda \equiv a \pmod p$ also. We therefore have a contradiction, and must conclude that if $\lambda = p^\alpha$ then $a \equiv 1 \pmod p$. And if $p = 2$, and $a \not\equiv 1 \pmod 4$, then $a \equiv 3 \pmod 4$. Let $\sigma = 2^{\alpha-1}$. Then $(a^\sigma - 1)/(a - 1) \equiv 0 \pmod{2^\alpha}$. We have therefore proved that if $\lambda = p^\alpha$, a must be of the form $1 + qp^\gamma$, where $p^\gamma > 2$, and $p \nmid q$.

Second, the converse may also be proved. We need the following lemma. Let x be an integer such that

$$x \equiv 1 \pmod{p^\alpha}, \quad x \not\equiv 1 \pmod{p^{\alpha+1}}.$$

Then,

$$x^p \equiv 1 \pmod{p^{\alpha+1}}, \quad x^p \not\equiv 1 \pmod{p^{\alpha+2}}.$$

Now by the condition of the lemma, $x = 1 + qp^\alpha$, where $p \nmid q$. Expanding by means of the binomial theorem, we have

$$x^p = 1 + qp^{\alpha+1}(1 + Q),$$

where Q is an integer. Thus $x^p = 1 + q'p^{\alpha+1}$, where $q' = q(1 + Q)$. Therefore $x^p \equiv 1 \pmod{p^{\alpha+1}}$.

But

$$Q = \binom{p}{2}qp^{\alpha-1} + \binom{p}{3}q^2p^{2\alpha-1} + \ldots + \binom{p}{p}q^{p-1}p^{(p-1)\alpha-1},$$

which is divisible by p because $(p - 1)\alpha > 1$ if, as supposed, $p^\alpha > 2$. Therefore $x^p \equiv 1 + qp^{\alpha+1} \pmod{p^{\alpha+2}}$, or $x^p \not\equiv 1 \pmod{p^{\alpha+2}}$.

By repeated application of the lemma, if $\rho = p^\beta$, then $a^\rho \equiv 1 \pmod{p^{\beta+\gamma}}$, and $a^\rho \not\equiv 1 \pmod{p^{\beta+\gamma+1}}$, for all $\beta \geq 0$. Therefore,

$$(a^\rho - 1)/(a-1) \equiv 0 \pmod{p^\beta}, \quad (a^\rho - 1)/(a-1) \not\equiv 0 \pmod{p^{\beta+1}}.$$

This is true if $\beta = \alpha$, $\rho = \lambda$, and hence $\lambda = p^\alpha$.

Finally, any integer can be expressed as a product of its prime factors, $m = p_1^\alpha p_2^\beta \ldots p_z^\zeta$. Consider first the case of just two prime factors, $m = \lambda_1 \lambda_2$, where $\lambda_1 = p_1^\alpha$ and $\lambda_2 = p_2^\beta$. If (x_n) is the sequence defined by $x_{n+1} = (ax_n + c) \bmod m$, and if $x'_n = x_n \bmod d$, where $d \mid m$, then (x'_n) is the sequence defined by $x'_{n+1} = (ax'_n + c) \bmod d$. Since λ_1 and λ_2 are both divisors of m, and $\gcd(\lambda_1, \lambda_2) = 1$, then if we define $y_n = x_n \bmod \lambda_1$ and $z_n = x_n \bmod \lambda_2$, for all $n \geq 0$, then $x_n = x_k$ if and only if $y_n = y_k$ and $z_n = z_k$.

Let $\lambda' = \mathrm{lcm}(\lambda_1, \lambda_2)$. We shall show that the period of (x_n) is $\lambda = \lambda'$. Since $x_n = x_{n+\lambda}$ for sufficiently large n, then $y_n = y_{n+\lambda}$, so $\lambda_1 \mid \lambda$, and also $z_n = z_{n+\lambda}$, so $\lambda_2 \mid \lambda$. But $y_n = y_{n+\lambda'}$, and $z_n = z_{n+\lambda'}$, therefore $x_n = x_{n+\lambda'}$ and $\lambda \leq \lambda'$. Therefore $\lambda = \lambda' = \mathrm{lcm}(\lambda_1, \lambda_2)$.

Drawing all these results together, the linear congruence sequence defined by the parameters m, a, c and x_0 will have the maximum period $\lambda = m$ if and only if:

    (i) c and m are coprime;
    (ii) $p \mid (a-1)$ for every prime $p \mid m$;
    (iii) $4 \mid (a-1)$ if $4 \mid m$.

We could simply have quoted these results without proof, but without the theory (which is of interest anyway) they seemed too much of a black box. Sequences A and B of the last section satisfy these conditions and have the maximum period, in this case 32; sequences C and D do not, and the periods are less than the maximum.

One final result concerning the length of the period we shall state without proof. If $c = 0$ we have seen that we cannot expect the period to be m, and moreover we know that if any term is zero all succeeding terms are zero. The period can therefore be at most $m - 1$. If $m = p_1^\alpha p_2^\beta \ldots p_z^\zeta$, then the period $\lambda(m)$ is given by:

$$\lambda(2) = 1, \; \lambda(4) = 2, \; \lambda(2^\alpha) = 2^{\alpha-2} \text{ if } \alpha > 2;$$
$$\lambda(p^\alpha) = p^{\alpha-1}(p-1), \text{ if } p > 2;$$
$$\lambda(p_1^\alpha p_2^\beta \ldots p_z^\zeta) = \mathrm{lcm}(\lambda(p_1^\alpha) \ldots \lambda(p_z^\zeta)).$$

Note that, for $m = 1, 2, 4, p^\alpha$ and $2p^\alpha$, where p is an odd prime, $\lambda = \phi(m)$, where ϕ is Euler's function.

7.4.3 Sequential potency

In the last section we found the conditions for the period to be as large as possible, a very important requirement in practice. However, we said nothing about the degree of randomness. Sequences A and B of section 7.4.1 both have maximum period, but whereas A seems fairly random, B does not. We need some measure of the randomness of successive terms of the sequence.

We have seen that if the period is the maximum m, we may take $x_0 = 0$ in order to obtain the closed formula

$$x_n = c[(a^n - 1)/(a - 1)] \bmod m.$$

If we set $b = a - 1$, so that $a^n - 1 = (b + 1)^n - 1$, we can expand by means of the binomial theorem to obtain

$$x_n = c[n + \binom{n}{2}b + \binom{n}{3}b^2 + \ldots + \binom{n}{n}b^{n-1}] \bmod m.$$

We consider only sequences of maximum period obeying the conditions found in 7.4.2 above. There must then be positive integers for which $b^s \equiv 0 \pmod{m}$. Let s be the smallest such integer; we call s the *potency* of $a \pmod m$. Then,

$$x_n = c[n + \binom{n}{2}b + \ldots + \binom{n}{s}b^{s-1}] \bmod m.$$

Let us simplify matters slightly by setting $c = 1$. If the potency is 1, then $a = 1$, and $x_n \equiv n \pmod{m}$. Successive pairs of the sequence, $(\xi_n, \eta_n) = (x_n, x_{n+1})$, lie on the line $\xi - \eta + 1 = 0$, which is far from random behaviour. If the potency is 2, we have $x_n \equiv n + \binom{n}{2}b \pmod{m}$, again hardly random. Successive pairs lie on the three lines $\xi - \eta + 1 = 0, \pm m/2$; while successive triples $(\xi_n, \eta_n, \zeta_n) = (x_n, x_{n+1}, x_{n+2})$ lie on the four planes $\xi - 2\eta + \zeta = d$, $d \pm m, d - 2m$, where, in general, $d = cb \bmod m$. With higher potency, successive multiples will lie on hyperplanes, which are the closer together the higher the potency. The closer together the hyperplanes the more nearly random is the sequence.

It can easily be seen that, if $m = p_1^\alpha p_2^\beta \ldots p_z^\zeta$ and if $a = 1 + kp_1^{\alpha'}p_2^{\beta'}\ldots p_z^{\zeta'}$, where k and m are coprime, then the potency is the smallest value of s such that $\alpha's \geq \alpha, \beta's \geq \beta, \ldots, \zeta's \geq \zeta$. Thus the potency of sequence A is 3, while that of B is only 2, and it will be recalled that A appeared to be much more random than B. In practice, potencies ranging from 4 to 6 are required if the sequence is to be reasonably random.

The concept of potency is placed upon a firmer foundation in the *spectral test*. Successive terms of the sequence form a *lattice* of points $(x_n, x_{n+1}, \ldots, x_{n+k-1})$ in k-space. These points lie on parallel hyperplanes, which, like trees in an orchard, may be chosen in many ways according to the "inclination" of the planes. Of all families of possible parallel hyperplanes, let $1/v_k$ be the maximum distance between adjacent planes. We may refer to v_k as the *accuracy* of the generator in k-space. Of course, it is a function of m and a, but is bounded in any event by the approximate relation $v_k \leq \gamma_k m^{1/k}$, where γ_k is independent of m and varies monotonically from 1.07 for $k = 2$ to 1.29 for $k = 6$. In terms of v one calculates the "volume" associated with the "radius" v_k. This is a rough measure of the number of points likely to be found within the accuracy. It is given by $\mu_k = \pi^{k/2}v_k^k/[(k/2)!m]$. The corresponding bounds for μ_k are as follows:

k	=	2	3	4	5	6
μ_k^{max}	=	3.63	5.92	9.87	14.89	23.87

These bounds are never met by congruence generators, but the best of them have $\mu_k > 1$ for all $k \in [2, 6]$. It is interesting to note that the generator with $m = 256$, $a = 137$, $c = 187$ has $\mu_2 = 3.63$, $\mu_3 = 2.69$, $\mu_4 = 3.78$, $\mu_5 = 1.81$, $\mu_6 = 1.29$, and so passes the spectral test extremely well, but of course it is of no practical use as a random number generator because its period $m = 256$ is much too small.

7.4.4 Practical generators

Linear congruence generators have been extensively investigated and exhaustively tested. The main requirements are that the multiplier a should be chosen to guarantee the maximum period m, and that they pass various tests of which the spectral test is the most searching. The value of c turns out to be unimportant as long as it is non-zero. It is often taken to be a prime number near to $(3 - \sqrt{3})m/6$ on somewhat disreputable grounds, but that is as good a choice as any. It is also necessary to avoid overflow in the computation of $am + c$, which argues for a fairly small value of a subject to potency being high. The onset of overflow also depends on the word size of the machine being used. We have used unsigned long integers of four bytes, which severely restricts the generators that may be used.

Three generators from the many which satisfy all these requirements (see the book by Press *et al.*) are as follows:

Label	m	a	c	Potency
A	714025	1366	150889	4
B	233280	9301	49297	6
C	1771875	2416	374441	5

Most user applications will call for relatively short sequences of random numbers, containing no more than a few thousand numbers, and probably much fewer. For such purposes all three generators should be adequate. Nevertheless we shall subject them to some tests in the next chapter.

One weakness all linear congruence generators share, however: the least significant digits are also the least random. For that reason it is well to retain only the first few digits. This may most readily be done through a generator of floating-point numbers uniformly distributed over the interval $0 \le x < 1$. Such a generator is easily derived from an integer generator:

```
inline float uniform(longword &s)
{ return float(next(s))/m; }
```

If we then wish to produce a sequence of random integers in the range $0 \le n < d$, where d is an `int`, all we need do is to call `int(d*uniform(s))`. The function `uniform` is of interest in its own right, because it is often desirable to have a generator of floating-point numbers, uniformly distributed. Do not forget, however, that it is only the first few digits of such numbers that have any pretence to being randomly distributed.

7.4.5 Shuffling

It is possible to increase the randomness of a sequence of numbers produced by a random number generator by shuffling them in a suitable way. Since shuffling need never *decrease* the degree of randomness it is usual to combine a random number generator with a shuffling procedure. Of course, nothing can change the number of *different* numbers produced by a generator, and it is doubtful whether even the periodicity, once it is established, is altered by shuffling; but we may hope to destroy whatever correlation there may be between the near successors of any number in a sequence. We shall illustrate what is involved by modifying the function which produces a uniform distribution of floating point numbers.

The idea is to fill a fixed array with consecutive numbers from the generator, and then to choose at random, using the generator, which of the numbers in the array to select as the next member of the sequence. The following routine, adapted from the routine **ran0** of Press *et al.*, carries out this task. It should be seeded, and re-seeded, with a negative long integer. For that reason the congruence generator must be redefined in terms of signed `long` instead of unsigned `longword`. And for *that* reason, of the three generators in 7.4.4 we were restricted to generator A, because otherwise we might have had overflow with a four-byte integer. Machines with larger word size do not suffer such severe restrictions.

```
const long m = 714025;
const long a = 1366;
const long c = 150889;

inline  long next(long &s) { return s = (a*s+c)%m; }

int potlen = 101;                   // arbitrary pot length

float uniform(long &s)
{       static long pot[potlen];       // shuffle pot
        static long temp;
        static bool start = FALSE;
        if (s < 0 || start == FALSE)   // fill pot
        {       start = TRUE;
                for (int i = 0; i < potlen; ++i)
                        pot[i] = s = next(s);
        }
        temp = next(s);                 // random element
        int j = int(float(potlen*temp)/m);
        temp = pot[j];                  // take element
        pot[j] = next(s);               // renew pot
        return float(temp)/m;
}
```

This is the form of the function `uniform` that we shall use in the remainder of this book. From `uniform` we can generate other forms of distribution, such as the Boltzmann distribution, exponentially distributed according to a "temperature" parameter, and the Gaussian distribution, normally distributed according to a variance. They are included in the header file `random.h` in Appendix D.

7.4.6 Example—Monte Carlo integration

There is a form of integration that depends upon sampling the integrand many times at random, and summing the samples suitable weighted. It can be very useful for evaluating multiple integrals which might otherwise prove intractable. We shall illustrate the method by computing the volume of the unit sphere.

The unit sphere is confined within the surface $x^2 + y^2 + z^2 = 1$. If we restrict x, y, z to their positive values, we are dealing with a one-eighth segment of the sphere. If we choose these coordinates uniformly distributed in the interval [0, 1], and test whether the corresponding point lies within the surface, we can weight that point as 1 if it does and 0 if it does not. Averaging the weights over a large number of random samples, and multiplying the result by eight, gives us an estimate of the volume of the unit sphere:

```
float volume(int n, long &s)
{       float w = 0.0;
        for (int i = 0; i < n; ++i)
        {       float x = uniform(s);
                float y = uniform(s);
                float z = uniform(s);
                if (x*x + y*y + z*z < 1) w += 1;
        }
        return 8.0*float(w)/n;
}
```

In a similar, but somewhat more complicated fashion, we could estimate, for example, the volume of the intersection of a sphere and a torus. However, in performing Monte Carlo integrations in several dimensions the limitations of the random number generator being used must not be forgotten. Sampling is not entirely random, and only a finite, if large, number of points are in fact being sampled. There is a considerable literature devoted to the subject, and generators have been specially designed to suit the Monte Carlo method.

8 Tests of randomness

There is really no way of telling whether a single number is random. Is zero random? One can hope only to say whether a *sequence* of numbers satisfies one or more well accepted tests of randomness. It is obviously important to test whether any particular sequence satisfies appropriate tests, or (more to the point) whether any particular *sequence generator* actually generates sequences which satisfy such tests.

We cannot go into all the intricacies of statistical testing. They are fully described elsewhere. But it is perhaps worthwhile to explain the basis of the more important methods. Many of them are based upon the χ^2 test, so called. For the sake of definiteness let us consider the throwing of a pair of dice. The scores can be 2 up to 12, but whereas there is only one way of getting 2 or 12, there are several ways of getting other scores. If the dice are unloaded, the probability of any particular pair of faces turning up is 1/36. There is 1 way of scoring 2, 2 ways of scoring 3, and so on; the probabilities p_s of the various scores s are therefore:

s =	2	3	4	5	6	7	8	9	10	11	12
p_s =	1/36	1/18	1/12	1/9	5/36	1/6	5/36	1/9	1/12	1/18	1/36

Now suppose we throw the dice $n = 144 \, (= 36 \times 4)$ times, and record the number of times a given score s occurs as n_s. The "expected" number would be $e_s = np_s$—expected in the limit as n becomes very large—and the comparison for a particular run is given below:

s =	2	3	4	5	6	7	8	9	10	11	12
n_s =	4	10	13	13	20	30	15	22	10	6	1
e_s =	4	8	12	16	20	24	20	16	12	8	4

Of course, *any* sequence of scores can turn up in practice, even 144 throws of 12. Such a score is not wrong: it is merely improbable, but it might happen. The χ^2 test tells us whether the sequence as a whole is "credible", or so improbable that one would have serious grounds to suspect that the dice were loaded, or the scores faked.

We define
$$\chi^2 = \sum_s (n_s - e_s)^2/e_s.$$

This is an obvious measure of the divergence of the observed from the expected. For the case in hand it gives $\chi^2 \approx 9.23$, and the question is whether this result is "credible" or not.

On the assumption that we are dealing with observations each of which is distributed according to a "Poisson distribution", as befits random data, it is possible to calculate the probability that $\chi^2 < X$, say. In the limit of a large number of observations, this χ^2-distribution can be expressed in terms of the incomplete gamma function, the parameters of which, in essence, are χ^2 and f, the number of "degrees of freedom"—in this case one less than the maximum number of different scores. The function is extensively tabulated, and we display some typical "percentiles" in the table below:

Some chi-squared percentiles

f	$p = 1\%$	$p = 5\%$	$p = 25\%$	$p = 50\%$	$p = 75\%$	$p = 95\%$	$p = 99\%$
1	0.0002	0.0039	0.102	0.455	1.32	3.84	6.64
3	0.115	0.352	1.21	2.37	4.11	7.82	11.34
9	2.09	3.33	5.90	8.34	11.39	16.92	21.67
10	2.56	3.94	6.74	9.34	12.55	18.3	23.2
30	15.0	18.5	24.5	29.3	34.8	43.8	50.9
100	69.2	77.1	89.1	98.3	108.1	123.2	134.7

From this table we find that the probability that $\chi^2 < 9.34$ is about 50% with 10 degrees of freedom. This must be accepted as highly satisfactory: there would have been only a 1% probability of χ^2 being below about 2.6 or above about 23.2. In the low case one might suspect that the observations had been tampered with to fit the expectations; in the high case, one might legitimately wonder whether the dice had been tampered with (especially if there was a preponderance of high scores!). At the very least one would repeat the experiment a few times to see whether extreme values of χ^2 persist.

It is easy but amusing to develop a dice throwing simulator. We need to generate a random sequence of integers uniformly distributed over the range [1,6]. We could use the routine uniform introduced in the last chapter, but it is better to respond directly to the spirit of the game with the function

```
inline int roll(longword &s)
{ return 1 + int(6*(float(next(s))/m)); }.
```

Then, if we store the probabilities for different scores in an array float prob[11], and the number of times each score appears in another array longword score[11], the program fragment

```
int    score[11] = { 0,0,0,0,0,0,0,0,0,0,0 };
for (longword i = 0; i < throws; ++i)
{      int u = roll(s);          // roll 1st dice
       int v = roll(s);          // roll 2nd dice
       int t = u + v;            // note score and
       ++score[t-2];             // store it
}
```

where `throws` is the number of times we throw a pair of dice, plays the game for us. All that remains is to write another program fragment to compute the value of chi-squared:

```
float X = 0.0;                    // chi-squared
for (int j = 0; j < 11; ++j)      // for each score
{       float e = throws*prob[j]; // expected numbers
        float t = score[j] - e;   // and differences
        X += t*t/e;
}
```

In this way, we may throw dice as often as we wish without tiring, and in doing so we provide a first test of the random number generator `next`.

Tests of this kind, but some of them very much more sophisticated, may be applied to the sequences produced by random number generators, and the results of such tests are the basis for deciding the merits of different generators. We discuss a few of them below.

8.1 Even distribution test

Suppose we produce a sequence of floating-point numbers u_j uniformly distributed in the range $0 \le u_j < 1$ using the procedures of the last chapter, and we wish to test the results for randomness. There are several things we can do, but the first is to perform a χ^2 test of the frequency with which each number appears in sub-ranges such as $0.30 \le u_j < 0.40$. We shall run the `uniform` generator 1000 times, sorting the resulting integers $k = \text{int}(100u)$ into 10 categories according as k lies in the ranges $[0, 9]$, $[10, 19]$, ..., $[90, 99]$. If the numbers are randomly distributed we expect there to be about 100 numbers in each category. We shall judge the significance of deviations from expectation against a χ^2 distribution. Moreover, we shall repeat the experiment 100 times on each generator, so that the variability of estimates of χ^2 can also be noted. The whole test program is reproduced below since it may be of some interest in its own right. The arrays `prob` and `perc` hold the probabilities and percentiles of the χ^2 distribution for 9 degrees of freedom; while `box` holds the number of integers in each category, and `chi` the range of values of χ^2 from different runs. The program can be used for any linear congruence generator simply by offering different values of m, a and c.

```
#include <standard.h>

longword    m, a, c;

inline      longword next(longword &s) { return s = ( a*s + c) % m; }
inline      float uniform(longword &s) { return float(next(s))/m; }

int   box[10], chi[8];
```

```
int    prob[8] = { 1, 5, 25, 50, 75, 95, 99, 100 };
float perc[8] = { 2.09, 3.33, 5.90, 8.34, 11.39, 16.92, 21.67, 100.0 };

void  normbox()
{      for (int i = 0; i < 10; ++i) box[i] = 0; }

main()
{      clear;
       statement("\nchi-squared test of random number generator\n");
       statement("\nchoose generator parameters:\n\n");
       input("m = ", m); input("a = ", a); input("c = ", c);
       clear;
       longword s;
       statement("\nchi-squared test of random number generator\n");
       cout << "\nm = " << m << ",  a = " << a << ",  c = " << c << "\n\n";
       for (int k = 0; k < 100; ++k)     // do 100 runs
       {      normbox();
              s = k;
              int i,j;
              for (i = 0; i < 1000; ++i)
              {      int k = int(100*uniform(s));
                     for (j = 0; j < 10; ++j)
                            if (10*j <= k && k < 10*(j+1)) ++box[j];
              }
              float X = 0;                 // to hold chi-squared
              for (i = 0; i < 10; ++i)
              {      float t = float(box[i] - 100);
                     X += t*t;
              }
              X /= 100;
              if (k == 0)                  // print out first run
              {      cout << "\ndistribution of single run\n\n";
                     for (j = 0; j < 10; ++j)
                            cout << box[j] << tab;
                     cout << "\nchi-squared = " << X << endl;
              }
              for (i = 0; i < 8; ++i) // develop distribution of chi-squared
                     if (X <= perc[i]) ++chi[i];
       }
       cout << "\ndistribution of chi-squared for 100 sequences\n";
       newline;
       for (int j = 0; j < 8; ++j)
              cout << "X <=  " << perc[j] << ":" << tab << chi[j] << endl;
}
```

Note that the category array box has to be initialized to zero contents on each run. Likewise, the "seed" s set to a new value, the choice of which is arbitrary; for simplicity we have taken it to be the running parameter k. We could have chosen anything else, such as next(s), but this would only have given a spurious appearance of greater randomness.

With generator A, there appeared on the screen:

```
m = 714025,  a = 1366,   c = 150889

distribution of single run

98     97     81     95     116    103    92     96     104    118

chi-squared = 10.84

distribution of chi-squared for 100 sequences

X <=    2.09:       2
X <=    3.33:       5
X <=    5.90:       25
X <=    8.34:       58
X <=   11.39:       82
X <=   16.92:       94
X <=   21.67:       100
X <=    100:        100
```

The program was run for each of the three sample generators, and the results obtained were:

A	2	5	25	58	82	94	100
B	0	2	18	31	64	97	100
C	2	4	22	51	77	94	100
exp	1	5	25	50	75	95	99

The last row, labelled "exp", gives the "expected" numbers of runs on the basis of a chi-squared distribution. There being 100 runs on each generator these expected numbers are just the probabilities stored in the array prob multiplied by 100. It will be seen that the performance is reasonable in each case, given that the numbers involved are not very large, but that the generators can be ordered according to increasing m, C being the best and B the worst. This is not very surprising; we are examining only the evenness of distribution over sequences of 100 numbers at a time, and correlations would not show up on such a test.

However, in this test we have been testing the sequence for the randomness of its first two digits only, because we have in fact been sorting the integers $k_j = \text{int}(100u_j)$, rather than the numbers u_j themselves, in order to put them into sub-ranges. To go further—to examine the randomness of the first three digits, say—we would have to compute the distribution of the integers $k_j = \text{int}(1000u_j)$ instead. We shall need 100 boxes, and to obtain about 100 numbers in each we shall need a sequence of 10 000 numbers. The appropriate χ^2 test is for 99 degrees of freedom, but 100 is near enough. Repeating that 100 times, we almost exhaust the capacity of the generator, which is limited to m numbers. We may expect trouble, and indeed we get it. This time the test on each generator gives the following distribution of the values of chi-squared:

A	0	7	45	79	97	100	100
B	58	94	100	100	100	100	100
C	1	8	46	87	96	100	100
exp	1	5	25	50	75	95	99

The results are hopeless, especially for generator B which has the smallest period. Great care must be exercised if randomness is required in the third decimal digit.

8.2 Serial correlation test

We may also examine a sequence for tangible correlation between successive terms—tangible to someone supposed not to know that a linear congruence generator is being used to produce them, for, of course, there is a very precise correlation between successive terms given by the generator formula. In particular, we may examine the correlation coefficient, as usually defined, between the sequence

$$(U_0, U_1, U_2, \ldots, U_{n-1})$$

of length n, and the cyclically shifted sequence

$$(U_q, U_{q+1}, \ldots, U_{n-1}, U_0, \ldots, U_{q-1}),$$

where q is a constant integer and $0 < q < n$. Then a range of correlation coefficients may be defined by

$$C_q = [n\sum_j U_j U_{q+j} - (\sum U_j)^2]/[n\sum U_j^2 - (\sum U_j)^2].$$

Their values tell us something about the apparent independence of successive terms of a sequence. In particular, the coefficients should all be *small*. More precisely, if the sequence of n terms is truly random, and if $\mu = -1/(n-1)$, $\sigma = (-\mu)\sqrt{[n(n-3)/(n+1)]}$, then we should have $\mu - 2\sigma < C < \mu + 2\sigma$ with a probability of about 95%. If we make many determinations of C, μ is their expected mean value, and σ their expected standard deviation. (Strictly, these quantities have been calculated for a normal, rather than a uniform, distribution, but there should not be too much difference.)

For example, a program may require the repeated computation of some function of q variables, each taking random values or having a random component. In that case, one should be interested in the value of C_q.

The following function computes the correlation coefficient C_1 of neighbouring pairs in a sequence of n unsigned integers in the range $[0, d-1]$ derived from a congruence sequence:

```
float corr(longword &s, word n)
{       float U = 0, UU = 0, UV = 0;
        float u = uniform(s);
        float w = u;                            // retain i = 0 term
        for (word i = 1; i < n; ++i)            // start at i = 1
        {       float v = uniform(s);
                U  += v;                        // Σv
                UU += v*v;                      // Σv²
                UV += u*v;                      // Σuv
                u = v;

        }
        U  += w;                                // include term
        UU += w*w;                              // with i = 0
        UV += u*w;
        return (n*UV - U*U)/(n*UU - U*U);
}
```

We have run this routine for each generator, taking 100 sequences of 1000 numbers each. The values of the correlation coefficient may be averaged over the 100 runs, and the standard deviation calculated. The results are as follows:

Theory	$\mu = -0.0010$	$\sigma = 0.0316$
A	0.0018	0.0285
B	−0.0029	0.0277
C	−0.0034	0.0286

Clearly, all three generators are satisfactory from the point of view of a serial correlation test. The test was run on uniformly distributed floating point numbers. Had it been run on integers in the range $0 \leq x_i < d$, the value of d would have had to be kept below about 250 unless the lack of randomness of decimal digits beyond the first two was to show itself.

It has sometimes been hoped that the correlation coefficient could be reduced by judicious choice of the additive parameter c in the congruence generator. Indeed, for the sequence of m integers *taken as a whole*, it can be shown that there are regions of c for which the coefficient is small, even zero. In practice, however, we are concerned with relatively short subsequences, and that being so there is no discernible effect of the value of c on the correlation coefficient.

8.3 The spectral test

In the spectral test, we generalize the serial test to consider, in principle, the whole set of correlations between any number of successive terms of a sequence. This may also be seen as an exploration of the concept of potency introduced in the last chapter. The treatment is again adapted from that given in Volume 2 of the book by Knuth.

Consider the k-dimensional vector

$$U^{(k)}{}_n = (u_n, u_{n+1}, \ldots, u_{n+k-1}),$$

where the numbers u_i are, as before, related to a congruence sequence of successive integers x_i of maximum period m by the relations:

$$u_n = x/m, \quad u_{n+1} = \mathrm{next}(x)/m, \quad u_{n+2} = \mathrm{next}(\mathrm{next}(x))/m, \ldots, \quad u_{n+k-1} = \mathrm{next}^{k-1}(x)/m,$$

and, as usual, $\mathrm{next}(x) = (ax + c) \bmod m$, so that $0 \le u_i < 1$. Note that:

$$\mathrm{next}^i(x) = (a^i x + (1 + a + \ldots + a^{i-1})c) \bmod m.$$

We are interested in the properties of the set $L^{(k)}$ of all vectors $\mathbf{U}^{(k)}{}_n$ such that $0 \le n < m$, or in set notation:

$$L^{(k)} = \{\ U^{(k)}{}_n \mid 0 \le n < m\ \}.$$

The vectors $U^{(k)}{}_n$ may be regarded as points in a k-dimensional unit cube, where they cluster on $(k-1)$-dimensional parallel hyperplanes.

If we imagine the unit cube extended indefinitely in all dimensions we can dispense for the most part with the modulo notation. This is a commonplace procedure in crystallography. Then we can write the set $L^{(k)}$ equivalently as:

$$L^{(k)} = \{x/m, (\mathrm{next}(x)/m) + \lambda_1, \ldots, (\mathrm{next}^{k-1}(x)/m) + \lambda_{k-1} \mid \text{integer } x, \lambda_1, \ldots, \lambda_{k-1}\}.$$

Define the following auxiliary k-vectors:

$$A_0 = (0, c, (1 + a)c, \ldots, (1 + a + \ldots + a^{k-2})c)/m$$
$$A_1 = (1, a, a^2, \ldots, a^{k-1})/m$$
$$A_2 = (0, 1, 0, \ldots, 0)$$
$$\ldots\ldots\ldots\ldots\ldots\ldots\ldots$$
$$A_k = (0, 0, 0, \ldots, 1)$$

Then our set may equivalently be written:

$$L^{(k)} = \{A_0 + y_1 A_1 + y_2 A_2 + \ldots + y_k A_k \mid \text{integer } y_1, y_2, \ldots, y_\kappa\}.$$

The points $(x_1, x_2, \ldots, x_k)$ of L with $0 \le x_i < 1$ are exactly the points of the set with which we began this discussion.

A further simplification is possible. The vector A_0 is a constant, effectively shifting the origin. We may therefore set it equal to the zero vector without affecting distances between the points of the set, remembering that the set now contains the point $(0, 0, \ldots, 0)$, however. The additive constant c of the congruence generator appears only in this vector. This therefore confirms our impression from the last section that the value of c is irrelevant to the correlations between successive terms. Setting $A_0 = 0$, the set L reduces to the *lattice L_0*:

$$L^{(k)}_0 = \{ \, y_1 A_1 + y_2 A_2 + \ldots + y_k A_k \mid \text{integer } y_1, \ldots, y_k \, \}.$$

Now a family of $(k-1)$-dimensional parallel hyperplanes may be defined by a non-zero vector $P = (p_1, p_2, \ldots, p_k)$ perpendicular (*ie*, orthogonal) to all of them. In that case, the set of all points on a given hyperplane may be written:

$$\{ \, (x_1, x_2, \ldots, x_k) \mid p_1 x_1 + p_2 x_2 + \ldots + p_k x_k = q \, \},$$

where q is a different constant for each hyperplane: the k-dimensional points $X = (x_1, x_2, \ldots, x_k)$ lie on hyperplanes defined by the scalar product $P.X = q$. Since neighbouring hyperplanes are separated by constant distance, and one of them contains the point $(0, 0, \ldots, 0)$, we may choose the vector P so that q is an integer. Then the perpendicular distance between neighbouring hyperplanes is simply:

$$\min_{X \neq 0} \{ \, \sqrt{(x_1^2 + \ldots + x_k^2)} \mid P.X = 1 \, \}$$

Equivalently, this distance is $1/\sqrt{(p_1^2 + \ldots + p_k^2)}$. Now the vector P is not unique: many families of hyperplanes may be drawn through the given points, varying in their inclination, just as there are many crystal planes. We want to find the family in which neighbouring hyperplanes are farthest apart. This is the distance $1/\nu_k$ of the last chapter.

Thus, $$\nu_k^2 = \min_{P \neq 0} \{ \, (p_1^2 + \ldots + p_k^2) \mid P.X = 1 \, \}.$$

P is the shortest vector to define a family of hyperplanes $\{ \, P.X = q \mid \text{integer } q \, \}$ containing all the points of L_0.

Since the vectors $(1, 0, \ldots, 0)$, $(0, 1, \ldots, 0)$, $\ldots$, $(0, 0, \ldots, 1)$ are all in L_0, it follows that $P.A$ is an integer for each of them, which implies that every p_i is an integer. Moreover, since $A_1 = (1, a, \ldots, a^{k-1})/m$ is in L_0, it follows that $(p_1 + ap_2 + \ldots + a^{k-1}p_k)/m$ is an integer, or $p_1 + ap_2 + \ldots + a^{k-1}p_k \equiv 0 \pmod{m}$.

Finally, therefore, we have an expression that we may use for computation:

$$\nu_k = \min_{P \neq 0} \{ \, \sqrt{(p_1^2 + p_2^2 + \ldots + p_k^2)} \mid p_1 + ap_2 + \ldots + a^{k-1}p_k \equiv 0 \pmod{m} \, \}.$$

where the p_i are integers, not all of which are zero.

8.3.1 Minimization process

In simple cases the formulae we have just derived suffice to calculate the minimum distances $1/\nu_k$. For example, if $m = 32$, $a = 13$, and $k = 2$, the set

$$\{ \, \text{integer } x, y \mid x + 13y \equiv 0 \pmod{32}, \, x^2 + y^2 < 200 \, \}$$

consists of the following values:

x	1	4	5	6	7	12	13
y	−5	12	7	2	−3	4	−1
x^2+y^2	26	160	74	40	58	160	170

to which must be added values in which the signs of x and y are both reversed. Clearly the minimum value of $x^2 + y^2$ is 26, so $v_2 = \sqrt{26} \approx 5.1$. However, a simple scanning procedure of this kind would be prohibitive for large values of m even with $k = 2$: it would in general be well-nigh impossible for $k > 2$. We badly need an algorithm.

Note, firstly, that we may get rid of the modulo condition by setting

$$p_1 = qm - ap_2 - \ldots - a^{k-1}p_k$$

for some integer q. The assignment $z_1 = q$, $z_j = p_j$, $j > 1$, then leads us to

$$v_k^2 = \min_{Z \neq 0} S(\mathbf{z}),$$

where $S(\mathbf{z}) = (mz_1 - az_2 - \ldots - a^{k-1}z_k)^2 + z_2^2 + \ldots + z_k^2$, and the z_j are integers.

This more explicit expression has the same minimum as before, but the values of the z_j which minimize it are not the same as the values of the p_j which minimize the original expression: more precisely, $z_1 \neq p_1$. In k-space, the "surface" $S = $ constant is an ellipsoid of which the lengths of the axes are determined by m and a—or at least it would be so if the z_j were not restricted to integer values. The function S provides us with a systematic method for maintaining the condition $x_1 + ax_2 + \ldots + a^{k-1}x_k \equiv 0 \pmod{m}$ during the minimization of $x_1^2 + x_2^2 + \ldots + x_k^2$.

We now define a $k \times k$ matrix,

$$A = (1/m) \begin{pmatrix} 1 & a & a^2 & \ldots & a^{k-1} \\ 0 & m & 0 & \ldots & 0 \\ 0 & 0 & m & \ldots & 0 \\ \ldots & \ldots & \ldots & \ldots & \ldots \\ 0 & 0 & 0 & \ldots & m \end{pmatrix}$$

the row vectors of which are exactly the vectors A_j introduced before. We also define a second $k \times k$ matrix,

$$X = \begin{pmatrix} m & 0 & 0 & \ldots & 0 \\ -a & 1 & 0 & \ldots & 0 \\ -a^2 & 0 & 1 & \ldots & 0 \\ \ldots & \ldots & \ldots & \ldots \\ -a^{k-1} & 0 & 0 & \ldots & 1 \end{pmatrix}$$

If we denote the scalar product of two row vectors by a dot, the rows of these two matrices satisfy the orthogonality relations:

$$X_i . A_j = \delta_{ij}, \quad 1 \le i, j \le k,$$

where δ is the Kroncikes delta: $\delta_{ij} = 1$ if $i = j$, 0 otherwise.

In terms of the matrix X we may express the integer vector $\mathbf{z}$ in the form

$$\mathbf{z} = z_1 X_1 + z_2 X_2 + \ldots + z_k X_k,$$

where the components $z_i = \mathbf{z} A_i$ because of the orthogonality relations between the matrices X and A. Written in terms of components, we have

$$\mathbf{z} = (\ (mz_1 - az_2 - \ldots - a^{k-1}z_k), z_2, \ldots, z_k\ ),$$

and the square of the length of $\mathbf{z}$ is therefore $\mathbf{z}^2 = S(\mathbf{z})$.

To restrict, if possible, the search for vectors which minimize S, and therefore the interplanar distance, we must find limits on the values of their components.

Now, $\qquad z_j = (z_1 X_1 + z_2 X_2 + \ldots + z X_k) . A_j,$

and so, by Cauchy's inequality,

$$z_j^2 = ((z_1 X_1 + \ldots + z_k X_k) . A_j)^2 \le (z_1 X_1 + \ldots + z_k X_k)^2 (A_j . A_j) = S(\mathbf{z})(A_j . A_j).$$

If $\mathbf{x}$ is the vector which minimizes $S(\mathbf{z})$, then *a fortiori* $x_j^2 \le S(\mathbf{z})(A_j . A_j)$. Any vector $\mathbf{z}$ provides us with limits on the components of the vector which minimizes S. We may in particular consider the limits obtained by taking $\mathbf{z}$ to be successively:

$$(1, 0, 0, \ldots, 0), (0, 1, 0, \ldots, 0), (0, 0, 1, \ldots, 0), \ldots, (0, 0, 0, \ldots, 1)$$

We then obtain $\qquad x_j^2 \le (X_i . X_i) . (A_j . A_j), \quad 1 \le i, j \le k,$

which is a set of k limits on each x_j, all of which may be used to reduce the search space. However, these limits may be very wide—as may be seen by expanding out the vector products in terms of m and a. The search for a vector to minimize the distance, though finite, would therefore be very long. We need to find a way of reducing the limits to something more practical.

Let $\mathbf{q}^{(j)} = (q_1, q_2, \ldots, q_{j-1}, 0, q_{j+1}, \ldots, q_k)$ be an integer k-vector, arbitrary except that the j-th element is chosen to be zero: the suffix j is to be kept fixed. Consider the following transformations:

$$A'_i = A_i - q^{(j)}_i A_j, \qquad x'_i = x_i - q^{(j)}_i x_j, \qquad X'_i = X_i, \qquad i \neq j,$$
$$A'_j = A_j, \qquad\qquad x'_j = x_j, \qquad\qquad X'_j = X_j + \sum_{i \neq j} q^{(j)}_i X_i.$$

The orthogonality relations are preserved under these transformations, for $X'_i . A'_j = m \, \delta_{ij}$, and so is the value of S; but if the vector $\mathbf{q}^{(j)}$ is chosen appropriately, the search limits will be reduced, perhaps greatly so. The essential task of the required algorithm, therefore, is to choose this vector in the best possible way, namely to make the quantities $A'_i . A'_i$ and $X'_i . X'_i$ as small as possible.

Now, if we were dealing with real numbers instead of integers, the minima would occur when $q^{(j)}_i = (A_i . A_j)/(A_j . A_j)$, which would make A'_i orthogonal to A_j. Since we are in fact dealing with integers, we must be content to choose the integers *nearest* to these values; these will also give the smallest values of $X'_j . X_j$. Exact orthogonality will not be achieved, but the scheme of transformations can be repeated until no further reduction takes place. We then have the smallest possible limits on the range of values of the x_j^2 which have to be considered in performing the final minimization search.

8.3.2 The case k = 2

In the simplest case when $k = 2$, a more direct approach is possible which makes no explicit reference to the matrices A and X. It is simplest to state the algorithm and then to establish that it works properly. The function `invsep2` which returns the value of v_2^2 may be written in terms of a recursive auxiliary function which has to be initialized to the starting values of the search:

```
inline float invsep2(long m, long a)
{ return auxinvsep2(m, a, -1, 0, 1.0+a*a); }
```

The auxiliary function `auxinvsep2` is then given by:

```
float auxinvsep2(long h, long k, long p, long t, float s)
{       long q = h/k;
        long x = h - q*k;
        long y = t - q*p;
        float r = x*x + y*y;
        if (r > s) return s;
        return auxinvsep2(k, x, y, p, r);
}
```

Here, the variable t represents the previous value of the variable p, as p represents the previous value of y.

It is a simple matter to include a running output in the function `auxinvsep`. For the interesting little generator with $m = 256$, $a = 137$ the computation proceeds as follows:

h	k	p	t	q	s
256	137	−1	0	1	18770
137	119	1	−1	1	14162
119	18	−2	1	6	328
18	11	13	−2	1	290
11	7	−15	13	1	274

nu2 = 16.552946

where the final number printed is the value of v_2. Notice that at each stage $h + at \equiv 0 \pmod{m}$ and $k + ap \equiv 0 \pmod{m}$. This observation allows us to prove by induction that at all stages $x + ay \equiv 0 \pmod{m}$, because h is the previous k, and t the previous p. The quantity $x^2 + y^2$ is stored in the variable s; since it is used only for comparison purposes we have chosen to represent it as a floating point number. The recursion stops when s reaches its minimum value, the square root of which is v_2.

The computation is extremely fast and usually requires very few steps. This is because at each stage h is replaced by k, and k by $h \% k$: the resemblance to Euclid's algorithm for the greatest common divisor is striking.

8.3.3 The case $k = 3$

In the general case where $k > 2$, we shall need to involve the matrices A and X explicitly. For purposes of exposition it is sufficient to consider the case of $k = 3$: it may easily be generalized to higher k. The handling of vectors and matrices is crude; we represent them as simple arrays:

```
typedef double vector[3];
typedef double matrix[3][3];
```

It is convenient to define a variable `long a2 = (a*a) % m;`. Then for the case of $k = 3$ the matrices X and A are given by:

```
matrix X = { m, 0,  0, -a, 1, 0, -a2, 0, 1 };
matrix A = { 1, a, a2,  0, m, 0,   0, 0, m };
```

Notice that we have multiplied A throughout by m to ensure that we are always dealing with integers. That does not affect the value of the transformation vector **q** because the factor m cancels out, but the limits have to be divided by m^2.

It is convenient to define three auxiliary functions. The first is an inline function to round a floating point number to the nearest integer value:

```
inline double round(double x) { return floor(x + 0.5); }
```

where the standard library function double floor(double x) returns the floating point value of the highest integer not greater than x, which is usually written $\lfloor x \rfloor$ in number theory. Because it is of general utility we have included the function round in the header file standard.h in Appendix A.

The second function computes the scalar product of two row vectors, not necessarily from the same matrix (it is needed for the evaluation of the expressions $(A_i . A_j)/(A_j . A_j)$ and $(X_i . X_i)(A_j . A_j)$:

```
double scalar(matrix u, matrix v, int i, int j)
{       double t = 0;
        for (int k = 0; k < 3; ++k)
        {       double uik = u[i][k], vjk = v[j][k];
                t += uik * vjk;
        }
        return t;
}
```

The third auxiliary function detects whether all the components of a vector are zero; it is written inline:

```
inline bool zero(vector q)
{ return (!q[0] && !q[1] && !q[2]) ? TRUE : FALSE; }
```

We can now write down a recursive function which attempts to find the square of the inverse distance for the case of $k = 3$. It is

```
double auxinvsep3(m, matrix X, matrix A, int j, double s)
{       static int c = 0;
        int j1 = (j+1)%3, j2 = (j+2)%3;     // to rotate q vector
        double Ajj  = scalar(A, A, j, j);   // construct q
        double Aij1 = scalar(A, A, j1, j);
        double Aij2 = scalar(A, A, j2, j);
        vector q = { 0, 0, 0 };
        q[j]  = 0;
        q[j1] = long(round(Aij1/Ajj));
        q[j2] = long(round(Aij2/Ajj));
        X[j][0]  += q[j1]*X[j1][0] + q[j2]*X[j2][0];
        X[j][1]  += q[j1]*X[j1][1] + q[j2]*X[j2][1];
        X[j][2]  += q[j1]*X[j1][2] + q[j2]*X[j2][2];
        A[j1][0] -= q[j1]*A[j][0];
        A[j1][1] -= q[j1]*A[j][1];
        A[j1][2] -= q[j1]*A[j][2];
        A[j2][0] -= q[j2]*A[j][0];
```

```
        A[j2][1] -= q[j2]*A[j][1];
        A[j2][2] -= q[j2]*A[j][2];
        double X20 = X[2][0], X21 = X[2][1], X22 = X[2][2];
        s = X20*X20 + X21*X21 + X22*X22;
        if (zero(q)) ++c; else c = 0;
        if (c == 3) return sqrt(s);
        return auxinvsep3(m, X, A, j1, s);
}
```

together with the initializer,

```
double invsep3(long m, matrix X, matrix A)
{       double X20 = X[2][0], X21 = X[2][1], X22 = X[2][2];
        double s3 = X20*X20 + X21*X21 + X22*X22;
        return auxinvsep3(m, X, A, 0, s3);

}
```

The core of the recursive function `auxinvsep3` is the repeated transformation of the matrices A and X by the vector $\mathbf{q}$, the latter calculated from the previous values of the matrix A. Again, we have to be careful with conversions since we are doing integer arithmetic. We may begin the process of transformation with any one of the vectors $\mathbf{q}^{(j)}$, the components of which are given by $\mathbf{q}^{(j)}{}_i = (A_i.A_j)/(A_j.A_j)$. Having chosen the initial j we then cycle repeatedly through the k values of j. This accounts for the appearance of j as one of the parameters of the function, and for its replacement by `j1 = (j+1)%3` in the recursive call. The objective is to reach a state where the vectors $\mathbf{q}^{(j)}$ remain zero for a complete cycle, so that no further transformation can take place. This state can be detected by a static integer variable that is set to zero whenever $\mathbf{q}$ differs from zero, but accumulates if zero transformations occur. When this variable reaches the value 3, the recursion ceases. In all the cases examined this takes place with less than 30 recursive calls.

The initializing function `invsep3` sets the initial value of s and sets an arbitrary choice of the initial value of j. We have chosen the latter to be `j = 0`.

However, `auxinvsep3` does not necessarily terminate at the minimum value of s, although it always reaches a small value. The important thing it achieves is the reduction of the search limits, given, as we saw in the last section, by $z_j^2 = S(\mathbf{z})(A_j.A_j)/m^2$ (the factor m^2 arising from the redefinition of the matrix A). We find the actual minimum by performing a systematic search over all values of $x_j \le z_j$, not forgetting that $S(-\mathbf{z}) = S(\mathbf{z})$.

Recalling that $S(\mathbf{z}) = (z_1 X_1 + \ldots + z_k X_k)^2$, it is easy to write a function returning S:

```
double S(vector x, matrix X)
{       double t = 0;
        for (int i = 0; i < 3; ++i)
        {       double s = 0;
                for (int j = 0; j < 3; ++j)
```

```
                    s += x[j]*X[j][i];
            t += s*s;
    }
    return t;
}
```

If we define a global vector **z** to hold the limits, we can also write a routine to calculate these limits in terms of the value of the matrix A and of s attained by `auxinvsep3`, namely:

$$z_i = \lfloor \sqrt{\lfloor (A_i . A_i)s/m^2 \rfloor} \rfloor.$$

A procedure for computing this global vector is therefore:

```
void  makezed(long m, matrix A, double s)
{      for (int i = 0; i < 3; ++i)
       {      double f = scalar(A, A, i, i);
              f /= m;
              f *= s;
              f /= m;
              f = sqrt(floor(f));
              z[i] = long(f);
       }
}
```

Note that we have avoided squaring m directly. Given the vector **z**, the search function may finally be written:

```
double search(vector x, matrix X)
{      double s = 1.0E+50;
       double r = 1.0E+50;
       long u, v;
       for (x[0] = 0; x[0] <= z[0]; ++x[0])
       {      u = !x[0] ? 0 : -z[1];
              for (x[1] = u; x[1] <= z[1]; ++x[1])
              {      v = !x[0] && !x[1] ? 0 : -z[2];
                     for (x[2] = v; x[2] <= z[2]; ++x[2])
                     {      r = S(x, X);
                            if (r > 0.1 && r < s) s = r; // exclude r = 0
                     }
              }
       }
       return s;
}
```

8.3.4 Results

It is merely tedious to generalize these functions to any value of k. The integer k has to be included among the list of parameters of every function, and replaces the particular value 3 used in `auxinvsep3`. The matrices A and X have to be extended correspondingly, declaring them by `typedef matrix[k][k]`. The following results are obtained for the lowest v_k for the set of congruence generators already considered and for the "little" generator:

	m	a	v_2	v_3	v_4
	256	137	16.6	5.5	3.7
A	714025	1366	654.2	75.6	27.5
B	233280	9301	385.6	52.7	14.9
C	1771875	2416	1197.5	57.9	26.9

Remembering that the maximum distances between neighbouring hyperplanes are the inverses $1/v_k$, we can easily see why, as observed in 8.2, correlations show up if more than a few decimal digits are significant in a sequence of numbers produced by the congruence generators examined here. These generators are severely limited by the requirement that they should not lead to overflow on a 32-bit machine; as it is, they operate near that limit. If random arithmetic is an important consideration—for Monte Carlo integration, for example, which we illustrated in the last chapter—it should be performed on a machine with a larger word size. If strict randomness is *not* an important consideration, in testing routines for matrix algebra, for example, these generators may be freely used—especially generator A, which is the one we have adopted in Appendix D which also employs shuffling (*cf.* 7.4.5).

9 Vectors and matrices

Many computations involve the solution of linear equations in several, perhaps many, variables, or can be transformed and approximated so as to do so. In this chapter we develop appropriate concepts for the solution of problems in linear algebra, especially the systematic definition of the classes `vector` and `matrix`. We shall regard a vector as a pointer to an array of floating point numbers, and a matrix as a pointer to an array of vectors. We shall be building directly on the work of chapter 5. We first define the class `vector`, then the class `matrix`. The chapter ends with a simple application to linear algebra.

9.1 The class vector

As in the class `string` (*cf* 5.2), we shall take the data members of the class `vector` to be a pointer to an array and the length of that array. We shall also need appropriate constructors as public members, and a destructor; the constructors should be able to create an initialized vector of given length, a vector converted from a given array, and a vector that is a copy of a given vector. To deal with these aspects first, we write:

```
class vector {
private:
      int size;
      double *vec;
      ........
public:
      vector(int);
      vector(const double*, int);
      vector(const vector&);
      ~vector();
      ........
};
```

Consider the constructors. They have to allocate memory for the correct number of floating point numbers; check that memory is actually available, printing out an error message if it is not; and then initialize the vector.

The first constructor creates a vector, the elements of which are initialized to zero, given only the length or size of the vector:

```
vector::vector(int n)
{       size = n;
        vec = new double [size];
        if (!vec)
        error("allocation failure in vector::vector(int)!");
        for (int i = 0; i < size; ++i) vec[i] = 0;
}
```

The second constructor creates a vector to hold a given array:

```
vector::vector(const double *a, int n)
{       size = n;
        vec = new double [size];
        if (!vec)
        error("allocation failure in vector::vector(double*, int)!");
        for (int i = 0; i < size; ++i) vec[i] = a[i];
}
```

The copy constructor, similarly, is:

```
vector::vector(const vector &v)
{       size = v.size;
        vec = new double [size];
        if (!vec)
        error("allocation failure in vector::vector(vector&)!");
        for (int i = 0; i < size; ++i) vec[i] = v.vec[i];
}
```

The destructor is so simple that it may be written inline:

```
inline vector::~vector() { delete vec; }
```

Notice that we have declared the given array a and the given vector v to be const because they are passed as a pointer and by reference, respectively (*cf* 2.5.4,5); otherwise, the constructor definitions could in principle change their values.

We now have to define a number of operators and auxiliary functions. The earlier work on the class string will continue to serve as a model. The complete class declaration is given in the header file vecmat.h of Appendix E, and we do not repeat it here. First, there is the indexing operator, which again may be written inline:

```
inline double& vector::operator[](int i)
{ return vec[range(i)]; }.
```

It returns a reference to a double, so that assignment is possible through the indexing operator, just as it is in the class `string`. Again, we have introduced a (private) function `range` to check that $0 \leq i <$ `size`, and to terminate the program with an error message if the check fails. As in 5.2.4, it will be defined:

```
inline int vector::range(int i)
{ return (i < 0 || i >= size) ? (error("..."), -1) : i; }
```

where the standard error function was defined in 3.3 and is to be found in the header file `standard.h` in Appendix A. An appropriate error message would be:

```
"index out of range in vector::operator[]".
```

When we wish to access the `i`th element of the vector `v`, whether for writing or for reading, we now simply have to write `v[i]`, which is pleasingly close to ordinary mathematical notation.

Then there is the unary plus operator, written inline as in the class `complex`:

```
inline vector& vector::operator+() { return *this; }
```

Unary minus we shall later define as a friend function. Next come the assignment operators `=`, `+=`, and `-=`. Here we define only the simple assignment operator; the other two definitions are similar:

```
vector& vector::operator=(const vector &v)
{      if (size != v.size)
       error("diff size in vector& vector::op=(const vector&)!");
       for (int i = 0; i < size; ++i) vec[i] = v.vec[i];
       return *this;
}
```

The assignment operators `*=`, and `/=` allow us to multiply or divide a vector by a double:

```
vector& vector::operator*=(double x)
{      for (int i = 0; i < size; ++i)
            vec[i] *= x;
       return *this;
}
```

Similarly for the division assignment, where, however, it is assumed that error trapping for division by zero is dealt with externally.

It is sometimes useful to be able to access the size of a vector directly. We define:

```
inline int vector::getsize() { return size; }
```

It is also useful to be able to interchange two elements of a vector:

```
void  vector::swap(int i, int j)
{       double tmp = vec[range(i)];
        vec[i] = vec[range(j)];
        vec[j] = tmp;
}
```

where the function range is again used to make external range checking unnecessary.

We define unary minus as a friend function. It must *not* return a reference, however, because (as explained in 2.5) it would be a reference to a local variable which goes out of scope after the function call is complete. The unary minus operator is defined to be:

```
vector operator-(const vector &v)
{       int n = v.size;
        vector u(n);
        for (int i = 0; i < n; ++i)
                u.vec[i] = -v.vec[i];
        return u;
}
```

Then come the usual arithmetic operators. As in the class complex studied in the last chapter, they are friend functions which also may not return a reference. The binary plus and minus operators are similar. For the plus operator, we have:

```
vector operator+(const vector &v1, const vector &v2)
{       int n = v1.size;
        if (v1.size != v2.size) error("...");
        vector v(n);
        for (int i = 0; i < n; ++i)
                v.vec[i] = v1.vec[i] + v2.vec[i];
        return v;
}
```

The multiplication and division operators allow multiplication or division by a floating point number. The multiplication operator has to be duplicated so that post-and pre-multiplication are both possible. We define post-multiplication in full:

```
vector operator*(const vector &v, double d)
{       vector vd = v;
        vd *= d;
        return vd;
}
```

Notice that we have avoided tiresome loop expressions by using on a local variable the multiplication assignment operator already defined. The pre-multiplication operator may now be defined inline:

```
inline vector operator*(double d, vector &v) { return v * d; }
```

The remaining multiplication operator will be dealt with in the next section, while the division operator may be written in the same way as post-multiplication.

The scalar product of two vectors, $\mathbf{u}.\mathbf{v} = \sum_i u_i v_i$, is simply:

```
double scalar(const vector &u, const vector &v)
(       double t = 0;
        int n = u.size;
        if (u.size != v.size) error("...");
        for (int i = 0; i < n; ++i)
                t += u.vec[i] * v.vec[i];
        return t;
}
```

Likewise, the euclidean norm of a vector, $\|\mathbf{v}\|_2 = \sqrt{(\sum_i v_i^2)}$, is:

```
double norm(const vector &v)                        // OVERFLOW!
{       int n = v.size;
        double t = 0.0;
        for (int i = 0; i < n; ++i)
        {       double vi = v.vec[i];
                t += vi * vi;
        }
        return sqrt(t);
}
```

However, this function is liable to overflow if the vector is large. It returns the square root of the sum of the squares of the elements v_i, but the sum of squares may be very large. Let t_i be the square root of the sum of the squares of the first i elements; then

$$t_{i+1} = (t_i^2 + v_i^2)^{1/2} = t_i(1 + (v_i/t_i)^2)^{1/2}.$$

A better definition is therefore:

```
double norm(const vector &v)                // BETTER!
{       int n = v.size;
        const double eps = 1.0E-15;         // small number
        double t = fabs(v[0]);
        for (int i = 1; i < n; ++i)
        {       double vi = fabs(v[i]);
                if (t < eps) t = vi;
                else
                {       double x = vi/t;
                        t *= sqrt(1 + x*x);
                }
        }
        return t;
}
```

where action has been taken to avoid dividing by values of t near zero.

We shall sometimes prefer to use the "infinite norm", or $\| \mathbf{v} \|_\infty = \text{Max} \, |v_i|$, which is equal to the magnitude of the element of greatest absolute value in the vector. This is given by:

```
double norminf(const vector &v)
{       int n = v.size;
        double t = 0.0;
        for (int i = 0; i < n; ++i)
        {       double vi = fabs(v.vec[i]);
                if (vi > t) t = vi;
        }
        return t;
}
```

Finally, there are the input and output operators. The output operator we shall define to be formatted, so that vectors of not more than eight elements can be displayed in consistent fashion on a single line of the screen, this being sufficient for our immediate purposes; but in large scale applications it may be desirable to alter the formats here offered. We shall require floating point numbers to be left adjusted and in fixed point notation; and we shall show them to 4 digits after the decimal point, within a field width of 10 characters. Using the class format introduced in 6.4.4, we therefore define the following format objects:

```
const long vml = ios::left + ios::fixed;
const int vmp = 4;
const int vmw = 10;

format vmfmt(vml, vmp, vmw), vmrst(0, 0, 0);
```

The output operator may then be written:

```
ostream& operator<<(ostream &s, const vector &v)
{      int n = v.size;
       for (int i = 0; i < n; ++i)
             s << vmfmt << v.vec[i];
       s << vmrst << "\n";
       return s;
}
```

We may now output a vector simply by writing cout << v;, or, using our standard output macro (*cf* Appendix A), output("Vector v = ", v);.

For input we might adopt the same approach; but to the extent that in this book we are concerned with input from the keyboard, we prefer to use a more direct method. We shall write:

```
istream& operator>>(istream &s, vector &v)
{      int n = v.size;
       cout << "enter " << n << " elements:\n";
       for (int i = 0; i < n; ++i)
       {      cout << "[" << i << "] = ";
              s >> v.vec[i];
       }
       return s;
}
```

Note that the input vector must *not* be declared const, otherwise input would be impossible! We may now seek input simply by writing cin >> v;, or using our standard input macro, input("Enter a vector:\n", v);. As a simple test of the input and output routines, we offer the following program:

```
#include <standard.h>
#include <vecmat.h>
void  main()
{      int n = 4;
       vector v(n);
       input("Enter a vector\n", v);
       newline;
       output("Vector is:\n ", v);
}
```

The screen will show something along the following lines:

```
Enter a vector
enter 4 elements:
v[0] = 1.234345
v[1] = 2.345456
```

```
v[2] = 3.456567
v[3] = 4.567678

Vector is:
1.2343     2.3455     3.4566     4.5677
```

The limitation to precision 4 in the formatted output operator means that the output values have been rounded to four decimal digits.

The input and output operators may be used in turn to test all other functions of the class vector. For example, we may enter two vectors u and v and print out the sum u + v. It is extremely important that all properties of a newly defined class should be thoroughly tested, especially if the class is likely to be of general application and therefore used in a variety of circumstances.

9.2 The class matrix

We shall regard a matrix as a vector of vectors, or rather as a pointer to an array of vectors. However, we shall often wish to refer to a particular element of a matrix. In ordinary mathematical notation we would refer to the element M_{ij}, say, of the matrix M. If we wish to use a formalism as similar to this as possible, we would hope to be able to write simply M[i][j]. This implies that the relevant data member of the class matrix should be vector **mat;, which means vector* *mat;, so that mat is a pointer to an array of pointers to vectors. The vector that is the ith row of the matrix is *mat[i], and the individual matrix elements are mat[i]->vec[j], and this expression is what we wish to have access to when we write M[i][j]. Remember that the data member of vector is itself a pointer to a double, so that **mat is a double.

Although much of our work will involve square matrices, this is not always the case. We must therefore allow for the possibility that the number of rows and the number of columns are different. This we do by having two data members referring to size; numrows is the number of rows and therefore the number of row vectors constituting the matrix, while numcols is the number of columns and therefore the size of each row vector.

One further preliminary remark: since the matrix has a pointer to a vector as a data member we should be able to refer directly to the members of the class vector. This we may do, as we saw in 5.1.3, if in defining the class vector we declare all the members of class matrix to be friends in the declaration of the class vector. A shorthand notation which achieves this very simply declares the whole class matrix to be a friend of vector, as follows:

```
class vector {          // declaration of vector
friend class matrix;    // matrix has access to vector
........
};
```

```
class matrix {              // declaration of matrix
........
};
```

The class `matrix` may now be developed, its members having full access to the members of class `vector`:

```
class matrix {
private:
        int numrows;
        int numcols;
        vector **mat;
        int range(int);            // row range check
public:
        matrix(int, int);          // rectangular matrix
        matrix(int);               // square matrix
        matrix(const matrix&);     // copy of matrix
        ~matrix();
        vector& operator[](int);
        ........
};
```

We shall first define the constructors. The constructor for a rectangular matrix is:

```
matrix::matrix(int nrows, int ncols)
{       numrows = nrows;  numcols = ncols;
        mat = new vector* [numrows];
        if (!mat) error("row alloc failure"
                    " in matrix::matrix(int, int)");
        for (int i = 0; i < numrows; ++i)
        {       mat[i] = new vector(numcols);
                if (!mat[i]) error("col alloc failure"
                        " in matrix::matrix(int, int)");
        }
}
```

We see that `mat` is defined as pointing to an array of pointers to vectors, `numrows` in number, each of which may be written `mat[i]` and in turn defined as pointing to a vector of size `numcols`, the elements of which are automatically initialized to zero by means of the constructor `vector::vector(int)` already defined. After each use of the operator `new`, the possibility arises that the intended memory allocation has failed; should this happen explanatory error messages are issued in the usual way before automatic termination of the program. Note also that the error messages, being too long to fit conveniently into a single line on the page, have been split into two: adjacent strings are automatically concatenated.

The square matrix constructor is similar, except that the first statement should be

```
numrows = numcols = n;.
```

The copy constructor takes the form:

```
matrix::matrix(const matrix &m)
{       numrows = m.numrows; numcols = m.numcols;
        mat = new vector* [numrows];
        if (!mat) error("row alloc failure"
                        " in matrix::matrix(matrix&)");
        for (int i = 0; i < numcols; ++i)
        {       mat[i] = new vector(numcols);
                if (!mat[i]) error("col alloc failure"
                                " in matrix::matrix(matrix&)");
        }
        for (i = 0; i < numrows; ++i) *mat[i] = *m.mat[i];
}
```

In the last statement, note that we have assigned to complete row vectors, vector assignment having already been defined. This simplifies the writing considerably.

The matrix destructor is a little more complicated than the vector destructor. We have to delete not only the pointer mat, but also the individual pointers mat[i]. This we should do in the opposite order to their creation:

```
matrix::~matrix()
{     for (int i = numrows; i > 0; --i)
            delete mat[i-1];
      delete mat;
}
```

Next, we define the indexing operator. If we refer back to the class declaration, we see that it has to return a complete row vector:

```
inline vector& matrix::operator[](int i)
{       return *mat[range(i)];          }
```

where the matrix::range function is given by

```
int matrix::range(int i)
{ return (i<0 || i>=numrows) ? (error("...")), -1) : i; }
```

and the error message might be "matrix row index out of range!".

If M is a matrix, the expression M[i] now refers to the ith (row) vector of M, while M[i][j] refers to the jth element of M[i]. We have therefore achieved our stated objective of being able to access the elements of the matrix M simply by writing M[i][j].

The assignment operators =, +=, −=, *=, / = follow in the class declaration of Appendix E. We define the first:

```
matrix& matrix::operator=(const matrix &m)
{    if (m.numrows != numrows || m.numcols != numcols)
     error("diff sizes in"
           " matrix& matrix::op=(const matrix&, const matrix&)!");
     for (int i = 0; i < numrows; ++i)
          *mat[i] = *m.mat[i];
     return *this;
}
```

The other assignment operators may be defined similarly. It is sometimes useful to be able to access the data members numrows and numcols. We define two functions accordingly, the analogues of the function getsize() in the class vector:

```
inline int matrix::getnumrows() { return numrows; }
inline int matrix::getnumcols() { return numcols; }
```

And in the case of a square matrix:

```
int   matrix::getsize()
{    if (numrows != numcols)
     error("getsize() requires square matrix");
     return numrows;
}
```

We may wish to interchange two row vectors:

```
void  matrix::swap(int i, int j)
{    vector *tmp = mat[range(i)];
     mat[i] = mat[range(j)];
     mat[j] = tmp;
}
```

Notice that we have chosen to interchange pointers to the row vectors concerned. (We shall not need to interchange *column* vectors.)

It is often necessary to transpose a matrix by interchanging rows and columns:

```
matrix  matrix::transpose()
{    int p = numrows, q = numcols;
     matrix mt(q, p);            // create transposed matrix
```

```
        for (int i = 0; i < q; ++i)
        {       for (int j = 0; j < p; ++j)
                        mt.mat[i]->vec[j] = mat[j]->vec[i];
        }
        return mt;
}
```

Unary plus and minus may be defined just as for the class **vector**. We leave them as an exercise for the reader. The binary addition friend operator can be simply written using the addition assignment already defined:

```
matrix operator+(const matrix &m1, const matrix &m2)
{       if (m1.numrows != m2.numrows || m1.numcols != m2.numcols)
        error("diff sizes in matrix op+(const matrix&, const matrix&)!");
        matrix mt = m1;
        mt += m2;
        return mt;
}
```

and similarly for the friend subtraction operator. We leave multiplication and division by a floating point number as an exercise for the reader; using the assignment operators already defined they are extremely simple.

The euclidean and infinite norms, as in the class **vector**, are defined as friends:

```
double norm(matrix &m)
{       double t = 0;
        int nr = m.numrows, nc = m.numcols;
        for (int i = 0; i < nr; ++i)
        {       for (int j = 0; j < nc; ++j)
                {       double mij = m[i][j];
                        t += mij * mij;
                }
        }
        return sqrt(t);
}
```

and

```
double norminf(matrix &m)
{       double t = 0;
        int nr = m.numrows, nc = m.numcols;
        for (int i = 0; i < nr; ++i)
        {       double s = 0;
```

```
              for (int j = 0; j < nc; ++j)
                    s += fabs(m.mat->vec[j]);
              if (s > t) t = s;
        }
        return t;
}
```

The function we have defined for the euclidean norm of a matrix, like its counterpart for a vector, is deficient if the matrix has large elements, because of the danger of overflow. We leave it to the reader to eliminate this deficiency, following the vector example.

There remain the input and output operators, which also are friend functions. We may build on what we did for the class `vector` and work entirely in terms of row vectors:

```
ostream& operator<<(ostream &s, const matrix &m)
{       int nr = m.numrows;
        for (int i = 0; i < nr; ++i)
              s << *m.mat[i];
        return s;
}
```

and

```
istream& operator>>(istream &s, matrix &m)
{       int nr = m.numrows;
        cout << "\nenter " << nr << "row vectors\n\n";
        for (int i = 0; i < nr; ++i)
        {       cout << "enter row vector " << i << "\n";
              s >> *m.mat[i];
        }
        return s;
}
```

9.3 Operations of linear algebra

Still further functions, friends of both classes, `vector` and `matrix`, need to be defined if we are to carry out the usual operations of linear algebra. First, however, it is necessary to make a distinction between row vectors and column vectors. It is a distinction of which the class `vector` as defined is innocent; the distinction involves the way that vectors combine with other vectors and with matrices. The vector $\mathbf{v} = \{v_0, v_1, \ldots, v_{n-1}\}$, written as a row vector, would be

$$(v_0 \quad v_1 \ldots v_{n-1})$$

whereas, written as a column vector, it would be

$$\begin{pmatrix} v_0 \\ v_1 \\ \cdot \\ \cdot \\ \cdot \\ v_{n-1} \end{pmatrix}$$

In mathematical notation, when we wish to emphasize that we are using a vector **v** as a row vector we shall write **v** ^{R}; when it is to be regarded as a column vector we shall write **v** ^{C}: in either case, the elements of the vector v are the same. The one is the transpose of the other: $v^R = (v^C)^T$, and *vice versa*. Similarly, we write M^T for the transpose of the matrix M. If the elements of the matrix M are m_{ij}, the elements of M^T are m_{ji}. Remember also that $(ab)^T = b^T a^T$ for both vectors and matrices. A matrix is regarded, as we have seen, as a column of row vectors. In writing products, the distinction between row vector and column vector must reflect itself in the order in which multiplications are carried out. Multiplication of vectors and matrices is not in general commutative.

In the next chapter we shall need to manipulate vectors and matrices to solve linear equations. Of recurring interest will be the pre-multiplication of a (column) vector by a matrix resulting in another (column) vector, and the transpose of this process. Also important is the multiplication of a (column) vector by a (row) vector resulting in a matrix. Three routines are presented to carry out these tasks, all declared as friend functions of *both* classes **vector** and **matrix**:

```
friend vector& operator*(matrix&, vector&);    // u^C = Mv^C
friend vector& operator*(vector&, matrix&);    // u^R = v^R M
friend matrix& operator*(vector&, vector&);    // M  = u^C v^R
```

The first routine returns a (column) vector obtained by multiplying a (column) vector by a matrix. It may be written as follows:

```
vector operator*(const matrix &m, const vector &v)
{       int nr = m.numrows;
        if (m.numcols != v.size)
        error("diff sizes in vector op*(const matrix&, const
vector&)!");
        vector u(nr);
        for (int i = 0; i < nr; ++i)
                u[i] = scalar(*m.mat[i], v);
        return u;
}
```

Note the use of the function **scalar** to avoid an additional explicit **for** loop.

The second routine, which is the transpose of the first, multiplies a row vector by a matrix to obtain a row vector:

```
vector operator*(const vector &v, const matrix &m)
{       int nr = m.numrows, nc = m.numcols;
        if (v.size != nr)
        error("diff sizes in vector op*(const vector&, const matrix&)!");
        vector u(nc);
        for (int i = 0; i < nc; ++i)
        {       double t = 0;
                for (int j = 0; j < nr; ++j)
                        t += v.vec[j] * m.mat[j]->vec[i];
                u.vec[i] = t;
        }
        return u;
}
```

The third routine, which multiplies a column vector by a row vector to produce a matrix, may likewise be written:

```
matrix operator*(const vector &u, const vector &v)
{       int nr = u.size, nc = v.size;
        matrix m(nr, nc);
        for (int i = 0; i < nr; ++i)
        {       for (int j = 0; j < nc; ++j)
                        m.mat[i][j] = u[i] * v[j];
        }
        return m;
}
```

These routines should be tested in many ways. A few of them are tested by the program:

```
main()
{       int n = 4;
        matrix m(n);
        vector v(n);
        input("\nenter matrix\n", m);
        input("\nenter vector\n", v);
        vector u = m * v;
        output("\nproduct vector is\n", u);
}
```

If we propose that the matrix M is

$$\begin{pmatrix} 1 & 2 & 3 & 4 \\ 2 & 3 & 4 & 5 \\ 3 & 4 & 5 & 6 \\ 4 & 5 & 6 & 7 \end{pmatrix}$$

and the (column) vector $\mathbf{v}$ is

$$(4 \quad 3 \quad 2 \quad 1)^{\mathrm{T}}$$

then the (column) vector $\mathbf{u} = M\mathbf{v}$ is:

$$(20 \quad 30 \quad 40 \quad 50)^{\mathrm{T}}.$$

Clearly, the pre-multiplication of a vector by a square matrix of the same dimension n in general requires n^2 multiplications and $n(n-1)$ additions. We consider the resulting speed of typical matrix manipulations in the course of the next chapter.

The complete header file `vecmat.h` is reproduced in Appendix E. It contains a great number of friend functions. This is the price one has to pay in C++ if one wishes to mimic as nearly as possible ordinary mathematical notation.

10 Direct solution of linear equations

We now have the apparatus to consider how to compute the solution of linear equations in an arbitrary number of unknowns. We shall denote by the symbols Eq_i the following set of n equations in n unknowns x_i with constant real coefficients a_{ij}:

$$\begin{aligned}
\mathrm{Eq}_0 &: \quad a_{00}x_0 + a_{01}x_1 + \ldots + a_{0\,n-1}x_{n-1} = b_0 \\
\mathrm{Eq}_1 &: \quad a_{10}x_0 + a_{11}x_1 + \ldots + a_{1\,n-1}x_{n-1} = b_1 \\
&\ldots\ldots\ldots\ldots\ldots\ldots\ldots\ldots\ldots\ldots\ldots\ldots\ldots\ldots\ldots \\
\mathrm{Eq}_{n-1} &: \quad a_{n-1\,0}x_0 + a_{n-1\,1}x_1 + \ldots + a_{n-1\,n-1}x_{n-1} = b_{n-1}
\end{aligned}$$

The given coefficients form a square matrix $A = (a_{ij})$, and the unknowns x_i and the given right-hand constants b_i are column vectors $\mathbf{x}$ and $\mathbf{b}$, respectively. In matrix notation,

$$\sum_j a_{ij}x_j = b_i, \quad \text{or} \quad A\mathbf{x} = \mathbf{b}.$$

A very important quantity associated with a square matrix is its *determinant*. Presently, we shall see how to compute the determinant economically. For the time being we merely note that it may be defined recursively. Let $A^{(p,q)}$ be the matrix obtained by striking out the pth row and the qth column of A. Then,

$$\det(A) = \sum_i (-1)^{i+j} a_{ij} \det(A^{(i,j)}),$$

for any column j; or alternatively,

$$\det(A) = \sum_j (-1)^{i+j} a_{ij} \det(A^{(i,j)}),$$

for any row i.

This is known as *Cramer's rule*. It should *not* be used to compute the determinants of matrices of more than a few rows and columns: it requires $n!$ multiplications to evaluate the determinant of an $n{\times}n$ matrix, which rapidly becomes prohibitive.

If A and B are $n{\times}n$ matrices, then $\det(AB) = \det(A)\det(B)$. Note also that if A^{T} is the transpose of A, obtained by interchanging rows with columns, then $\det(A^{\mathrm{T}}) = \det(A)$.

We state without proof the well-known

Theorem

The following statements are equivalent:

(i) $\det(A) \neq 0$;

(ii)   The system of linear equations $A\mathbf{x} = \mathbf{b}$ has a unique solution for every vector $\mathbf{b}$;

(iii) $A\mathbf{x} = \mathbf{0}$ has the unique solution $\mathbf{x} = \mathbf{0}$, where $\mathbf{0}$ is the column vector every element of which is zero;

(iv) A has a unique inverse A^{-1} such that $AA^{-1} = A^{-1}A = I$, where I is the identity matrix, the only non-zero elements of which are diagonal and equal to unity.

In this chapter we study direct methods, where the solution of the above equations is obtained in a single step; we pay particular attention to the speed of computation. In Chapter 12 we consider iterative methods.

10.1 Gaussian elimination

Three operations are allowed on a system of linear equations. Firstly, a multiple of one equation may be added to another equation; this we shall denote symbolically by $\mathrm{Eq}_i \mathrel{+}= \lambda \mathrm{Eq}_j$. Second, two equations may be exchanged, which we denote by $\mathrm{Eqswap}(i,j)$. Third, an equation may be multiplied through by a constant, which we denote by $\mathrm{Eq}_i \mathrel{*}= \mu$. After carrying out any of these operations we shall continue to use the symbols Eq_i, although the coefficients will have changed. These operations will allow us, in certain circumstances, to compute the solution to the system of equations.

10.1.1 Algorithms

The simplest method is to eliminate x_0 from all Eq_i with $i > 0$ by using only the first of these three operations; x_1 from all the resulting equations Eq_i with $i > 1$; and in general x_j from all Eq_i with $i > j$. To eliminate x_0 from Eq_1, for example, we replace it with $\mathrm{Eq}_1 \mathrel{+}= \lambda \mathrm{Eq}_0$, setting $\lambda = -a_{10}/a_{00}$. At a certain stage, the equations will take the form:

$$
\begin{array}{llllllllll}
\mathrm{Eq}_0 & : & a_{00}x_0 & + & a_{01}x_1 & +\ldots+ & a_{0j}x_j & +\ldots+ & a_{0\,n-1}x_{n-1} & = & b_0 \\
\mathrm{Eq}_1 & : & & & a_{11}x_1 & +\ldots+ & a_{1j}x_j & +\ldots+ & a_{1\,n-1}x_{n-1} & = & b_1 \\
\multicolumn{11}{c}{\dotfill} \\
\mathrm{Eq}_j & : & & & & & a_{jj}x_j & +\ldots+ & a_{j\,n-1}x_{n-1} & = & b_j \\
\multicolumn{11}{c}{\dotfill} \\
\mathrm{Eq}_{n-1} & : & & & & & a_{n-1\,j}x_j & +\ldots+ & a_{n-1\,n-1}x_{n-1} & = & b_{n-1}
\end{array}
$$

and we now wish to eliminate x_j from all Eq_k with $k > j$. In "pseudo-code" we do this by repeating the plus assignment:

```
for (int k = j+1; k < n; ++k)
{       λ = -a_kj/a_jj;
        Eq_k += λEq_j;
}
```

Complete triangulation is achieved if we repeat this process, starting with $k = 1$. The result is a triangular matrix of coefficients a_{ij} with all elements zero below (*ie*, to the left of) the principal diagonal. The process is known as *gaussian elimination*. The following procedure is written in real code, where in addition an error is declared if any of the diagonal elements should become zero:

```
void   triangulate(matrix &a, vector &b)
//     make matrix a upper triangular
{      int n = a.getsize();
       for (int i=0; i<n-1; i++)
       {      double diag = a[i][i];
              if (fabs(diag) < gauss0_toosmall)
                     error("diagonal too small in triangulate()");
              for (int j=i+1; j<n; j++)
              {      double mult = a[j][i]/diag;
                     a[j] -= mult * a[i];
                     b[j] -= mult * b[i];
              }
       }
}
```

For example, if we apply the procedure `triangulate` to the following equations,

$$\begin{pmatrix} 1 & 2 & 3 & 4 \\ 2 & 3 & 4 & 6 \\ 3 & 4 & 2 & 5 \\ 4 & 6 & 5 & 7 \end{pmatrix} \begin{pmatrix} x_0 \\ x_1 \\ x_2 \\ x_3 \end{pmatrix} = \begin{pmatrix} 8 \\ 6 \\ 4 \\ 2 \end{pmatrix},$$

we obtain

$$\begin{pmatrix} 1 & 2 & 3 & 4 \\ 0 & -1 & -2 & -2 \\ 0 & 0 & -3 & -3 \\ 0 & 0 & 0 & -2 \end{pmatrix} \begin{pmatrix} x_0 \\ x_1 \\ x_2 \\ x_3 \end{pmatrix} = \begin{pmatrix} 8 \\ -10 \\ 0 \\ -10 \end{pmatrix}.$$

It is then a simple matter to find the values of x_i that satisfy the original equations; one merely has to backsubstitute in the triangular system just obtained. Starting with the last equation, Eq_{n-1}, which contains the single unknown x_{n-1}, we have $x_{n-1} = b_{n-1}/a_{n-1\,n-1}$, where the elements of A and **b** are now those resulting from triangulation. This value may be substituted into Eq_{n-2} in order to calculate x_{n-2}; and so on. Again, it is necessary to beware of zero divisors:

```
void   backsubst(matrix &a, vector &x, vector &b)
//     given triangular a, solve for x given b
{      int n = a.getsize();
```

```
for (int i=n-1; i>=0; i--)
{      double diag = a[i][i];
       if (fabs(diag) < gauss0_toosmall)
             error("diagonal too small in backsubst()");
       x[i] = (b[i] - scalar(a[i], x))/diag;
}
}
```

Applying `backsubst` to the equations resulting from `triangulate` in the above example, we obtain:

$$\mathbf{x} = (-17 \quad 10 \quad -5 \quad 5)^{\mathrm{T}}.$$

Finally, the two procedures may be combined into one, which we call `gauss0`:

```
void  gauss0(matrix &m, vector &x, vector &b)
{      triangulate(m, b);
       backsubst(m, x, b);
}
```

10.1.2 Speed of computation

We shall now estimate the speed of the function `gauss0` by counting the number of multiplications and divisions involved; other arithmetic operations will be ignored, although there are other processes for which there are so many additions and subtractions, and so few multiplications and divisions, that this convention might prove misleading (*cf* 3.10).

We shall often need the following well-known results:

$$\sum_{i=0}^{i<N} 1 = N, \qquad \sum_{i=0}^{i<N} i = N(N-1)/2, \qquad \sum_{i=0}^{i<N} i^2 = N(N-1)(2N-1)/6.$$

Consider first the function `triangulate`. The definition of `mult` requires 1 division. The multiplication assignment to the vector `a[j]` in the `j` loop requires n multiplications. The multiply assignment to `b[j]` requires 1 further multiplication. There are thus $n + 2$ multiplications or divisions in the inner, `j`, loop. This loop is repeated $n-1-i$ times. The outer, `i`, loop runs over $i = 0, \ldots, n-2$. The total number of multiplications and divisions in `triangulate` is therefore:

$$\sum_{i=0}^{i<n-1} (n+2)(n-1-i) = (n+2)[(n-1)\sum_{i=0}^{i<n-1} 1 - \sum_{i=0}^{i<n-1} i]$$

$$= (n+2)[(n-1)(n-1) - (n-1)(n-2)/2]$$

$$= n(n-1)(n+2)/2$$

$$= n^3/2 + O(n^2).$$

The function `backsubst`, on the other hand, because of the function `scalar`, has n multiplications and 1 division, making $n + 1$ multiplications or divisions altogether, per iteration. There is only a single loop, and it runs over $i = 0, \ldots, n-1$. The total number of multiplications and divisions in `backsubst` is therefore:

$$(n + 1) \sum_{i=0}^{i<n} 1 = n(n + 1).$$

Adding together the contributions from both functions, we find that `gauss0` requires

$$\tfrac{1}{2}n^2(n + 3) = n^3/2 + O(n^2).$$

multiplications and divisions. The number of additions and subtractions is also $O(n^3)$. The time taken to solve a set of n linear equations therefore increases sharply with n. For large n, `backsubst`, being only quadratic in n, does not add appreciably to the number of operations and therefore to the time of computation.

10.2 Refinement

The function `gauss0` is too simple for many applications. It may fail because one of the divisors `diag` becomes zero or small in the process of triangulation; and many of its operations consist of addition or multiplication of those elements which have already been reduced to zero. It is therefore both unpredictable and inefficient. We must refine the function in order to eliminate these features as far as possible.

10.2.1 Pivoting

The second procedure available for the solution of a system of linear equations is the interchange of two complete equations, which we have denoted symbolically by Eqswap(i, j). The order in which the equations are presented makes no difference to their solution, but it can make a great deal of difference to the possibility of *computing* the solution. Returning to the procedure `gauss0`, we observe that division by a zero diagonal element has to be avoided. Indeed, division by a sufficiently small diagonal element is enough to lose precision. Small or zero divisors can easily occur during the execution of the function `gauss0`, and we should therefore seek to reduce their incidence as much as possible. This may be done by rearranging the order of the equations during the course of the computation.

To take a very simple example, suppose we try to use `gauss0` to compute the solution of the pair of equations,

$$ax + by = c$$
$$dx + ey = f$$

where $a = 0$. The function gauss0 fails immediately because the first divisor is zero. If we interchange the equations, however, the first divisor becomes d, which we suppose not to be zero, so gauss0 succeeds.

To avoid occurrences of zero divisors of this trivial kind, and in order to preserve precision, we should obviously try to arrange that divisors are as large as possible. This will be our criterion in deciding how best to arrange the equations at each step of the process. Consider again the function triangulate. Instead of taking the diagonal divisor in the first loop to be a[i][i], let us find a rearrangement of that part of the matrix not so far triangulated which makes this element as large as possible. Instead of the statement

```
double diag = a[i][i];
```

we shall instead write

```
double diag = pivot(a, b, i);
```

leaving the rest of the function triangulate unchanged. The function pivot is intended to return the largest available diagonal element obtained by rearranging the equations from Eq$_i$ downwards; those above the pivotal equation Eq$_i$ have already been dealt with and their arrangement clearly must not be disturbed again. Those below it have zero for their first $i-1$ elements. We therefore seek the element of largest magnitude, from row i down, in column i. Having located it in Eq$_j$, say, we interchange Eq$_i$ and Eq$_j$ to bring this largest available element into diagonal position. In pseudo-code, the function pivot should perform the following operation:

$$\text{int } j = i;$$
$$\text{findmaxelem}(i, j);$$
$$\text{if } (j > i) \text{ Eqswap}(i, j);$$

This operation is performed by the function pivot if its definition is:

```
double pivot(matrix &a, vector &b, int i)
//      row i to have largest element from lower col i as diag
{       int n = a.getsize();
        int j = i;                              // row variable
        double t = 0;
        for (int k=i; k<n; ++k)                 // find max elem
        {       double aki = fabs(a[k][i]);
                if (aki > t) { t = aki;   j = k; }
        }
        if (j > i)                              // swap equations
        {       a.swap(i, j);
                b.swap(i, j);
        }
        return a[i][i];                         // return diagonal
}
```

10.2.2 The case of the vanishing determinant

The resulting redefinition of `triangulate` will avoid many of the potential failures of the original version. However, it may be that there is a zero diagonal element with all the elements in the lower column also zero. In that case our revised procedure will be to no avail; a zero diagonal element cannot be avoided. This is not an imperfection of the computation procedure: it is because $\det(A) = 0$.

The following theorem may readily be proved by expanding the determinant about its first column as in Cramer's rule:

> If $M = (m_{ij})$ is a triangular matrix (upper or lower), or as a special case is diagonal, the determinant of M is given by $\det(M) = \prod_i m_{ii}$.

Thus, in a triangular matrix, if any diagonal element is zero, so is the determinant. If $\det(A) = 0$, there is no unique solution to the system of equations, and the method must fail. To illustrate this, suppose that triangulation leads to the following system of equations:

$$\begin{pmatrix} 1 & 2 & 3 & 4 \\ 0 & 0 & 2 & -1 \\ 0 & 0 & -1 & 1 \\ 0 & 0 & 0 & 1 \end{pmatrix} \begin{pmatrix} x_0 \\ x_1 \\ x_2 \\ x_3 \end{pmatrix} = \begin{pmatrix} 10 \\ -4 \\ 3 \\ 2 \end{pmatrix}.$$

Clearly the determinant vanishes, because the second diagonal element is zero: the rows and columns of the matrix are not linearly independent. We can attempt to solve the equations by back substitution. From the last equation, $x_3 = 2$. Substituting this value into the next equation gives $x_2 = -1$. The next equation is satisfied identically, reducing to $-4 = -4$. The first equation then reduces to $x_0 + 2x_1 = 5$, or $x_1 = (5 - x_0)/2$, where x_0 is arbitrary: there is a continuum of solutions. If, on the other hand, the right-hand constant of the second equation were -5, say, instead of -4, then this equation would be inconsistent with the last two, and there would be *no* solution. Either way, there would be no *unique* solution to the system of equations, because the determinant vanishes.

10.2.3 Minimizing the arithmetic

In the function `triangulate`, there is the statement `a[j] -= mult * a[i];` within the j loop, where `j > i`. But the row vector `a[i]` has already been triangulated, so its first i−1 elements are zero, as are the corresponding elements of `a[j]`. To multiply zero elements by `mult`, and then to subtract them, is wasteful computer effort. We shall therefore replace this statement by the less elegant but more efficient statement:

```
for (int k = i+1; k < n; ++k)
      a[j][k] -= mult*a[i][k];
```

The function `backsubst` of the procedure `gauss0` is similarly inefficient. It contains the statement

```
x[i] = (b[i] - scalar(a[i], x))/diag;
```

but only some of the elements occurring in the scalar product are non-zero. We shall replace it with

```
x[i] = (b[i] - dotprod(a[i], x, i+1, n-1))/diag;
```

where the function `dotprod` also avoids unnecessary arithmetic:

```
double dotprod(vector &u, vector &v, int k1, int k2)
//      sum u[i]*v[i], i = k1...k2
{       double sum = 0;
        for (int i = k1; i <= k2; ++i) sum += u[i] * v[i];
        return sum;
}
```

10.2.4 Refined gauss

We may now draw these considerations together into a revised procedure for the solution of linear equations by gaussian elimination. The function `triangulate` becomes

```
void   triangulate(matrix &a, vector &b)                // IMPROVED
//     with pivoting
{      int n = a.getsize();
       for (int i = 0; i < n-1; ++i)
       {      double diag = pivot(a, b, i);
              if (fabs(diag) < gauss_toosmall)
                     error("zero determinant!");
              for (int j = i + 1; j < n; ++j)
              {      double mult = a[j][i]/diag;
                     for (int k = i+1; k < n; ++k)
                            a[j][k] -= mult*a[i][k];
                     b[j] -= mult * b[i];
              }
       }
}
```

and `backsubst` becomes

```
void   backsubst(matrix &a, vector &x, vector &b)    // IMPROVED
{      int n = a.getsize();
       for (int i = n - 1; i >= 0; i--)
       x[i] = (b[i]-dotprod(a[i],x,i+1,n-1))/a[i][i];
}
```

The refined gaussian procedure is then given by

```
void   gauss(matrix &a, vector &x, vector &b)
{      triangulate(a, b);
       backsubst(a, x, b);
}
```

where the improved versions of `triangulate` and `backsubst` are implied.

10.2.5 Improved speed

The refined `gauss` procedure is still cubic in n, but there is an improvement in speed nevertheless, which is due to more efficient arithmetic. In the function `triangulate`, the inner, j, loop now contains $n - i + 1$ multiplications or divisions, and this has to be iterated $n - i - 1$ times to give

$$\sum_{j=i+1}^{j<n} (n - i + 1) = (n - i)^2 - 1$$

multiplications and divisions. The total number of such operations in `triangulate` is therefore:

$$\sum_{i=0}^{i<n-1} [(n - i)^2 - 1] = n^3/3 + O(n^2).$$

The function `backsubst` still has only $O(n^2)$ multiplications and divisions. Thus, the total for the function `gauss` is $n^3/3 + O(n^2)$, which for large n is 50% faster than the simple function `gauss0`.

10.2.6 Scaling

One further operation remains to us. It may be that the coefficients of the matrix a differ widely in magnitude, and there is a danger that precision will suffer. In that case, the equations can each be multiplied through by a constant chosen to make the matrix coefficients more nearly of the same order of magnitude. How this is done is arbitrary, but a commonplace strategy is to calculate a *scaling vector* with elements $\mu_i = \|a_i\|_\infty^{-1}$, where a_i is the ith row of the matrix A. In each equation, the largest coefficient will then be ± 1. Better scaling methods exist, however, which avoid the use of so many costly multiplications, and we shall pursue the matter no further.

10.3 Matrix decomposition

Let us now reconsider the process `gauss0` from the point of view of what is happening to the matrix of coefficients when we perform the successive operations $\mathrm{Eq}_i \mathrel{+}= \lambda \mathrm{Eq}_j$. We shall for the moment assume there will be no zero divisors.

Set $\lambda_{i0} = -a_{i0}/a_{00}$, and consider the matrix

$$
L_1 = \begin{pmatrix} 1 & 0 \ldots \ldots 0 \\ \lambda_{10} & 1 \ldots \ldots 0 \\ \cdots\cdots\cdots\cdots \\ \lambda_{n-10} & 0 \ldots \ldots 1 \end{pmatrix}.
$$

The inverse L_1^{-1} of the matrix L_1, satisfying $L_1^{-1}L_1 = L_1 L_1^{-1} = I$, where I is the identity matrix, is given by

$$
L_1^{-1} = \begin{pmatrix} 1 & 0 \ldots \ldots 0 \\ -\lambda_{10} & 1 \ldots \ldots 0 \\ \cdots\cdots\cdots\cdots \\ -\lambda_{n-10} & 0 \ldots \ldots 1 \end{pmatrix},
$$

as may readily be seen by direct matrix multiplication. Then define:

$$
A_1 \equiv L_1 A = \begin{pmatrix} a_{00} & a_{01} & \ldots & a_{0\,n-1} \\ 0 & a^{(1)}_{11} & \ldots & a^{(1)}_{0\,n-1} \\ \cdots\cdots\cdots\cdots\cdots\cdots \\ 0 & a^{(1)}_{n-1\,1} & \ldots & a^{(1)}_{n-1\,n-1} \end{pmatrix},
$$

where $a^{(1)}_{ij} = a_{ij} + \lambda_{i0} a_{0j}$ $(i, j \neq 0)$. This is the first step of the gaussian elimination process. If we repeat it we eventually obtain, as we know, an upper triangular matrix

$$
U \equiv A_{n-1} = L_{n-1} L_{n-2} \ldots L_1 A.
$$

Let us define $L^{-1} = L_{n-1} L_{n-2} \ldots L_1$, so that $L = L_1^{-1} L_2^{-1} \ldots L_{n-1}^{-1}$. Then $U = L^{-1} A$, and $LU = A$.

It may readily be shown that L takes the form

$$
L = \begin{pmatrix}
1 & 0 & 0\ldots\ldots 0 \\
\lambda_{10} & 1 & 0\ldots\ldots 0 \\
\lambda_{20} & \lambda_{20} & 1\ldots\ldots 0 \\
\cdots\cdots\cdots\cdots\cdots\cdots \\
\lambda_{n-10} & \lambda_{n-11} & \lambda_{n-12}\ldots\ldots 1
\end{pmatrix}.
$$

The question is, whether *any* square $n \times n$ matrix may be directly *decomposed* (or factored) into the product of a lower and an upper triangular matrix, without *overtly* using the gaussian process. Each triangular matrix contains in general $\tfrac{1}{2}n(n + 1)$ non-zero elements (although some of them also may *happen* to be zero). That is altogether n elements more than the n^2 contained by the original matrix. It is customary to remove the redundancy by setting the diagonal elements of L equal to unity (in the Doolittle method, so-called), or the diagonal elements of U equal to unity (the Crout method), or setting the diagonal elements of L equal to those of U (the Choleski method, which is sometimes used for symmetric matrices). The Doolittle method is essentially that which results from gaussian elimination. However, there is no need to assume gaussian elimination in order to obtain it.

If, in the Doolittle method, we set

$$
LU = \begin{pmatrix}
1 & 0 & 0 & 0 \\
l_{10} & 1 & 0 & 0 \\
l_{20} & l_{21} & 1 & 0 \\
l_{30} & l_{31} & l_{32} & 1
\end{pmatrix}
\begin{pmatrix}
u_{00} & u_{01} & u_{02} & u_{03} \\
0 & u_{11} & u_{12} & u_{13} \\
0 & 0 & u_{22} & u_{23} \\
0 & 0 & 0 & u_{33}
\end{pmatrix}
= \begin{pmatrix}
1 & 2 & 3 & 4 \\
2 & 3 & 4 & 6 \\
3 & 4 & 2 & 5 \\
4 & 6 & 5 & 7
\end{pmatrix} = A,
$$

then, by successive substitution, we find

$$
L_D = \begin{pmatrix}
1 & 0 & 0 & 0 \\
2 & 1 & 0 & 0 \\
3 & 2 & 1 & 0 \\
4 & 2 & 1 & 1
\end{pmatrix}, \quad
U_D = \begin{pmatrix}
1 & 2 & 3 & 4 \\
0 & -1 & -2 & -2 \\
0 & 0 & -3 & -3 \\
0 & 0 & 0 & -2
\end{pmatrix}.
$$

Similarly, in the Crout method, we find

$$
L_C = \begin{pmatrix}
1 & 0 & 0 & 0 \\
2 & -1 & 0 & 0 \\
3 & -2 & -3 & 0 \\
4 & -2 & -3 & -2
\end{pmatrix}, \quad
U_C = \begin{pmatrix}
1 & 2 & 3 & 4 \\
0 & 1 & 2 & 2 \\
0 & 0 & 1 & 1 \\
0 & 0 & 0 & 1
\end{pmatrix}.
$$

Now that L and U are both known, the original equations $Ax = \mathbf{b}$ may be decomposed into the *two* triangular systems of equations

$$
Ly = \mathbf{b}, \qquad Ux = y,
$$

where we have introduced the auxiliary vector variable **y**, the value of which is to be found by solving the first triangular set by direct substitution. Once **y** is known, **x** may be found in the same fashion from the second set.

Taking **b** = (8 6 4 2)^T as before, we obtain:

$$\text{Doolittle} \ldots \mathbf{y} = (8 \ -10 \ \ 0 \ -10)^T$$
$$\text{Crout} \ldots \ldots \mathbf{y} = (8 \ \ \ 10 \ \ 0 \ \ \ \ 5)^T$$

and in both cases, $$\mathbf{x} = (-17 \ \ 10 \ -5 \ \ 5)^T$$

just as we found using the function `gauss0`. Notice that both L and U may be stored compactly within a *single* matrix. This is because there is no need to store diagonal elements the values of which have been set equal to unity.

The compacted Doolittle matrix we shall denote by A^D, and the Crout matrix by A^C. In the present case, they are given by:

$$A^D = \begin{pmatrix} 1 & 2 & 3 & 4 \\ 2 & -1 & -2 & -2 \\ 3 & 2 & -3 & -3 \\ 4 & 2 & 1 & -2 \end{pmatrix}, \quad A^C = \begin{pmatrix} 1 & 2 & 3 & 4 \\ 2 & -1 & 2 & 2 \\ 3 & -2 & -3 & -1 \\ 4 & -2 & -3 & -2 \end{pmatrix}$$

These are the matrices that are to be computed directly. Notice that once the vector **b** has been used to compute **y** it is no longer needed; it may be used to store the solution **x**. A single vector template may be used to store first **b**, then **y**, then **x**. Moreover, since all we have to do is to make orderly substitutions, a single matrix template will serve to store first the matrix A of coefficients (a_{ij}), then the Doolittle or Crout matrix. The work is very compact.

10.4 Crout algorithm

We must now program the above processes. We shall develop a routine for the Crout method; Doolittle is similar, but it will be left as an exercise for the reader. First, we present a routine equivalent to the simple process `gauss0`, which assumes there will be no zero divisors. Then we shall show how to include pivoting.

10.4.1 Simple Crout

We shall need a function to return the lesser of two integers:

```
inline  int min(int a, int b) { return a < b ? a : b; }
```

The first thing we have to do is to transform each matrix element m_{ij} from its initial value a_{ij} to its final value l_{ij} or u_{ij}, as the case may be, always remembering that in Crout the diagonal elements u_{ii} are unity, but unexpressed in m. The following function performs this task:

```
double revise(matrix &m, int i, int j)
{      double t = m[i][j];
       for (int k = 0; k < min(i, j); ++k)
             t -= m[i][k] * m[k][j];
       if (j > i)
       {      double diag = m[i][i] ;
              if (fabs(diag) < crout_toosmall)
                    error("diag near zero in revise()");
              t /= diag;
       }
       return t;
}
```

The elements m_{ij} with $j \leq i$ are the elements l_{ij}, those with $j > i$ are the elements u_{ij} which have to be divided by the diagonal elements l_{ii}.

In terms of the auxiliary function **revise** just defined we may easily write a procedure to perform the decomposition $A = LU$. It is:

```
void   decompose(matrix &m)
{      int n = m.getsize();
       for (int i = 0; i < n; ++i)
       {      for (int j = 0; j < n; ++j)
                    m[i][j] = revise(m, i, j);
       }
}
```

We now have to perform the forward substitution implied by $L\mathbf{y} = \mathbf{b}$, followed by the backward substitution implied by $U\mathbf{x} = \mathbf{y}$:

```
void   substitute(matrix &m, vector &v)              // v = b
{      int n = m.getsize();
       for (int i = 0; i < n; ++i)                   // Ly = b
       {      double t = v[i];
              for (int j = 0; j < i; ++j)
                    t -= m[i][j] * v[j];
              double diag = m[i][i];
              if (fabs(diag) < crout_toosmall)
                    error("diag near zero in substitute()");
              t /= diag;
              v[i] = t;                              // v = y
       }
       for (i = n-1; i >= 0; i--)                    // Ux = y
       {      double t = v[i];
              for (int j = i+1; j < n; j++)
```

```
                        t -= m[i][j] * v[j];
            v[i] = t;                                    // v = x
        }
    }
}
```

Finally, we combine these functions into the single function crout0:

```
void   crout0(matrix &m, vector &v)
//     input m is matrix A of coefficients
//     input v is b, output is x
{      decompose(m);
       substitute(m, v);
}
```

Notice that, when crout0 is used to solve the equations $Ax = b$, the matrix A, once decomposed, may be used over and over again with different values of the given vector **b**. The function decompose does not depend upon the vector variable. For this reason the functions decompose and substitute should themselves be generally available in any header file as well as their combination crout0. This is an important feature worth preserving when pivoting is introduced.

10.4.2 Speed

We may again estimate the time of computation by counting the number of multiplications involved in the function crout0. Since combinatorics are rather tedious we prefer to compute this number in what follows. All we have to do is to rewrite the functions in such a way that each multiplication is replaced by an increment of an integer variable. We are concerned only with the function decompose, because that is clearly cubic in n; the function substitute is only quadratic, and so does not add appreciably to the amount of computation for large n.

Let us therefore define a function count which returns the number of multiplications in decompose:

```
int    count(int n)
{      int c = 0;
       for (int i = 0; i < n; ++i)
       {      for (int j = 0; j < n; ++j)
              {      for (int k = 0; k < min(i, j); ++k) ++c;
                     if (j > i) ++c;
              }
       }
       return c;
}
```

Running this routine for a few low values of n readily shows that the number of multiplications in decompose is $(n^3 - n)/3 \approx n^3/3$ for large n. This is to be compared with $n^3/2$ for the function `triangulate` in `gauss0`, which did not perform the arithmetic quite so efficiently, and with $n^3/3$ in `gauss`, which did.

10.4.3 Crout with pivoting

We must now introduce pivoting to avoid, as far as possible, zero divisors—just as we did to go from `gauss0` to `gauss`. First, however, note that it is easy to modify the function `decompose` so that it returns the value of the determinant of the matrix of coefficients.

We rely upon the fact that $\det(A) = \det(LU) = \det(L)\det(U)$. But $\det(U) = \prod_i u_{ii} = 1$, because U is triangular and in the Crout method its diagonal elements are unity. Therefore $\det(A) = \det(L) = \prod_i l_{ii}$. The following modification of `decompose` therefore returns the value of the determinant:

```
double decompose(matrix &m)                        // REVISED
{       int n = m.getsize();
        double det = 1;
        for (int i = 0; i < n; ++i)
        {       for (int j = 0; j < n; ++j)
                        m[i][j] = revise(m, i, j);
                det *= m[i][i];
        }
        return det;
}
```

Now recall from our experience of gaussian elimination that the purpose of pivoting will be to reduce as far as possible the occurrence of zero divisors in the function `revise` (*cf* 10.4.1). We have to interchange the rows of the matrix A and of the (column) vector **b** so that chance occurrences of zero diagonal elements are if possible avoided. There are two complications. Firstly, if any two rows of the matrix are interchanged the determinant changes sign. Second, if the function which is to play the part of `pivot` in the earlier routine `gauss` receives the vector variable as a parameter, then so does the function `decompose`, and this we wish to avoid. In `gauss`, however, the vector variable is involved in `pivot` only so that its rows may be interchanged in unison with the rows of the matrix. All that we require is a record of the row permutations invoked by `pivot`, so that at a later stage we may apply the same permutation of rows to the vector variable.

Let us define a column vector $\mathbf{p} = (0 \ \ 1 \ \ 2 \dots n{-}1)^{\mathrm{T}}$ with the aid of the function

```
vector index(int n)
{       vector v(n);
        for (int i = 0; i < n; ++i)
                v[i] = double(i);
        return v;
}
```

If we in
the matrix
Thus, let p
overall si

```
int
{
```

```
}
```

```
}
```

An introduction to nume

```
void permute(vector &v, vector &p)
{
    int n = v.getsize();
    vector u(n);
    for (int i = 0; i < n; ++i)
        u[i] = v[int(p[i])];
    v = u;
}
```

Finally, we may assemble all the abov

```
double crout(matrix &m,
{
    int n = m.getsize
    if (v.getsize()
        error("
    vector p =
    double det
    permute(
    substit
    retur
}
```

Apar
nant,
co
h

```
{
    in
    if (p.getsize()
        error("diff sizes in u
    double det = 1;
    for (int i = 0; i < n; ++i)
    {   int sgn = pivot(m, p, i);
        double piv = m[i][i];
        det *= sgn*piv;
        for (int j = i+1; j < n; ++j) m[i][j] /= piv;
        for (int j = i+1; j < n; ++j)
            if (double mult = m[j][i] != 0)
                for (int k = i+1; k < n; ++k)
                    m[j][k] -= mult*m[i][k];
    }
    return det;
}
```

The function substitute is unchanged, but it must be applied to the permuted version of the vector **b**, the elements of which are obtained by applying the permutation of the indices recorded in the vector **p**:

functions into the final `crout` routine:

```
                    vector &v)
                    ();
                !=  n)
            diff  sizes  in  crout(...)!");
            ndex(n);
            =  decompose(m,  p);
        v,  p);
        ute(m,  v);
        n  det;
```

from additional multiplications required in **decompose** to compute the determi-
the number of multiplications is essentially unaffected, because **pivot** only requires
parisons. The time of computation remains $n^3/3$ for large n. The function **decompose**,
owever, is independent of the given vector **b**. If several computations are required for different vectors **b**, but the same matrix A, the advantage is in favour of **crout** rather than **gauss**;
but they are otherwise equivalent.

10.5 The inverse matrix

We may now use the routines **decompose** and **substitute** to compute the inverse of a
matrix, which exists provided the determinant is non-zero.

We should begin by programming the identity matrix. Since the constructor
matrix::matrix(int) constructs a matrix all the elements of which are initialized to
zero, it is necessary only to define the diagonal elements:

```
matrix identity(int n)
{       matrix m(n);                    // initialized to zero
        for (int i = 0; i < n; ++i)
                m[i][i] = 1;
        return m;
}
```

Let B be a matrix such that $AB = I$, and denote the *columns* of B by B_{*j} and of I by I_{*j}. Then $AB_{*j} = I_{*j}$, so that the procedure `crout` may be used to calculate the matrix B, column by column. The solution will, of course, be represented by *row* vectors; therefore the resulting matrix must be transposed to obtain B, which is the desired inverse. We may therefore write:

```
matrix inverse(matrix &m)
{       int n = m.getsize();
        vector p = index(n);
        matrix t = identity(n);
        double det = decompose(m, p);
        if (fabs(det) < crout_toosmall)
            error("zero determinant in inverse(matrix&)!");
        // treat col vectors as row vectors
        for (int i = 0; i < n; ++i)
        {       permute(t[i], p);
                substitute(m, t[i]);
        }
        // transpose to col vectors
        matrix inv = t.transpose();
        return inv;
}
```

If we apply `inverse` to the matrix:

$$A = \begin{pmatrix} 1 & 2 & 3 & 4 \\ 2 & 3 & 4 & 6 \\ 3 & 4 & 2 & 5 \\ 4 & 6 & 5 & 7 \end{pmatrix},$$

the inverse is given by

$$A^{-1} = (1/6) \begin{pmatrix} -23 & 16 & -5 & 3 \\ 16 & -14 & 4 & 0 \\ -5 & 4 & -5 & 3 \\ 3 & 0 & 3 & -3 \end{pmatrix}.$$

Provided $\det(A) \neq 0$, it is possible to write the solution of the equations $Ax = \mathbf{b}$ as $\mathbf{x} = A^{-1}\mathbf{b}$. This is permissible for formal purposes, but it should never be used as a practical way of solving linear equations; it requires far too many arithmetic operations. However, there are occasions when it is necessary to work with the inverse matrix, and the routine `inverse` is then useful.

Given two $n \times n$ matrices A and B, it is possible to solve the matrix equation $AX = B$ for the unknown matrix X in the same way.

10.6 Tridiagonal equations

Some matrices contain *systematic* zeros. In that case, special treatments should be introduced wherever possible to reduce unnecessary computing effort. Sometimes the problem can be dealt with by defining a special class derived from the general class `matrix`. There are, for example, many situations in mathematics where the solution of a system of tridiagonal equations is required; that is to say, systems of linear equations in which the matrix of coefficients consists only of the diagonal elements and the immediately adjacent elements. An example of such a system of equations is

$$\begin{pmatrix} 4 & 1 & 0 & 0 & 0 \\ 1 & 4 & 1 & 0 & 0 \\ 0 & 1 & 4 & 1 & 0 \\ 0 & 0 & 1 & 4 & 1 \\ 0 & 0 & 0 & 1 & 4 \end{pmatrix} \begin{pmatrix} x_0 \\ x_1 \\ x_2 \\ x_3 \\ x_4 \end{pmatrix} = \begin{pmatrix} 6 \\ 11 \\ 12 \\ 11 \\ 6 \end{pmatrix},$$

the solution of which is easily seen to be $\mathbf{x} = (1\ 2\ 2\ 2\ 1)^{\mathrm{T}}$. If the solution is sought by gaussian elimination or matrix decomposition there will be many arithmetic operations involving nothing more than multiplication by, or addition or subtraction of, zero. In the case of a large tridiagonal matrix, most of the operations would be of this kind. Instead of the full elimination process for a square matrix, all that is necessary is to eliminate the lower diagonal, and then backsubstitute to find the solution. In this way the unnecessary operations can be avoided.

Moreover, a lot of memory is used up recording the presence of these zeros if the usual $n \times n$ matrix representation is used. Each equation in a tridiagonal system requires just *four* coefficients, including the constant on the right-hand side. The essential data can therefore be contained in an $n \times 4$ matrix. This suggests that we create a data type consisting of $n \times 4$ matrices. We shall do this by derivation from the class `matrix`. Since there will be no "additional" data members, we can use a structure, but we must remember that the data members of `matrix` were declared to be `private`. That means that they are inaccessible to the derived structure, and must be obtained using the public index function of `matrix`:

```
struct tridiag : public matrix {
    tridiag(int n) : matrix(n, 4) {}
    ~tridiag() {}
    int getnum() { return matrix::getnumrows(); }
};
```

The alternative approach is to redefine the class `matrix` so that its data members are declared to be `protected` and therefore accessible directly to the derived structure. We shall not do that here; the method we have adopted has the advantage of transparency. As a result, however, `getnum` and the input and output functions have to be defined in terms of the public access functions of `matrix`.

In order to identify matrix elements transparently we shall introduce the anonymous enumeration

```
enum {L, D, U, B};
```

If we now declare equ to be a system of n tridiagonal equations

```
tridiag equ(n);
```

the subdiagonal element of equation i may be accessed as equ[i][L], and so on. In terms of equ the solution algorithm now becomes

```
void  tridiagonal(tridiag &equ)
{       int n = equ.getnum();
        for (int k = 1; k < n; ++k)        // elim lower diagonal
        {       double diag = equ[k-1][D];
                if (fabs(diag) < tridiag_toosmall)
                        error("diag too small in elimination");
                double mult = equ[k][L]/diag;
                equ[k][L] = 0;
                equ[k][D] -= equ[k-1][U] * mult;
                equ[k][B] -= equ[k-1][B] * mult;
        }
        for (k = n-1; k >= 0; --k)          // then back substitute
        {       double diag = equ[k][D];
                if (fabs(diag) < tridiag_toosmall)
                        error("diag too small in back substitution");
                if (k != n-1) equ[k][B] -= equ[k][U]*equ[k+1][B];
                equ[k][B] /= diag;
        }
}
```

where we have introduced the global constant `tridiag_toosmall` lest a diagonal element be, or become, too small. In applications where tridiagonal equations occur zero diagonals do not usually appear, so that avoiding them does not normally lead to termination of the program.

It is convenient to redefine input in terms of the enumeration, and to format the output appropriately, using the class format introduced in Chapter 6. The new operators may be written:

```
ostream& operator<<(ostream &s, tridiag &equ)
{       int n = equ.getnum();
        const long tril = ios::left + ios::fixed;
        format trifmt(tril, 4, 10);
```

```
        for (int i = 0; i < n; ++i)
              s << trifmt << equ[i][B];
        return s;
}

istream& operator>>(istream &s, tridiag &equ)
{       int n = equ.getnum();
        cout << "Enter " << n << " equations:\n";
        for (int i = 0; i < n; ++i)
        {     cout << "L" << i << " = ";
              s >> equ[i][L];
              cout << "D" << i << " = ";
              s >> equ[i][D];
              cout << "U" << i << " = ";
              s >> equ[i][U];
              cout << "B" << i << " = ";
              s >> equ[i][B];
        }
        return s;
}
```

The result of taking into account the systematic zeros of a tridiagonal system of equations is an enormous saving of computational effort. Gaussian elimination or Crout decomposition followed by substitution involves, as we have seen, $n^3/3$ multiplications or divisions. The procedure `triagonal` involves just $5n - 4$. This is a very efficient procedure which will be used whenever tridiagonal equations occur. The header file `tridiag.h` may be written:

```
#ifndef TRIDIAG
#define TRIDIAG

        #include <vecmat.h>
        extern const double tridiag_toosmall;
        enum    {L, D, U, B};
        struct tridiag : matrix { ... };
        void    tridiagonal(tridiag&);
        ostream& operator<<(0stream&, tridiag&);
        istream& operator>>(istream&, tridiag&);

#endif  // TRIDIAG
```

where it is intended that the structure should be included in full as defined above.

10.7 Modular programming

Before leaving the subject of linear equations, we shall use the simple procedure gauss0 to illustrate the modular approach to programming. Most practical programs will involve a number of common procedures, of which gauss0 is an example. Rather than encode the procedure afresh each time it is required, it is better to compile it into a separate module which may be linked to any other common procedures required by a given program. The essence of the process was described in 5.5.

The collection of functions declared in the header file standard.h, which is displayed in Appendix A, can be compiled into a separate module, or object file—which, under the operating system DOS, would be called standard.obj. Similarly, the collection of functions declared in the header file vecmat.h, displayed in Appendix E, would be compiled into a module vecmat.obj. None of these modules would contain a function main(); by themselves they are therefore not executable. They become executable only when linked to a file which contains a function main(), and there can only be one such file in any given program.

In order to refer correctly and consistently to gauss0, we should first write a header file for it, which will be included in any module that uses it. The header file we have provided will be called gauss0.h, and written:

```
#ifndef GAUSS0
#define GAUSS0

        #include <standard.h>
        #include <vecmat.h>

        extern const double gauss0_toosmall;
        void gauss0(matrix&, vector&, vector&);

#endif        // GAUSS0
```

As we explained in 1.7.3, the whole header file is made conditional, so that if it has already been included in any program, through having been included in a module forming part of that program, it will not be included repeatedly. There follows a list of header files to be included in the module gauss0 because the modules they refer to are required in order to compile it. They are standard.h, to be found in Appendix A, and vecmat.h. There is then the indication that the constant gauss0_toosmall, which occurs in the functions triangulate and backsubst which constitute the procedure gauss0, is to be defined elsewhere: it will be defined as a global variable in the user program. The statement beginning extern is not really a declaration; it reserves no place in memory; it merely warns the compiler to look for a proper declaration of the variable elsewhere, and in the meantime tells it the type of the variable concerned. Note that we have given this variable a name related to the module in which it occurs. This is because in a complex program there may be many variables playing a similar role, and with a similar name, and confusion may easily result if they are not distinguished in a systematic fashion.

Finally, there is a declaration of the function `gauss0`. Notice that the functions `triangulate` and `backsubst` are *not* declared. They are private to the module `gauss0`, will not be used separately from it, and therefore should not be declared. Modules allow us to hide functions and variables required only locally to the module, and in particular allow us to use undeclared identifiers in other modules without the risk of confusion or ambiguity.

The `gauss0` module itself includes the header file `gauss0.h`, and otherwise consists of the definitions of the function `triangulate`, `backsubst` and `gauss0`. As we have already seen, it contains no function `main()`. It is compiled to the object file `gauss0.obj`. When wishing to use the procedure `gauss0` in a program all that is necessary is to include the header file, provide a definition of the constant `gauss0_toosmall`, and link in the module `gauss0.obj` when compiling the program.

For example, the following program may be used to test the procedure `gauss0`:

```
#include <gauss0.h>
#include <standard.h>
#include <vecmat.h>

const double gauss0_toosmall = 3.0E-7;

main()
{       clear;
        int n;
        input("Enter number of variables: ", n);
        matrix m(n);
        vector x(n), b(n);
        input("\nEnter matrix m\n", m);
        input("\nEnter vector b\n", b);
        gauss0(m, x, b);
        output("\nSolution vector x is:\n", x);
}
```

Notice that we have included the header files `standard.h` and `vecmat.h`, even though they are already included *via* the header file `gauss0.h`. This does no harm since we have made inclusion conditional upon each header file not having been included already; but it has the advantage of reminding us which modules to link together to obtain an executable program. Illustrated for the simple case of `gauss0`, this is the procedure we shall adopt when developing useful modules more generally.

11 Errors in matrix manipulation

Computing speed is one thing; accuracy is another. We now have to try to estimate the errors involved in matrix operations. We shall need convenient and mutually consistent measures of magnitude of vectors and matrices which correspond to the modulus of a scalar quantity. We introduced the concept of a *norm* in Chapter 9. We must now formalize that concept, and learn how to apply it to matrix manipulation in order to estimate errors. The results we shall obtain will prove to be somewhat pessimistic, but they will serve at least to indicate situations where large errors are to be expected, and if possible avoided.

11.1 Norms

A *vector norm* $\|\mathbf{x}\|$ is a *function* of a vector $\mathbf{x}$ which has the following properties:

  (i) $\|\mathbf{x}\| \geq 0$, equality holding only if $\mathbf{x} = \mathbf{0}$;

  (ii) $\|\alpha\mathbf{x}\| = |\alpha|\,\|\mathbf{x}\|$, for all scalars α and vectors $\mathbf{x}$;

  (iii) $\|\mathbf{x} + \mathbf{y}\| \leq \|\mathbf{x}\| + \|\mathbf{y}\|$, the "triangle inequality".

The last may also be written

$$\|\mathbf{x} - \mathbf{z}\| \leq \|\mathbf{x} - \mathbf{y}\| + \|\mathbf{y} - \mathbf{z}\|,$$

and in the further form $\big|\,\|\mathbf{x}\| - \|\mathbf{y}\|\,\big| \leq \|\mathbf{x} - \mathbf{y}\|,$

which is known as the "reverse triangle inequality".

  Note that all these properties are possessed by the length of a vector as usually defined in 2- or 3-dimensional euclidean space, namely the square root of the sum of the squares of its components. An appropriate generalization that also possesses these properties is

$$\|\mathbf{x}\|_p = \left(\textstyle\sum_i |x_i|^p\right)^{1/p}, \; p \geq 1,$$

which reduces to the euclidean length when $p = 2$. There is no reason why p should be an integer, but the three most common norms are those having $p = 1, 2$ and ∞, the last being interpreted as the limit of $\|\mathbf{x}\|_p$ as $p \to \infty$. In these cases, therefore,

$$\|\mathbf{x}\|_1 = \Sigma_i |x_i|, \quad \|\mathbf{x}\|_2 = (\Sigma_i |x_i|^2)^{1/2}, \quad \|\mathbf{x}\|_\infty = \mathrm{Max}_i(|x_i|).$$

Thus, if $\mathbf{x} = (1 \ {-2} \ \ 3)$, then $\|\mathbf{x}\|_1 = 6$, $\|\mathbf{x}\|_2 = \sqrt{14} \approx 3.7$, $\|\mathbf{x}\|_\infty = 3$.

Sometimes, we shall speak of the norm $\|\cdot\|$, or more specifically, $\|\cdot\|_p$, without the argument, in the same way that we might speak of the function f as shorthand for $f(x)$.

We now have to consider matrices. The choice is limited by the fact that $A\mathbf{x}$ is a vector and therefore has vector norms as already defined. A matrix norm should if possible be consistent with this requirement. We define a *natural matrix norm*, or the *matrix norm* $\|A\|$ *induced by the vector norm* $\|\mathbf{x}\|$ as

$$\|A\| \equiv \mathrm{Max}(\|A\mathbf{x}\|), \text{ over all } \mathbf{x} \text{ such that } \|\mathbf{x}\| = 1.$$

So defined, a matrix norm possesses the properties (i)–(iii) possessed by a vector norm, together with two extra properties. In all:

(i) $\|A\| \geq 0$, equality holding only if A is null;

(ii) $\|\alpha A\| = |\alpha| \, \|A\|$, for all scalars α and matrices A;

(iii) $\|A + B\| \leq \|A\| + \|B\|$, the "triangle inequality";

(iv) $\|AB\| \leq \|A\| \, \|B\|$;

(v) $\|A\mathbf{x}\| \leq \|A\| \, \|\mathbf{x}\|$, for all vectors $\mathbf{x}$ and matrices A.

Again, there are alternative ways of writing (iii).

The matrix norm induced by the vector norm $\|\mathbf{x}\|_p$ is written $\|A\|_p$. In the cases $p = 1$ and $p = \infty$, we have:

$$\|A\|_1 = \mathrm{Max}_j(\Sigma_i |a_{ij}|), \text{ where the sum is over a column;}$$

$$\|A\|_\infty = \mathrm{Max}_i(\Sigma_j |a_{ij}|), \text{ where the sum is over a row.}$$

Naturally, for a *symmetric* matrix in which $a_{ij} = a_{ji}$, $\|A\|_1 = \|A\|_\infty$. For a diagonal matrix D, with diagonal elements d_i, $\|D\|_1 = \|D\|_\infty = \mathrm{Max}_i|d_i|$; both norms equal the modulus of the diagonal element of greatest magnitude. In particular, both norms are unity for the unit matrix: $\|I\| = 1$.

The case of $p = 2$ is the most important in practice, but more difficult to establish. Firstly, however, note that if we treat $A\mathbf{x}$ as a vector, and make use of the Cauchy–Schwartz inequality, we get

$$\|A\mathbf{x}\|_2 = [\Sigma_i|\Sigma_j a_{ij}x_j|^2]^{1/2} \leq [\Sigma_i\{\Sigma_j|a_{ij}|^2\}\{\Sigma_j|x_j|^2\}]^{1/2} \leq \|A\|_E\|\mathbf{x}\|_2,$$

where $\|A\|_E = \{\Sigma_i\Sigma_j|a_{ij}|^2\}^{1/2}$ may be called the euclidean, or sometimes Frobenius, norm. It possesses all the above properties for a matrix norm, but it is *not* the norm induced by the vector norm $\|\mathbf{x}\|_2$. Applied to the unit matrix in n dimensions, its value is not unity as it would be for a natural norm; instead, this norm is $\|I\|_E = \sqrt{n}$. As this example suggests, the euclidean norm may be somewhat large, and this results from treating a (square) matrix of dimension n as if it were a vector of dimension n^2.

The modulus of the eigenvalue of greatest magnitude of a matrix A is known as the *spectral radius* $\rho(A)$ of the matrix. It can be shown, moreover, that if A is real, the matrix norm induced by the euclidean vector norm is

$$\|A\|_2 = \sqrt{\rho(A^T A)},$$

where the suffix T as usual denotes the transposed matrix. If A is also *symmetric* (the case that will concern us most), then:

$$\|A\|_2 = \text{Max}_i \, |\lambda_i| = \rho(A) \quad \text{(real symmetric } A\text{)}.$$

We shall consider how to compute the eigenvalues of matrices in Chapter 13. For the present it is sufficient to know that if there are n non-zero vectors $\mathbf{x}_i$ and scalars λ_i, which together satisfy the equations $A\mathbf{x}_i = \lambda_i \mathbf{x}_i$, where A is a square $n \times n$ matrix, then the constants λ_i are called the *eigenvalues* of the matrix A, and $\mathbf{x}_i$ are the corresponding *eigenvectors*. The eigenvalues are the solutions of the *characteristic equation* $\det(A - \lambda I) = 0$.

If the matrix is diagonal the eigenvalues are equal to the diagonal elements, and the spectral radius is simply the largest of the absolute values of the diagonal elements. In particular, the spectral radius of the unit matrix is unity: $\|I\|_2 = \rho(I) = 1$. To take a less trivial example, the matrix

$$A = \begin{pmatrix} 1 & 3 \\ 3 & 2 \end{pmatrix},$$

for which the characteristic equation is $\lambda^2 - 3\lambda - 7 = 0$, has norms $\|A\|_1 = \|A\|_\infty = 5$, $\|A\|_2 = 4.54$, $\|A\|_E = 4.80$. Note that $\|A\|_2$ is the smallest of these.

The importance of the spectral radius is that for *any* matrix norm, $\rho(A) \leq \|A\|$. For if λ is *any* eigenvalue of a real symmetric matrix A, and $\mathbf{x} \neq \mathbf{0}$ a corresponding eigenvector, then $\lambda \mathbf{x} = A\mathbf{x}$. Taking norms, we have

$$|\lambda| \, \|\mathbf{x}\| = \|\lambda \mathbf{x}\| = \|A\mathbf{x}\| \leq \|A\| \, \|\mathbf{x}\|, \text{ so that } |\lambda| \leq \|A\|.$$

But this is true for *any* eigenvalue λ, and therefore for that with largest modulus. Hence $\rho(A) \equiv \text{Max} \, |\lambda_i| \leq \|A\|$. Thus, $\rho(A)$ is a lower bound for *any* matrix norm. It may further be shown that for any $\varepsilon > 0$, no matter how small, there exists a natural norm such that:

$$\rho(A) < \|A\| < \rho(A) + \varepsilon.$$

Therefore, $\rho(A)$ is the *greatest lower bound* for the natural norms of A.

11.2 Convergence

The spectral radius is also important because it determines whether successive products of a matrix converge. Let A be a real symmetric matrix, and consider the sequence of matrices:

$$A, A^2 = AA, \ldots, \ A^k = AA^{k-1} = A^{k-1}A.$$

Let $\mathbf{x}$ be an eigenvector of A of eigenvalue λ. Then,

$$A^2\mathbf{x} = A(\lambda\mathbf{x}) = \lambda(A\mathbf{x}) = \lambda^2\mathbf{x},$$

and in general, $A^k\mathbf{x} = \lambda^k\mathbf{x}$, where k is any positive integer. The right-hand side of this equation converges to zero as $k \to \infty$ if and only if $|\lambda| < 1$. If every eigenvalue of A has this property, and in particular the maximum modulus eigenvalue, then $\lim_{k \to \infty} A^k = O$, where O is the null matrix, if and only if $\rho(A) < 1$. If this condition is satisfied, we say that A is a *convergent matrix*. This is the generalization to real symmetric matrices of the similar relation for scalars, namely, $\lim_{k \to \infty} t^k = 0$, if and only if $|t| < 1$.

Now let A be a real symmetric matrix with $\rho(A) < 1$, and suppose there exists a non-zero vector $\mathbf{x}$ such that $(I - A)\mathbf{x} = \mathbf{0}$. Then $A\mathbf{x} = I\mathbf{x} = \mathbf{x}$, so that A has the eigenvalue unity, in contradiction to the condition that $\rho(A) < 1$. Hence, $\mathbf{x} = \mathbf{0}$, and $(I - A)$ is a non-singular matrix with $\det(I - A) \neq 0$. According to the theorem of chapter 10, $(I - A)^{-1}$ therefore exists if $\rho(A) < 1$.

We shall need a limit on the norm of $(I - A)^{-1}$. Now, for any matrix A:

$$(I - A)(I + A^2 + \ldots + A^k) = I - A^{k+1}.$$

If there is any norm for which $\|A\| < 1$, then $\rho(A) \leq \|A\| < 1$, so that $(I - A)^{-1}$ exists and we can pre-multiply each side by this factor to obtain:

$$I + A + A^2 + \ldots + A^k = (I - A)^{-1}(I - A^{k+1}).$$

But, since $\rho(A) < 1$, $A^{k+1} \to O$ as $k \to \infty$. Therefore:

$$(I - A)^{-1} = \lim_{k \to \infty}(I + A + A^2 + \ldots + A^k).$$

If we now take norms on both sides, we have:

$$\|(I - A)^{-1}\| = \|I + A + A^2 + \ldots\|$$
$$\leq \|I\| + \|A\| + \|A^2\| + \ldots$$
$$\leq 1 + \|A\| + \|A\|^2 + \ldots$$
$$= 1/(1 - \|A\|), \text{ since } \|A\| < 1.$$

This is the result we sought.

11.3 Error estimation

We use these formal results to obtain information about the errors generated in computing the solution of the linear system $A\mathbf{x} = \mathbf{b}$, where $\det(A) \neq 0$. The treatment is standard (see, for example, the book by K. E. Atkinson). Suppose first that there is a small perturbation $\delta\mathbf{b}$ in the value of $\mathbf{b}$. Keeping A unchanged, the resulting perturbation $\delta\mathbf{x}$ of the solution is given by

$$A(\mathbf{x} + \delta\mathbf{x}) = (\mathbf{b} + \delta\mathbf{b}),$$

so that $\qquad\qquad A\delta\mathbf{x} = \delta\mathbf{b}, \text{ and } \delta\mathbf{x} = A^{-1}\delta\mathbf{b}.$

Taking norms, we have $\|\delta\mathbf{x}\| \leq \|A^{-1}\| \, \|\delta\mathbf{b}\|$. Values of the *relative* perturbations are of most interest. Dividing both sides by $\|\mathbf{x}\|$, we obtain:

$$\|\delta\mathbf{x}\| / \|\mathbf{x}\| \leq \|A\| \, \|A^{-1}\| \, \|\delta\mathbf{b}\| / \|A\| \, \|\mathbf{x}\|$$

$$\leq K(A)\|\delta\mathbf{b}\| / \|\mathbf{b}\|,$$

where $\qquad\qquad K(A) = \|A\| \, \|A^{-1}\|$

is called the *condition number* of A with respect to the norm $\|\cdot\|$. If K is large, it may be impossible to preserve accuracy in the calculation of $\mathbf{x}$ because perturbations due to rounding errors and the like may grow rapidly.

Let us now perturb A keeping $\mathbf{b}$ fixed. The perturbation of $\mathbf{x}$ is given by

$$(A + \delta A)(\mathbf{x} + \delta\mathbf{x}) = \mathbf{b},$$

or $\qquad\qquad (A + \delta A)\delta\mathbf{x} + (\delta A)\mathbf{x} = \mathbf{0}.$

Again, we are interested in relative changes. We therefore rewrite this equation as

$$A(I + A^{-1}\delta A)\delta\mathbf{x} + (\delta A)\mathbf{x} = \mathbf{0},$$

and since A is non-singular,

$$(I + A^{-1}\delta A)\delta\mathbf{x} + (A^{-1}\delta A)\mathbf{x} = \mathbf{0}.$$

We must now assume that the perturbation δA is chosen such that

$$\|A^{-1}\delta A\| \leq K(A)\|\delta A\| / \|A\| < 1.$$

Then we may write

$$\|\delta\mathbf{x}\| / \|\mathbf{x}\| \leq \|(I + A^{-1}\delta A)^{-1}A^{-1}\delta A\|,$$

$$\leq \|A\| \, \|(I + A^{-1})^{-1}\| \, \|A^{-1}\| \, \|\delta A\| / \|A\|,$$

or in terms of the condition number K,

$$\frac{\|\delta \mathbf{x}\|}{\|\mathbf{x}\|} \leq \frac{K(A)}{1 - K(A)\|\delta A\| / \|A\|} \cdot \frac{\|\delta A\|}{\|A\|}.$$

Finally, if we allow both A and $\mathbf{b}$ to be perturbed, the resulting perturbation satisfies the inequality:

$$\frac{\|\delta \mathbf{x}\|}{\|\mathbf{x}\|} \leq \frac{K(A)}{1 - K(A)\|\delta A\| / \|A\|} \left(\frac{\|\delta A\|}{\|A\|} + \frac{\|\delta \mathbf{b}\|}{\|\mathbf{b}\|} \right).$$

Let us consider the magnitude of the condition number. Its minimum value is unity, for

$$1 = \|I\| = \|AA^{-1}\| \leq \|A\| \, \|A^{-1}\| = K(A),$$

and $K(I) = 1$. Now, if λ is an eigenvalue of A with eigenvector $\mathbf{x}$, so that $A\mathbf{x} = \lambda\mathbf{x}$, then λ^{-1} is an eigenvalue of A^{-1}, also with eigenvector $\mathbf{x}$, because $\mathbf{x} = A^{-1}\lambda\mathbf{x}$, or $A^{-1}\mathbf{x} = \lambda^{-1}\mathbf{x}$. Therefore, the eigenvalues of A^{-1} are the reciprocals of the eigenvalues of A. Since $\rho(A) = \lambda_{\max} \equiv \mathrm{Max}(|\lambda_i|)$, it follows that $\rho(A^{-1}) = \lambda_{\min}^{-1} \equiv \mathrm{Max}(|\lambda_i^{-1}|)$. Therefore, $K_2(A) \equiv \rho(A)\rho(A^{-1}) = \lambda_{\max}/\lambda_{\min}$. Defined relative to the norm $\|\cdot\|_2$, we see that $K_2(A)$ is a measure of the spread of eigenvalues of the matrix A. Since for any norm $\rho(A) \leq \|A\|$, it follows that $\|A^{-1}\|^{-1} \leq |\lambda| \leq \|A\|$ for all eigenvalues λ.

We are not yet in a position to calculate eigenvalues and therefore the spectral radius. Instead, we shall calculate the infinite norm, obtained by summing the moduli of the elements in each row; the infinite norm is the maximum such sum. For the 4×4 matrix

$$A = \begin{pmatrix} 1 & 2 & 3 & 4 \\ 2 & 3 & 4 & 6 \\ 3 & 4 & 2 & 5 \\ 4 & 6 & 5 & 7 \end{pmatrix},$$

which we have often used before, the inverse of which we computed in 8.5, we find $\|A\|_\infty = 22$ and $\|A^{-1}\|_\infty = 47/6$. Therefore $K_\infty(A) \approx 172$. This means that it is likely that in a computation, such as Gaussian elimination, we shall lose a couple of places of decimals, which is acceptable, especially if we are working to double precision. We say that the matrix is *well-conditioned* if K is not too large.

On the other hand, consider the Hilbert matrix $H^{(n)}$, the elements of which are given by

$$H_{ij} = 1/(i + j + 1), \quad 0 \leq i, j < n.$$

The inverse matrix has elements which are integers and are known analytically:

$$(H^{-1})_{ij} = [\prod_{0 \le k < n}(i+k+1)(j+k+1)]/[(i+j+1)\prod_{0 \le k(\ne i)<n}(i-k)\prod_{0 \le k(\ne j)<n}(j-k)].$$

In the case of $n = 4$, we have:

$$H = \begin{pmatrix} 1 & 1/2 & 1/3 & 1/4 \\ 1/2 & 1/3 & 1/4 & 1/5 \\ 1/3 & 1/4 & 1/5 & 1/6 \\ 1/4 & 1/5 & 1/6 & 1/7 \end{pmatrix}, \quad H^{-1} = \begin{pmatrix} 16 & -120 & 240 & -140 \\ -120 & 1200 & -2700 & 1680 \\ 240 & -2700 & 6480 & -4200 \\ -140 & 1680 & -4200 & 2800 \end{pmatrix}.$$

so that $\|H\|_\infty = 25/12$, $\|H^{-1}\|_\infty = 13\,620$, and $K_\infty(H) = 28\,375$. It would have been much larger had we chosen $n > 4$. For $n = 5$, $K_\infty(H) = 943\,656$. Thus, the Hilbert matrix is *ill-conditioned*, and attempts to compute it by standard methods are likely to be very inaccurate for all but the smallest values of n.

It is rather difficult to apply the above analysis to Gaussian elimination. It has, however, been done (see the book by K. E. Atkinson), and we shall quote the result. The slightly inaccurate result of the triangulation process is such that:

$$LU = A + \delta A, \text{ where } \|\delta A\|_\infty / \|A\|_\infty \le f(n)10^{1-t} \text{Max} |a^{(k)}_{ij}| / \|A\|_\infty.$$

Here, it is assumed that we are working in t-decimal arithmetic, and that $a^{(k)}_{ij}$ is the element with the largest modulus at any stage k of the triangulation process. In the worst case $f(n) \le 1.01(n^3 + 3n^2)$, but usually with pivoting it is found that $f(n) \approx n$.

11.3.1 Random perturbations

Rather than carry the analysis any further, we shall simulate computational errors by applying small random variations to the matrix A. As an example, we therefore add to the above matrix A a random matrix δA, the elements of which are chosen from a gaussian distribution (*cf* Appendix D) with zero mean and given variance:

```
#include <random.h>

matrix ranmat(int n, double var, long &s)
{       matrix m(n);
        for (int i = 0; i < n; ++i)
        {       for (int j = 0; j < n; ++j)
                        m[i][j] = gaussian(s, var);
        }
        return m;
}
```

Taking a variance var $= 10^{-5}$, a typical perturbating matrix is

$$\delta A = \begin{pmatrix} .0021 & .0037 & .0035 & -.0114 \\ .0027 & -.0031 & -.0016 & .0062 \\ .0042 & .0009 & .0010 & .0002 \\ -.0023 & -.0012 & -.0014 & .0010 \end{pmatrix},$$

for which $\|\delta A\|_\infty = 0.02075$ and $\|\delta A\|/\|A\| = 0.000943$. Note that for a gaussian distribution of matrix elements the mean norm is

$$\langle\|\delta A\|_\infty\rangle \geq n\langle\|\delta a_{ij}\|\rangle = 2n(\text{var}/\pi)^{1/2} \approx 0.0143.$$

Using $\mathbf{b} = (8 \ \ 6 \ \ 4 \ \ 2)^{\mathrm{T}}$, the perturbed solution obtained using the routine `gauss` is

$$\mathbf{x} + \delta\mathbf{x} = (-16.5557 \ \ 9.6758 \ \ -4.9248 \ \ 4.9808)^{\mathrm{T}}$$

instead of the unperturbed solution

$$\mathbf{x} = (-17 \ \ 10 \ \ -5 \ \ 5)^{\mathrm{T}}.$$

The differences are not negligible.

From these results, $\|\delta\mathbf{x}\|/\|\mathbf{x}\| = 0.0261$. The ratio $(\|\delta\mathbf{x}\|/\|\mathbf{x}\|)/(\|\delta A\|/\|A\|) \approx 28$. We have already found that $K(A) \approx 172$, so that $K(A)/(1 - K(A)\|\delta A\|/\|A\|) \approx 205$. The inequality in the estimated error in the solution is observed with a handsome factor of 7.5 in hand.

Of course, we have been using infinite norms; had we used the norm $\|\cdot\|_2$ this factor would not have been so great. Even so, the best we can conclude is that the solution of $A\mathbf{x} = \mathbf{b}$ will be accurate if the condition number $K(A)$ is not too great; it *may* be accurate even if the condition number is quite large.

11.4 Iterative refinement

The rounding errors introduced by finite arithmetic may be reduced by an iterative process. Rather than working in single precision (`float` instead of `double`), we shall simulate the effect of single precision by randomly perturbing the matrix. However, for the moment let $A\mathbf{x} = \mathbf{b}$, and let B be the approximate value of A obtained by gaussian elimination (or otherwise). Let $\mathbf{y}$ be the solution of $B\mathbf{y} = \mathbf{r}$, where $\mathbf{r} = \mathbf{b} - A\mathbf{y}$ is the *residual*, calculated using the accurate matrix A. (In practice, A and $\mathbf{b}$ would be given in double precision, B in single, but $\mathbf{r}$ must be calculated in double precision if increased accuracy is desired. Let $\mathbf{e}$ be the solution of the approximate equations $B\mathbf{e} = \mathbf{r}$. Then we assert that $\mathbf{y} + \mathbf{e}$ is a better approximation to the accurate solution $\mathbf{x}$ than $\mathbf{y}$ is. This serves as the basis of an iterative scheme, the convergence of which we now demonstrate.

We have $A\mathbf{x} = \mathbf{b}$, so formally, if A is non-singular, $\mathbf{x} = A^{-1}\mathbf{b}$. Similarly, $\mathbf{y}^{(0)} \equiv \mathbf{y} = B^{-1}\mathbf{b} = (B^{-1}A)\mathbf{x}$. Define the matrix $R = B^{-1}A$. Then $\mathbf{y}^{(0)} = R\mathbf{x}$, and we can calculate $\mathbf{r}^{(0)} = \mathbf{b} - A\mathbf{y}^{(0)}$, and in principle $\mathbf{e}^{(0)} = B^{-1}\mathbf{r}^{(0)} = R(\mathbf{x} - \mathbf{y}^{(0)})$. Define $\mathbf{y}^{(1)} = \mathbf{y}^{(0)} + \mathbf{e}^{(0)}$, leading to a new $\mathbf{r}^{(1)} = \mathbf{b} - A\mathbf{y}^{(1)}$, and so on. In general, we find:

$$\mathbf{y}^{(k+1)} = \mathbf{y}^{(k)} + \mathbf{e}^{(k)} = [I - (I - R)^{k+2}]\mathbf{x}.$$

If $\|I - R\|_2 < 1$, the sequence $\mathbf{y}^{(k)}$ tends to the precise solution $\mathbf{x}$. Let us represent the difference between the accurate A and the inaccurate B by the expression $B = A + \delta A$. Then:

$$I - R = I - B^{-1}A = B^{-1}\delta A = A^{-1}(I - A^{-1}\delta A)^{-1}\delta A,$$

so that for any natural norm,

$$\|I - R\| = \|A^{-1}(I - A^{-1}\delta A)^{-1}\delta A\| \leq \frac{K(A)}{1 - K(A)\|\delta A\|/\|A\|} \cdot \frac{\|\delta A\|}{\|A\|}.$$

Let us again consider the system of equations

$$\begin{pmatrix} 1 & 2 & 3 & 4 \\ 2 & 3 & 4 & 6 \\ 3 & 4 & 2 & 5 \\ 4 & 6 & 5 & 7 \end{pmatrix} \begin{pmatrix} x_0 \\ x_1 \\ x_2 \\ x_3 \end{pmatrix} = \begin{pmatrix} 8 \\ 6 \\ 4 \\ 2 \end{pmatrix}$$

and simulate the errors δA by a random matrix of variance 10^{-5}. In a typical case we find $\|\delta A\| \approx 0.015$, so that $\|\delta A\|/\|A\| \approx 0.00068$ and $\|I - R\| \leq 0.13$. The successive approximations to the correct solution are as follows, with corresponding values of $\|\mathbf{x} - \mathbf{y}\|$ in parentheses:

−16.7745	9.8524	−5.0641	5.0300	(0.2255)
−16.9948	9.9963	−4.9993	4.9996	(0.0053)
−16.9999	9.9999	−5.0000	5.0000	(9.7E−5)
−17.0000	10.0000	−5.0000	5.0000	(2.0E−6)

This technique is sometimes useful for improving a result obtained inaccurately because the matrix is insufficiently well-conditioned. It can also be useful in compensating for an inaccurate LU decomposition, even though the problem may be well-conditioned.

12 Iterative solutions of systems of equations

We turn now from direct methods of solution to iterative methods, considering linear systems first. The system of linear equations $A\mathbf{x} = \mathbf{b}$ may be transformed into a different system reminiscent of the transformation of a non-linear equation $f(x) = 0$ into a fixed point relation $x = g(x)$, for which we found in 4.5 that an iterative method for a root might be written $x^{(k+1)} = g(x^{(k)})$. We therefore begin by re-arranging the system of equations as follows:

$$x_0 = [b_0 - a_{01}\, x_1 - a_{02}\, x_2 - \ldots - a_{0n-1} x_{n-1}]/a_{00} \ ,$$

$$x_1 = [b_1 - a_{10}\, x_0 - a_{12}\, x_2 - \ldots - a_{1n-1} x_{n-1}]/a_{11} \ ,$$

$$\ldots\ldots\ldots\ldots\ldots\ldots\ldots\ldots\ldots\ldots\ldots\ldots\ldots\ldots\ldots\ldots\ldots\ldots$$

$$x_{n-1} = [b_{n-1} - a_{n-1\,1}\, x_1 - a_{n-1\,2}\, x_2 - \ldots - a_{n-1\,n-2} x_{n-2}]/a_{n-1\,n-1} \ ,$$

This suggests a number of iterative schemes, the theoretical treatment of which follows the book by K. E. Atkinson.

12.1 Gauss–Jacobi iteration

The simplest scheme is that named after Gauss and Jacobi:

$$x_i^{(k+1)} = [b_i - \sum_{j\neq i} a_{ij} x_j^{(k)}]/a_{ii}.$$

We start with an approximate solution, if known; otherwise, we can start with $x_i^{(0)} = (b_i/a_{ii})$, or even one step further back with $\mathbf{x}^{(0)} = (0, \ldots, 0)^\mathrm{T}$. We iterate until the modulus of *each* of the differences $x_i^{(k+1)} - x_i^{(k)}$ is less than a prescribed tolerance; the last iterate is the solution to within that tolerance. It is, of course, assumed that none of the diagonal elements is zero. This will usually be so when iteration is the chosen method; if it is not, the equations may be rearranged, as in pivoting (*cf* 10.2.1), to avoid the problem as far as possible. The following routine achieves the stated task:

```
state jacobi(matrix &m, vector &x, vector &b)
{       int n = m.getsize();
        int iter = 0;
        state s = ITERATING;
        while (s == ITERATING)
        {       vector oldx = x;
                for (int i = 0; i < n; ++i)
                {       double bi = b[i];
                        double mi = m[i][i];
                        if (fabs(mi) < toosmall)
                                return NEARZERO;
                        for (int j = 0; j < n; ++j)
                        {       if (j != i)
                                        bi  -= m[i][j] * oldx[j];
                        }
                        x[i] = bi/mi;
                }
                if (norminf(x - oldx) < tolerance)
                        s = SUCCESS;
                if (++iter == maxiter)
                        s = WONTSTOP;
                // put running output here
        }
        return s;
}
```

Note that we need a further vector variable `oldx` to retain the previous value of **x**, and the terminating condition is neatly written in terms of `norminf(x - oldx)`.

As we shall see presently, the Gauss–Jacobi procedure transforms the system of equations $A\mathbf{x} = \mathbf{b}$ into an iterative scheme:

$$\mathbf{x}^{(k+1)} = B\mathbf{x}^{(k)} + \mathbf{c}.$$

Let us consider the convergence of this sequence $\mathbf{x}^{(k)}$ to a limit $\mathbf{x}$, if it exists. We have

$$\mathbf{x} \equiv \lim_{k \to \infty} \mathbf{x}^{(k)} = \lim_{k \to \infty} B^k \mathbf{x}^{(0)} + \lim_{k \to \infty} (\textstyle\sum_j B^j)\mathbf{c},$$

$$= (I - B)^{-1}\mathbf{c},$$

provided $\rho(B) < 1$, so that $B^k \to O$ and $(I - B)^{-1}$ exists. Conversely, let the sequence $\mathbf{x}^{(k)} \to \mathbf{x}$ for any $\mathbf{x}^{(0)}$, then $\mathbf{x} - \mathbf{x}^{(k)} = B^k(\mathbf{x} - \mathbf{x}^{(0)})$; and for any vector $\mathbf{x}^{(0)}$

$$\lim_{k \to \infty} B^k(\mathbf{x} - \mathbf{x}^{(0)}) = \lim_{k \to \infty} (\mathbf{x} - \mathbf{x}^{(k)}) = \mathbf{0},$$

which implies that $\rho(B) < 1$.

It follows, for any natural norm $\|B\| < 1$ that

$$\|\mathbf{x} - \mathbf{x}^{(k)}\| \le \|B\|^k \|\mathbf{x} - \mathbf{x}^{(0)}\|$$

$$\le [\|B\|^k/(1 - \|B\|)] \|\mathbf{x}^{(1)} - \mathbf{x}^{(0)}\|,$$

so that the iteration converges linearly in that norm.

For formal purposes, we can always write A as the sum of a non-zero diagonal matrix, a strictly lower triangular matrix (its diagonal elements being zero), and a strictly upper triangular matrix:

$$A = L + D + U.$$

Then the Gauss-Jacobi iteration matrix is $B_{GJ} = -D^{-1}(L + U)$, while $\mathbf{c}_{GJ} = D^{-1}\mathbf{b}$. To see this, note that:

$$x_i^{(k+1)} = [b_i - \sum_{j \ne i} a_{ij} x_j^{(k)}]/a_{ii}$$

$$= [b_i - \sum_{j<i} a_{ij} x_j^{(k)} - \sum_{j>i} a_{ij} x_j^{(k)}]/a_{ii}$$

$$= [b_i - \sum_j l_{ij} x_j^{(k)} - \sum_j u_{ij} x_j^{(k)}]/d_{ii}.$$

Multiplying through by d_{ii} immediately gives

$$D\mathbf{x}^{(k+1)} = -(L + U)\mathbf{x}^{(k)} + \mathbf{b},$$

and the result follows.

In terms of the initial matrix A, we have $B_{ij} = a_{ij}/a_{ii}$, $i \ne j$; $B_{ii} = 0$. Therefore:

$$\mu \equiv \|B\|_\infty = \text{Max}_i (\sum_{j \ne i} |a_{ij}|)/|a_{ii}|.$$

If $\mu < 1$, we say that the matrix A is *strictly diagonally dominant*. If A is strictly diagonally dominant the iteration converges, because $\rho(B) \le \mu < 1$. It may, however, converge even if A is not strictly diagonally dominant; the condition is sufficient but not necessary. Notice that it is enough for the condition to fail that any one of the rows of A should sum in modulus to more than unity. The following routine tests for this condition:

```
bool   sdd(matrix &m)
// strictly diagonally dominant
{      int n = m.getsize();
       for (int i = 0; i < n; ++i)
       {      double t = 0.0;
```

```
                    for (int j = 0; j < n; ++j)
                    {       if (j != i)
                                    t += fabs(m[i][j]);
                    }
                    if (t >= fabs(m[i][i]))
                            return FALSE;
            }
            return TRUE;
}
```

12.2 Gauss–Seidel iteration

The Gauss–Jacobi method does not make full use of the information available. If we recall the iterative scheme,

$$x_i^{(k+1)} = [b_i - \sum_{j \neq i} a_{ij} x_j^{(k)}]/a_{ii},$$

we note that on the right-hand side the components x_j with $j < i$ have already been updated, but their updated values are not used until all have been calculated. This is an unnecessary, indeed wasteful, restriction. It is corrected in the Gauss–Seidel method,

$$x_i^{(k+1)} = [b_i - \sum_{j<i} a_{ij} x_j^{(k+1)} - \sum_{j>i} a_{ij} x_j^{(k)}]/a_{ii},$$

for which the iteration matrix is $B_{GS} = -(D + L)^{-1} U$, while $\mathbf{c}_{GS} = (D + L)^{-1}\mathbf{b}$.

To see this, write as before,

$$x_i^{(k+1)} = [b_i - \sum_j l_{ij} x_j^{(k+1)} - \sum_j u_{ij} x_j^{(k)}]/d_{ii},$$

so that

$$d_{ii} x_i^{(k+1)} + \sum_j l_{ij} x_j^{(k+1)} = b_i - \sum_j {}_{ij} x_j^{(k)},$$

or

$$(D + L)\mathbf{x}^{(k+1)} = \mathbf{b} - U\mathbf{x}^{(k)},$$

from which the result follows.

Again, strict diagonal dominance guarantees convergence, but the rate of convergence is usually improved. Let $\alpha_i = \sum_{j<i} |a_{ij}/a_{ii}|$, and $\beta_i = \sum_{j>i} |a_{ij}/a_{ii}|$, and define

$$\mu \equiv \|B\|_\infty = \text{Max}_i (\alpha_i + \beta_i), \quad \eta = \text{Max}_i [\beta_i/(1 - \alpha_i)].$$

The iteration converges linearly at the rate $\eta \leq \mu < 1$. For if we define the error at each iteration to be $\mathbf{e}^{(k)} = \mathbf{x} - \mathbf{x}^{(k)}$, then

$$e_i^{(k+1)} = -\sum_{j<i} a_{ij} e_i^{(k+1)}/a_{ii} - \sum_{j>i} a_{ij} e_i^{(k)}/a_{ii},$$

from which it follows that

$$|e_i^{(k+1)}| \le \alpha_i \|\mathbf{e}^{(k+1)}\|_\infty + \beta_i \|\mathbf{e}^{(k)}\|_\infty.$$

Let m be the subscript for which $\|\mathbf{e}^{(k+1)}\| = |e_m^{(k+1)}|$; then

$$\|\mathbf{e}^{(k+1)}\|_\infty \le \alpha_m \|\mathbf{e}^{(k+1)}\|_\infty + \beta_m \|\mathbf{e}^{(k)}\|_\infty,$$

or

$$\|\mathbf{e}^{(k+1)}\|_\infty \le \eta \|\mathbf{e}^{(k)}\|_\infty,$$

so the sequence converges linearly with rate η. It remains only to show that $\eta \le \mu$. Since for every i,

$$(\alpha_i + \beta_i) - \frac{\beta_i}{1-\alpha_i} = \frac{\alpha_i[1-(\alpha_i+\beta_i)]}{1-\alpha_i} \ge \frac{\alpha_i}{1-\alpha_i} \cdot (1-\mu) \ge 0,$$

we deduce that $\eta \le \mu < 1$. The following function incorporates these improvements:

```
state seidel(matrix &m, vector &x, vector &b)
{       state s = ITERATING;
        int iter = 0, n = m.getsize();
        while (s == ITERATING)
        {       vector oldx = x;
                for (int i = 0; i < n; ++i)
                {       double bi = b[i], mi = m[i][i];
                        if (fabs(mi) < toosmall)
                                return NEARZERO;
                        for (int j = 0; j < i; ++j)
                                bi -= m[i][j] * x[j];
                        for (j = i+1; j < n; ++j)
                                bi -= m[i][j] * oldx[j];
                        x[i] = bi/mi;
                }
                if (norminf(x - oldx) < tolerance)
                        s = SUCCESS;
                if (++iter == maxiter)
                        s = WONTSTOP;
                // put running output here
        }
        return s;
}
```

12.3 Jacobi and Seidel compared

The particular system $Ax = \mathbf{b}$ we studied by direct methods in chapter 10 is certainly not diagonally dominant, and iteration will fail. Consider instead the system

$$
\begin{pmatrix}
7 & 1 & -1 & 2 \\
1 & 8 & 0 & -2 \\
-1 & 0 & 4 & -1 \\
2 & -2 & -1 & 6
\end{pmatrix}
\begin{pmatrix}
x_0 \\
x_1 \\
x_2 \\
x_3
\end{pmatrix}
=
\begin{pmatrix}
9 \\
7 \\
2 \\
5
\end{pmatrix}
$$

which has the exact solution $\mathbf{x} = (1 \ \ 1 \ \ 1 \ \ 1)^T$, and is strictly diagonally dominant. We find $\mu = 5/6$, and $\eta = 4/7$. The ratio $\mu/\eta \approx 1.5$ is a rough estimate of the increased rate of convergence of Gauss–Seidel over Gauss–Jacobi.

The following results are obtained from the the Gauss–Jacobi and Gauss–Seidel routines, respectively:

Iteration	Gauss–Jacobi			
0	0.0000	0.0000	0.0000	0.0000
1	1.2857	0.8750	0.5000	0.8333
2	0.9940	0.9266	1.0298	0.7798
3	1.0782	0.9457	0.9435	0.9812
4	1.0051	0.9855	1.0148	0.9464
5	1.0195	0.9860	0.9879	0.9960
6	1.0014	0.9966	1.0039	0.9868
7	1.0048	0.9965	0.9971	0.9990
8	1.0004	0.9992	1.0010	0.9967
9	1.0012	0.9991	0.9993	0.9998
10	1.0001	0.9998	1.0002	0.9992
11	1.0003	0.9998	0.9998	0.9999
12	1.0000	0.9999	1.0001	0.9998
13	1.0001	0.9999	1.0000	1.0000
14	1.0000	1.0000	1.0000	1.0000

Iteration	Gauss–Seidel			
0	0.0000	0.0000	0.0000	0.0000
1	1.2858	0.7143	0.8214	0.7798
2	1.0782	0.9352	0.9645	0.9464
3	1.0195	0.9842	0.9915	0.9868
4	1.0048	0.9961	0.9979	0.9967
5	1.0012	0.9990	0.9995	0.9992
6	1.0003	0.9998	0.9999	0.9998
7	1.0001	0.9999	1.0000	1.0000
8	1.0000	1.0000	1.0000	1.0000

The second converges at nearly twice the rate of the first, in reasonable accord with the expected factor of 1.5.

12.4 Successive over-relaxation

It is interesting to enquire whether the rate of convergence can be further improved. It can, by biasing the iteration formula still more in favour of updated values. We introduce an auxiliary vector $\mathbf{z}$ and an "acceleration parameter" ω.

Let
$$\mathbf{z}^{(k+1)} = D^{-1}[\mathbf{b} - L\mathbf{x}^{(k+1)} - U\mathbf{x}^{(k)}],$$

and
$$\mathbf{x}^{(k+1)} = \omega\mathbf{z}^{(k+1)} + (1 - \omega)\mathbf{x}^{(k)}.$$

If $\omega = 1$ we retain the Gauss–Seidel method, but if $\omega > 1$ the rate of convergence may be much improved. Note that

$$(I + \omega D^{-1}L)\mathbf{x}^{(k+1)} = \omega D^{-1}\mathbf{b} + [(1 - \omega)I - \omega D^{-1}U]\mathbf{x}^{(k)},$$

so that the error behaves as
$$\mathbf{e}^{(k+1)} = M(\omega)\mathbf{e}^{(k)},$$

where the matrix M is given by

$$M(\omega) = (I + \omega D^{-1}L)^{-1}[(1 - \omega)I - \omega D^{-1}U].$$

The parameter ω should in principle be chosen to minimize $\rho(M(\omega))$ if convergence is to be as fast as possible. Usually, however, it is found by trial and error.

The method of successive over-relaxation may be obtained most simply by modifying the Gauss–Seidel routine: merely follow the statement $x[i] = bi/mi$; with the statements

```
x[i] *= w;
x[i] += (1 - w)*oldx[i];
```

The parameter w must of course be part of the signature of the routine, which then becomes

```
state sor(matrix &m, vector &x, vector &b, double w);
```

Using the above system of equations, eight iterations are required to achieve convergence to four decimals for both $\omega = 1.00$ and $\omega = 1.25$. The minimum is five at $\omega = 1.10$. More dramatic reductions might be expected for a matrix not quite diagonally dominant. For example, the system of equations

$$\begin{pmatrix} 7 & 3 & -1 & 2 \\ 3 & 8 & 1 & -4 \\ -1 & 1 & 4 & -1 \\ 2 & -4 & -1 & 6 \end{pmatrix} \begin{pmatrix} x_0 \\ x_1 \\ x_2 \\ x_3 \end{pmatrix} = \begin{pmatrix} 11 \\ 8 \\ 3 \\ 3 \end{pmatrix}$$

has $\mu = \|B\|_\infty = 7/6$. With $\omega = 1$ (the Gauss–Seidel case) 37 iterations are required to achieve 4–decimal accuracy; the minimum occurs at $\omega \approx 1.4$, with just 12 iterations.

12.5 Speed of computation

As we found in chapter 10, the direct methods of elimination with pivoting, and decomposition, require for large n approximately $n^3/3$ multiplications or divisions, which (we have assumed, cf 3.10) are indicative of the speed of computation. Each iteration in the methods described in this chapter requires just n^2 such operations, so to determine the total number of operations we must multiply by the number m of iterations. Direct methods will therefore be faster than iterative methods if $n^3/3 < mn^2$, or $n < 3m$. Unless a very accurate initial guess is made of the solution, the number of iterations is likely to be several, to say the least, more likely several tens. Therefore, direct methods are likely to be superior unless n is quite large. There are matrices of special form, however, for which the number of multiplications or divisions may be sharply reduced. These are *sparse matrices*, many of the elements of which are zero, for which the number of multiplications required for each iteration may be only linear in n. Iteration may then be the faster method of computation. We dealt with the special case of tridiagonal matrices in chapter 10.

12.6 Roots of systems of non-linear equations

We learned how to extract roots of non-linear equations with one unknown in chapter 4. We left aside *systems* of non-linear equations in more than one variable because their solution calls for the methods developed in this chapter.

We wish to find by iteration the simultaneous roots of a system of equations, which in general may be written

$$f_0(x_0, x_1, \ldots, x_{n-1}) = 0$$
$$f_1(x_0, x_1, \ldots, x_{n-1}) = 0$$
$$\cdots\cdots\cdots\cdots\cdots\cdots$$
$$f_{n-1}(x_0, x_1, \ldots, x_{n-1}) = 0$$

given a sufficiently close initial approximation. We shall sketch only one method, the extension of Newton's algorithm to three variables. A more general treatment would take us too far from our main purpose.

We shall consider in particular the non-linear system:

$$x^2 + y^2 + z^2 = 2$$

$$x + y + z = 0$$

$$x(y + z) + 1 = 0$$

which has the solution $(1 \quad 0 \quad -1)^T$, as may be seen by inspection.

The straightforward extension of Taylor's theorem to three variables gives, to first order:

$$f_i(\xi,\eta,\zeta) = f_i(x_0,y_0,z_0) + (\xi-x_0)\partial f_i(x_0,y_0,z_0)/\partial x$$

$$+ (\eta-y_0)\partial f_i(x_0,y_0,z_0)/\partial y$$

$$+ (\zeta-z_0)\partial f_i(x_0,y_0,z_0)/\partial y$$

$$+ \text{higher order terms,}$$

for $i = 0, 1, 2$.

Let us denote by $\mathbf{w}$ the vector $(x, y, z)^T$, and by $\mathbf{f}$ the vector $(f_0(\mathbf{w}), f_1(\mathbf{w}), f_2(\mathbf{w}))^T$. Then, if $\mathbf{a} \equiv (\xi, \eta, \zeta)^T$ is a root of $\mathbf{f}(\mathbf{w}) = 0$, we have approximately:

$$0 \approx \mathbf{f}(\mathbf{w}^{(0)}) + J(\mathbf{w}^{(0)})(\mathbf{a} - \mathbf{w}^{(0)}),$$

where J is the *Jacobean matrix*:

$$J(\mathbf{w}) = \begin{pmatrix} \partial f_0(\mathbf{w})/\partial x & \partial f_0(\mathbf{w})/\partial y & \partial f_0(\mathbf{w})/\partial z \\ \partial f_1(\mathbf{w})/\partial x & \partial f_1(\mathbf{w})/\partial y & \partial f_1(\mathbf{w})/\partial z \\ \partial f_2(\mathbf{w})/\partial x & \partial f_2(\mathbf{w})/\partial y & \partial f_2(\mathbf{w})/\partial z \end{pmatrix}.$$

Solving for **a**, $\qquad \mathbf{a} \approx \mathbf{w}^{(0)} - J(\mathbf{w}^{(0)})^{-1}\mathbf{f}(\mathbf{w}^{(0)}) \equiv \mathbf{w}^{(1)},$

which we assert is a better approximation to **a** than is $\mathbf{w}^{(0)}$; and in general,

$$\mathbf{w}^{(k+1)} = \mathbf{w}^{(k)} - J(\mathbf{w}^{(k)})^{-1}\mathbf{f}(\mathbf{w}^{(k)}).$$

In practice we do not invert the Jacobean matrix; instead we use any standard procedure, such as Crout decomposition, to solve the linear equations

$$J(\mathbf{w}^{(k)})\delta\mathbf{w}^{(k)} = -\mathbf{f}(\mathbf{w}^{(k)})$$

for the differences $\qquad\qquad \delta\mathbf{w}^{(k)} = \mathbf{w}^{(k+1)} - \mathbf{w}^{(k)}.$

This is Newton's method applied to the roots of simultaneous non-linear equations.

In order to deal with the chosen example we shall need a number of auxiliary functions. First, we must program the three given equations, and a routine to evaluate them as a vector at given values of the variables:

```
inline double F0(vector &w)
{ return w[0]*w[0] +w[1]*w[1] + w[2]*w[2] - 2.0; }

inline double F1(vector &w)
{ return w[0] + w[1] + w[2]; }

inline double F2(vector &w)
{ return w[0]*(w[1]) + w[2]) + 1.0; }
```

together with

```
void   makevec(vector &F, vector &w)
{      F[0] = F0(w);
       F[1] = F1(w);
       F[2] = F2(w);
}
```

In a precisely similar fashion we must also assemble the Jacobean matrix, using a function the corresponding signature of which is

```
void jacobean(matrix &J, vector &w);
```

where each matrix element must be calculated analytically as $J_{ij}(\mathbf{w}) = \partial f_i(\mathbf{w})/\partial w_j$. The following routine then carries out Newton's algorithm for the non-linear system:

```
state newtsys(vector &x)
{       int n = x.getsize();
        state s = ITERATING;
        while (s == ITERATING)
        {       int iter = 0;
                // put running output here
                vector F(n);
                makevec(F, x);
                matrix J(n);
                jacobean(J, x);
                vector y = -F;
                crout(J, y);
                x += y;
                if (norminf(dx) < tolerance)
                        s = SUCCESS;
                if (++iter == maxiter)
                        s = WONTSTOP;
        }
        return s;
}
```

Convergence is quadratic as befits Newton's method (although we shall not prove this), and the following results are obtained if as starting approximation we set $x = (0.75 \ 0.5 \ -0.5)^T$ and tolerance = 5.0E–7:

Iteration	Approximation		
0	0.75	0.5	–0.5
1	1.333333	–0.135417	–1.197917
2	1.041667	–0.019914	–1.021752
3	1.000833	–0.000416	–1.000418
4	1	0	–1

Of course, the starting point must be chosen circumspectly not to be too far from the true solution, as with all applications of Newton's method. In 4.7, where we were concerned with the roots of single non-linear equations, we remarked that it may sometimes be helpful to precede the quadratic Newton's method with a few iterations of the linear bisection method because that is much less sensitive to the choice of starting point. Likewise, in the case of several simultaneous equations it may be advantageous to begin with a few iterations of a less efficient, linear, method. The *method of steepest descent* may be used for this purpose, although it may converge too slowly to be worth pursuing beyond a starting approximation for Newton's method.

In the method of steepest descent, which is described more fully in the book by Burden and Faires, and with considerable elaboration by Press *et al.*, we seek a root of the set $\mathbf{f}(\mathbf{w}) = 0$ of simultaneous equations by first forming the sum of squares:

$$g(\mathbf{w}) = |\mathbf{f}(\mathbf{w})|^2 = \sum_{i=1}^{n} [f_i(x_1,...,x_n)]^2.$$

The solutions of $\mathbf{f}(\mathbf{w}) = 0$ are then the solutions of $g(\mathbf{w}) = 0$, and since $g(\mathbf{w}) \geq 0$, they may be found by seeking the minima of $g(\mathbf{w})$. In general this would be too difficult; but if we can approach a solution in linear steps, we may at any rate arrange that each such step towards the minimum is the "steepest" possible. If $\nabla g(\mathbf{w})$ is the gradient of g at the point $\mathbf{w}$, then a suitable sequence of steps is given by

$$\mathbf{w}^{(k+1)} = \mathbf{w}^{(k)} - \alpha \nabla g(\mathbf{w}^{(k)}),$$

where α is some parameter; and therefore at each step we wish to minimize the function

$$h(\alpha) = g(\mathbf{w} - \alpha \nabla g(\mathbf{w}))$$

with respect to α, in order that the step down from $\mathbf{w}$ is as steep as possible.

The minimization of $h(\alpha)$ is again too difficult in general, but we may seek an approximation to it in the manner of the Brent method for a single function (*cf* 4.11), namely by finding a quadratic approximation, and taking the minimum of that. Let us call the three points α_1, α_2 and α_3. At the end of the process we expect the minimum to be very close to $\alpha = 0$, so let us take $\alpha_1 = 0$. If we set $\alpha_3 = 1$, that represents rather a large step, unless the gradient is very small, possibly one which takes us beyond the minimum. Let $g_1 = g(\mathbf{w})$, and $g_3 = g(\mathbf{w} - \alpha_3 \nabla g(\mathbf{w}))$. Divide α_3 by 2 repeatedly as long as $g_3 > g_1$, then take $\alpha_2 = \alpha_3/2$. If $g_2 = g(\mathbf{w} - \alpha_2 \nabla g(\mathbf{w}))$, we then have three points (α_i, g_i) to which a quadratic may be fitted, and an approximate minimum found at α_0, say. If $g_0 = g(\mathbf{w} - \alpha_0 \nabla g(\mathbf{w})) < g_3$, we shall assume that $\alpha = \alpha_0$ gives us a sufficient approximation to the steepest descent from $\mathbf{w}$; if not we take $\alpha = \alpha_3$. In either case, the process is repeated recursively at successive points $\mathbf{w}^{(k)}$.

Clearly, we first of all need some way of obtaining the gradient. If the functions f_i are simple we may calculate g and ∇g analytically. In general that will not be so. We shall use the second-order midpoint method (*cf* 16.1) to compute it:

```
vector grad(func f, vector &x, double h)
//      second-order midpoint differential
{       int n = x.getsize();
        vector g(n);
        double h2 = 2.0*h;
        for (int i = 0; i < n; ++i)
        {       x[i] += h;
                double f1 = f(x); // f(..., x[i]+h, ...)
```

```
                    x[i] -= h2;
                    double f2 = f(x);    // f(..., x[i]-h, ...)
                    g[i] = (f1 - f2)/h2;
                    x[i] += h;           // restore x[i]
            }
        return g;
}
```

Then the following function very simply carries out the process just described:

```
vector steepest(state &s, func g, vector &x)
{       static int iter = 0;
//      put internal output here
        vector z = grad(g, x, tol);
        double z0 = norm(z);
        if (z0 < toosmall) { s = NEARZERO; return x; }
        z /= z0;                     // unit vector
        double a1 = 0.0, a3 = 1.0;
        double g1 = g(x);
        double g3 = g(x - a3*z);
        while (g3 > g1)
        {       a3 /= 2.0;
                g3 = g(x - a3*z);
        }
        double a2 = a3/2;
        double g2 = g(x - a2*z);
        double p = (a2 - a1)*(g2 - g3);      // a1 = 0
        double q = (a2 - a3)*(g2 - g1);
        double r = (a2 - a1)*p - (a2 - a3)*q;
        double t = p - q;
        r /= (2.0*t);
        double a0 = a2 - r;                 // minimum of parabola
        double g0 = g(x - a0*z);
        double a = (g0 < g3 ? a0 : a3);
        if (a < tol) { s = SUCCESS; return x; }
        if (iter++ == maxiter) { s = WONTSTOP; return x; }
        return steepest(s, g, x-a*z);
}
```

Of course, the value of the state variable s must be initialized to ITERATING when steepest is called. Applied to the same system of three equations, the convergence is very slow compared with the second-order Newton method, although still adequate for a first approximation. The first few iterations are reproduced below for the same starting conditions:

Iteration	Approximation		
0	0.75	0.5	−0.5
1	0.897444	0.373606	−1.047675
2	0.890086	0.240095	−1.154650
3	0.870816	0.206531	−1.113337
4	0.907067	0.168479	−1.126558
5	0.900293	0.150919	−1.093599

It is interesting to note that the sequence $\mathbf{w}^{(k)}$ pursues a zig-zag path down the "valley" represented by the function $g(\mathbf{w})$. This follows from the fact that each step terminates at or near the minimum in that direction, so each new gradient must be perpendicular to the old one. Since we have not sought the exact minimum at each step, but only a quadratic approximation to it, the extent to which successive unit vectors are indeed orthogonal is an indication of how good the approximation actually is. We reproduce below the unit vectors corresponding to the steps above:

Iteration	Unit vector		
0	−0.2537	0.2175	0.9425
1	0.0430	0.7797	0.6247
2	0.3404	0.5929	−0.7298
3	−0.6689	0.7022	0.2440
4	0.1785	0.4627	−0.8684
5	−0.7778	0.6037	0.1751

13 Matrix eigenvalue problems

The set of n linear equations represented by the single matrix equation

$$A\mathbf{x} = \lambda\mathbf{x},$$

where A is a square $n \times n$ matrix and λ is a constant, permits of solutions other than the trivial solution $\mathbf{x} = \mathbf{0}$ if and only if

$$p_A(\lambda) \equiv \det(A - \lambda I) = 0,$$

where I is the unit matrix. This is the *characteristic equation* of the matrix A. If we expand the determinant by Cramer's rule (*cf* chapter 10) we see that p_A is a polynomial of degree n in λ. The characteristic equation therefore possesses n roots λ_i, not all necessarily distinct from each other, which are the *eigenvalues* of the matrix A. Correspondingly, there are n *eigenvectors* $\mathbf{x}^{(i)}$, not necessarily all distinct, which satisfy the relations

$$A\mathbf{x}^{(i)} = \lambda_i\mathbf{x}^{(i)}.$$

Note that the eigenvectors are undetermined to a multiplicative factor. This may be removed by a consistent choice of normalization; and it is customary to set $\|\mathbf{x}^{(i)}\|_\infty = 1$ or $\|\mathbf{x}^{(i)}\|_2 = 1$ depending on the ease of computation and the use to which the results are to be put.

A set of vectors $\mathbf{x}^{(k)}$ is said to be *linearly independent* if the equation $\Sigma_k c_k\mathbf{x}^{(k)} = \mathbf{0}$ has no solution other than $\mathbf{c} = \mathbf{0}$; otherwise they are linearly dependent. If the eigenvalues λ_k of a matrix A are distinct the corresponding eigenvectors $\mathbf{x}^{(k)}$ are linearly independent. Suppose the contrary were true, then there would exist coefficients c_k such that $\Sigma_k c_k\mathbf{x}^{(k)} = \mathbf{0}$. Multiplying through by the matrix A, we have $\Sigma_k c_k A\mathbf{x}^{(k)} = \Sigma_k c_k\lambda_k\mathbf{x}^{(k)} = \mathbf{0}$, which—since the λ_k are distinct— is a contradiction unless $\mathbf{c} = \mathbf{0}$.

For example, the characteristic equation of the real symmetric matrix

$$A = \begin{pmatrix} 2 & 1 & 0 \\ 1 & 3 & 1 \\ 0 & 1 & 2 \end{pmatrix}$$

is $\lambda^3 - 7\lambda^2 + 14\lambda - 8 = 0$, the roots of which are $\lambda_1 = 1$, $\lambda_2 = 2$, $\lambda_3 = 4$. These are three distinct eigenvalues, and it may easily be verified that the corresponding distinct eigenvectors (apart from a factor) are

$$\mathbf{x}^{(1)} = \begin{pmatrix} 1 \\ -1 \\ 1 \end{pmatrix}, \qquad \mathbf{x}^{(2)} = \begin{pmatrix} 1 \\ 0 \\ -1 \end{pmatrix}, \qquad \mathbf{x}^{(3)} = \begin{pmatrix} 1 \\ 2 \\ 1 \end{pmatrix},$$

and that they are linearly independent.

If $\mathbf{x}^{(k)}$ are a set of n linearly independent n-dimensional vectors with, as here, real coefficients, then *any* such vector may be expressed as a linear combination of the $\mathbf{x}^{(k)}$, which are therefore said to form a *basis* for the space of all such vectors.

In this chapter we shall consider several methods for computing eigenvalues and eigenvectors. But first we give some additional definitions and general theory.

13.1 General theory

A set of vectors $\mathbf{u}^{(i)}$ is said to be *orthogonal* if the scalar products $\mathbf{u}^{(i)}.\mathbf{u}^{(j)} \equiv \Sigma_k u^{(i)}{}_k u^{(j)}{}_k = 0$, $i \neq j$. It is said to be *orthonormal* if in addition $\|\mathbf{u}^{(i)}\|_2 \equiv \sqrt{(\mathbf{u}^{(i)}.\mathbf{u}^{(i)})} = 1$ for each i. The above eigenvectors $\mathbf{x}^{(i)}$ form an orthogonal set which is not orthonormal; however, the set $\{ \mathbf{x}^{(1)}/\sqrt{3}, \mathbf{x}^{(2)}/\sqrt{2}, \mathbf{x}^{(3)}/\sqrt{6} \}$ *is* orthonormal.

A square matrix P is said to be orthogonal if $P^{\mathrm{T}}P = I$, which implies that $P^{-1} = P^{\mathrm{T}}$, where as usual the suffix T denotes the transposed matrix.

If $\det(A) = 0$ the square matrix A is said to be *singular*; otherwise it is *non-singular*. If S is any non-singular matrix, then the matrix $B = S^{-1}AS$ has the same set of eigenvalues as A:

$$\det(B - \lambda I) = \det(S^{-1}AS - \lambda I) = \det(S^{-1}(A - \lambda I)S)$$

$$= \det(S^{-1})\det(A - \lambda I)\det(S)$$

$$= \det(A - \lambda I),$$

where we have used the fact that $\det(XY) = \det(X)\det(Y)$. We say that A and B are *similar*, and the relationship between them, $\mathbf{S}: B = S^{-1}AS$, is a *similarity transformation*. The characteristic equation is invariant under a similarity transformation: $p_B(\lambda) \equiv p_A(\lambda)$.

It is useful to take the analysis of the characteristic equation one step further. If we expand the determinant of a $n \times n$ matrix A by Cramer's rule, we obtain:

$$p_A(\lambda) = (-\lambda)^n + (-\lambda)^{n-1}\mathrm{trace}(A) + \ldots + \det(A).$$

where $\mathrm{trace}(A) = \Sigma_k a_{kk}$ is the sum of the diagonal elements of A. It follows that $\mathrm{trace}(A)$ is invariant under a similarity transformation, as are all the other coefficients of the characteristic equation.

Moreover, if $\mathbf{x}$ is an eigenvector of a matrix A corresponding to the eigenvalue λ, and B is related to A by a similarity transformation $\mathbf{S}$, then the eigenvector of B corresponding to the eigenvalue λ is $S^{-1}\mathbf{x}$.

If A is a real *symmetric* $n \times n$ matrix, it can be shown that there exists an orthogonal matrix P such that the similar matrix $D = P^{-1}AP$ is diagonal, the diagonal elements being the eigenvalues of A: and we write $D = \mathrm{diag}(\lambda_1, \ldots, \lambda_n)$. That being so, there exist n eigenvectors of A that form an orthonormal set. For if $D = P^{-1}AP$ then $AP = PD$. Let $\mathbf{p}^{(k)}$ be the k-th column of P. Then $A\mathbf{p}^{(k)} = \lambda_k\mathbf{p}^{(k)}$, and $\mathbf{p}^{(k)}$ is an eigenvector of A corresponding to the eigenvalue λ_k. There are n such eigenvectors and if they are normalized so that $\|\mathbf{p}^{(k)}\|_2 = 1$ they form an orthonormal set. Moreover, the eigenvalues of A are all real, for if $A\mathbf{p}^{(k)} = \lambda_k\mathbf{p}^{(k)}$, then $\mathbf{p}^{(k)\mathrm{T}}A\mathbf{p}^{(k)} = \lambda_k\|\mathbf{p}^{(k)}\|_2^2 = \lambda_k$. But the left-hand side is real, therefore λ_k is real also.

For example, if A is the above matrix, then the matrix P, the columns of which are the orthonormal eigenvectors of A, is

$$P = (1/\sqrt{6}) \begin{pmatrix} \sqrt{2} & \sqrt{3} & 1 \\ -\sqrt{2} & 0 & 2 \\ \sqrt{2} & -\sqrt{3} & 1 \end{pmatrix},$$

and it may readily be verified that

$$P^{-1}AP = \begin{pmatrix} 1 & 0 & 0 \\ 0 & 2 & 0 \\ 0 & 0 & 4 \end{pmatrix}.$$

We have seen that $\mathrm{trace}(A)$ is invariant under a similarity transformation. It follows that if A can be diagonalized by a similarity transformation, $D = P^{-1}AP$, then $\mathrm{trace}(A) = \mathrm{trace}(D)$. But the diagonal elements of D are the eigenvalues of A. Therefore,

$$\mathrm{trace}(A) = \Sigma_k \lambda_k.$$

The trace of a symmetric matrix is the sum of its eigenvalues. It may readily be seen that this is true for the matrix studied above, where $\mathrm{trace}(A) = 7 = \mathrm{trace}(P^{-1}AP)$.

13.1.1 Locating eigenvalues

It is sometimes useful to know roughly where the eigenvalues of a matrix A are to be found. The crudest estimate is given by $|\lambda| \le \|A\|$, where λ is any eigenvalue and $\|\cdot\|$ is any norm. We can do somewhat better by means of

Gershgorin's theorem:

         If the $n \times n$ matrix $A = (a_{ij})$, let $r_i = \Sigma_{j \ne i}|a_{ij}|$; and in the complex z-plane define the disk

$$C_i: |z - a_{ii}| \le r_i$$

as the set of points bounded by the circle of radius r_i centred on a_{ii}. There are n such disks, not necessarily distinct. Then, if λ is any eigenvalue of A, it lies on one of the disks C_i. Moreover, if m disks overlap but are disjoint from the remaining $n - m$ disks, then the union of the overlapping disks contains m eigenvalues, counting them according to their multiplicity as roots of the characteristic equation of the matrix A.

For if λ is an eigenvalue and $\mathbf{x}$ the corresponding eigenvector, so that $A\mathbf{x} = \lambda\mathbf{x}$, let x_k be a component of $\mathbf{x}$ of greatest magnitude: $|x_k| = \text{Max}_i |x_i| = \|\mathbf{x}\|_\infty$. Then $\Sigma_j a_{kj}x_j = \lambda x_k$. Rearranging terms and taking norms, we have successively:

$$(\lambda - a_{kk})x_k = \Sigma_{j \neq k} a_{kj}x_j,$$

$$|\lambda - a_{kk}|\,|x_k| \leq \Sigma_{j \neq k} |a_{kj}|\,|x_j|,$$

$$|\lambda - a_{kk}|\,\|\mathbf{x}\|_\infty \leq r_k\|\mathbf{x}\|_\infty.$$

Cancelling the common factor $\|\mathbf{x}\|_\infty$, it follows that $\lambda \in C_k$, which proves the first part of the theorem.

The second part may be obtained by continuity arguments, which we shall merely sketch. Let D be a diagonal matrix, the diagonal elements of which coincide with those of A, and let $O = A - D$ be the off-diagonal part of A. Further, let ε be a parameter such that $0 \leq \varepsilon \leq 1$, and define the parametrized matrix

$$A(\varepsilon) = D + \varepsilon O.$$

Correspondingly, let $\lambda_k(\varepsilon)$ be the eigenvalues of $A(\varepsilon)$. Clearly, the eigenvalues of $A(0) = D$ are the diagonal elements $\lambda_k(0) = a_{kk}$, while the eigenvalues of $A(1) = A$ are the original eigenvalues $\lambda_k(1) = \lambda_k$. In general, the eigenvalues of $A(\varepsilon)$ are the roots of the characteristic equation $p_{A(\varepsilon)}(\lambda) \equiv \det(A(\varepsilon) - \lambda I) = 0$. Now the coefficients of the polynomial p are continuous functions of ε; and the roots of any polynomial are continuous functions of its coefficients. Therefore the eigenvalues of $A(\varepsilon)$ are continuous functions of ε. But we have already proved that each eigenvalue $\lambda_k(\varepsilon)$ of $A(\varepsilon)$ lies on the disk $C_k(\varepsilon)$: $|z - a_{kk}| \leq \varepsilon r_k$, with r_i defined as before. Therefore, if $n - m$ of the disks $C_k = C_k(1)$ form a connected set disjoint from the rest, that set must contain the m eigenvalues $\lambda_k = \lambda_k(1)$ for which $\lambda_k(0) = a_{kk}$.

Consider, for example, the symmetric matrix

$$\begin{pmatrix} 1 & 1 & 0 \\ 1 & 3 & 1 \\ 0 & 1 & 7 \end{pmatrix}.$$

The Gershgorin disks are C_1: $|z - 1| \leq 1$, C_2: $|z - 3| \leq 2$, C_3: $|z - 7| \leq 1$; C_1 and C_2 overlap forming the set $C_1 \cup C_2$, while C_3 is disjoint. Since the eigenvalues must in this case be real, this implies that $0 \leq \lambda_1, \lambda_2 \leq 5$, $6 \leq \lambda_3 \leq 8$. The characteristic equation is

$$\lambda^3 - 11\lambda^2 + 29\lambda - 13 = 0,$$

and the actual roots are $\lambda_1 \approx 0.56$, $\lambda_2 \approx 3.19$, $\lambda_3 \approx 7.25$, which accords with the theorem.

13.1.2 Eigenvalue stability and errors of computation

In this section we show that the eigenvalues of a real symmetric matrix are stable against perturbations of the matrix. This result is not in general true of non-symmetric matrices, and it may therefore be difficult to obtain reliable estimates of their eigenvalues.

Suppose we have a real symmetric matrix A with eigenvalues λ_i. Let λ be an eigenvalue of the perturbed matrix $A + E$, and $\mathbf{x}$ the corresponding eigenvector, so that

$$(\lambda I - A)\mathbf{x} = E\mathbf{x}.$$

Further, let P be the orthogonal matrix which diagonalizes A, so that $P^{-1}AP = D$, where D is the diagonal matrix, the elements of which are the eigenvalues λ_i. Then, successively:

$$(\lambda I - PDP^{-1})\mathbf{x} = E\mathbf{x},$$

$$P(\lambda I - D)(P^{-1}\mathbf{x}) = EPP^{-1}\mathbf{x},$$

$$(\lambda I - D)(P^{-1}\mathbf{x}) = (P^{-1}EP)(P^{-1}\mathbf{x}),$$

$$(P^{-1}\mathbf{x}) = (\lambda I - D)^{-1}(P^{-1}EP)(P^{-1}\mathbf{x}),$$

$$\|P^{-1}\mathbf{x}\| \leq \|(\lambda I - D)^{-1}(P^{-1}EP)\| \, \|P^{-1}\mathbf{x}\|,$$

$$1 \leq \|(\lambda I - D)^{-1}\| \, \|(P^{-1}EP)\|.$$

We shall assume that λ differs, however slightly, from all the eigenvalues λ_i of A, so that

$$\det(\lambda I - D) \neq 0,$$

otherwise there is nothing to prove. Then,

$$(\lambda I - D)^{-1} = \operatorname{diag}[(\lambda - \lambda_1)^{-1}, \ldots, (\lambda - \lambda_n)^{-1}].$$

It follows that
$$1 \leq \operatorname{Max}_i |\lambda - \lambda_i|^{-1} \|P^{-1}\| \, \|E\| \, \|P\|,$$

or
$$\operatorname{Min}_i |\lambda - \lambda_i| \leq K(P)\|E\|,$$

where $K(P) = \|P^{-1}\| \, \|P\| \geq 1$ is the condition number of the matrix P (cf 11.3). This result holds good even if A is non-symmetric, provided it can be diagonalized by a similarity transformation, which will often be the case.

However, if A is symmetric as we have supposed, then P is an orthogonal matrix, and

$$\|P\|_2 = \sqrt{\rho(P^{\mathrm{T}}P)} = \sqrt{\rho(P^{-1}P)} = \sqrt{\rho(I)} = 1.$$

Therefore $$\mathrm{Min}_i \, |\lambda - \lambda_i| \leq \|E\|_2.$$

Since this result applies to *any* eigenvalue of the perturbed matrix, we can say that the eigenvalues of a real symmetric matrix are stable against perturbations of the matrix elements due, for example, to rounding errors. If there is a wide spread of eigenvalues, however, the *relative* errors may exceed acceptable limits for some of them. Double precision is normally used in the computation of eigenvalues.

Similarly, we may set bounds to the error resulting from an eigenvalue computation. For suppose λ is an approximate eigenvalue of the matrix A and $\mathbf{x}$ is the corresponding approximate eigenvector. Denote the exact eigenvalues of A by λ_i. Then we may compute the "discrepancy vector",

$$\mathbf{e} = A\mathbf{x} - \lambda\mathbf{x}$$

(working to higher precision if necessary), and show as before that

$$\mathrm{Min}_i \, |\lambda - \lambda_i| \leq K(P)\|\mathbf{e}\|,$$

or in the case of a symmetric matrix,

$$\mathrm{Min}_i \, |\lambda - \lambda_i| \leq \|\mathbf{e}\|_2.$$

Although we shall not concern ourselves much with non–symmetric matrices, it is worth pointing out that the condition number K may then be quite large because the diagonalizing matrix P is not in general orthogonal. This can lead to great difficulty in computing the eigenvalues of such matrices, including slower convergence of iterative methods.

13.2 The power method

The first method we shall consider finds only the eigenvalue of maximum modulus (the *dominant eigenvalue*) of a matrix, not necessarily symmetric, and the corresponding eigenvector. Although of limited application, we have already seen the need to know this eigenvalue in order to calculate the spectral radius, which is so significant in determining rates of convergence and computational errors.

We assume that there is a set $\mathbf{x}^{(i)}$, where $1 \leq i \leq n$, of eigenvectors of the matrix A corresponding to the eigenvalues λ_i, and we order them so that $|\lambda_1| > |\lambda_2| \geq \ldots \geq |\lambda_n|$. Notice that to avoid complication we have assumed that the dominant eigenvalue λ_1 is distinct from the others. We also assume that $\|\mathbf{x}^{(i)}\|_\infty = 1$.

The power method is an iterative technique for finding the dominant eigenvalue. First, we guess an initial approximation $\mathbf{y}_0$ to the eigenvector. It is convenient to choose the normalization $\|\mathbf{y}_0\|_\infty = 1$. Second, suppose that this initial vector is expressed as a linear combination of the exact eigenvectors: $\mathbf{y}_0 = \sum_k \alpha_k \mathbf{x}^{(k)}$. Let $\mathbf{z} = A\mathbf{y}_0 = \sum_k \alpha_k A\mathbf{x}^{(k)} = \sum_k \alpha_k \lambda_k \mathbf{x}^{(k)}$, and let $\mu_1 = \|\mathbf{z}\|_\infty$. If we now define the first iterate as $\mathbf{y}_1 = \mathbf{z}/\mu_1$, we shall have $\|\mathbf{y}_1\|_\infty = 1$. Eliminating the auxiliary variable $\mathbf{z}$, we have

$$\mathbf{y}_1 = [\textstyle\sum_k \alpha_k \lambda_k \mathbf{x}^{(k)}]/\mu_1.$$

Similarly, the second iterate will be

$$\mathbf{y}_2 = [\textstyle\sum_k \alpha_k \lambda_k^2 \mathbf{x}^{(k)}]/(\mu_1\mu_2),$$

where μ_2 is defined in a similar fashion; and in general,

$$\mathbf{y}_s = [\textstyle\sum_k \alpha_k \lambda_k^s \mathbf{x}^{(k)}]/(\mu_1\mu_2 \ldots \mu_s).$$

Since λ_1 is the eigenvalue of maximum modulus the term with $k = 1$ eventually dominates in this sum of powers as $s \to \infty$, and therefore $\mathbf{y}_s \to \alpha_1 \lambda_1^s \mathbf{x}^{(1)}/(\mu_1\mu_2 \ldots \mu_s)$. But we have chosen $\|\mathbf{y}_s\|_\infty = \|\mathbf{x}^{(1)}\|_\infty = 1$. Therefore $\mathbf{y}_s \to \mathbf{x}^{(1)}$ as $s \to \infty$, and so $\mu_1\mu_2 \ldots \mu_s \to \alpha_1 \lambda_1^s$; similarly $\mu_1\mu_2 \ldots \mu_{s+1} \to \alpha_1 \lambda_1^{s+1}$. Dividing, we finally obtain $\mu_{s+1} \to \lambda_1$ as $s \to \infty$.

A little more care shows that $\lim_{s\to\infty}[\|\mathbf{y}_{s+1} - \mathbf{x}^{(1)}\|_\infty / \|\mathbf{y}_s - \mathbf{x}^{(1)}\|_\infty] = |\lambda_2/\lambda_1| < 1$. The sequence converges to the dominant eigenvector linearly, at a rate determined by the eigenvalue nearest in magnitude to the dominant eigenvalue. This is the basis of the power method. Notice that it applies to non-symmetric as well as symmetric matrices.

The following function computes the dominant eigenvalue p and corresponding eigenvector $\mathbf{x}$ of an arbitrary real matrix A. The eigenvector is calculated to within a preset error denoted by `tolerance`. It is convenient to choose $\mathbf{x} = (1,1,\ldots,1)^T$ as the starting approximation, and easy to define an auxiliary function

```
int maxmod(vector &x);
```

which returns the index of the element of maximum modulus of the vector $\mathbf{x}$. Then the function we seek may be written:

```
state power(matrix &A, vector &x, double &p)
{      state s = ITERATING;
       int n = A.getsize();
       int iter = 0;
       p = 0.0;
       for (int i = 0; i < n; ++i)          // initialize x
              x[i] = 1;
       while (s == ITERATING)
       {      vector z = A*x;
              int q = maxmod(x);            // index of max mod
              p = z[q];                     // approx eigenvalue
              if (norminf(z) < toosmall)    // p = 0
                     s = NEARZERO;
              else
              {      q = maxmod(z);
                     z /= z[q];             // normalize z
                     double e = norminf(x - z);
                     x = z;                 // update x
                     if (++iter == maxiter)
                            s = WONTSTOP;
                     if (e < tolerance)
                            s = SUCCESS;
              }
       }
       return s;
}
```

We have again chosen a function which returns the final value of a variable of the enumerated type `state`, defined in the header file `standard.h`. The constants `tolerance`, `toosmall` and `maxiter` must as usual be provided in the user program. The function `norminf` is declared in the header file `vecmat.h` of Appendix E, and defined in 9.1.

The non-symmetric matrix

$$\begin{pmatrix} -4 & 14 & 0 \\ -5 & 13 & 0 \\ -1 & 0 & 2 \end{pmatrix}$$

has the characteristic equation $(\lambda - 2)(\lambda^2 - 9\lambda + 18) = 0$, with roots $\lambda_1 = 6, \lambda_2 = 3, \lambda_3 = 2$. Thus, the dominant eigenvalue is $\lambda_1 = 6$. If suitable internal output is included, and if we set `tolerance = 5.10^{-6}`, the following results are obtained:

Iteration	λ_1	Eigenvector		
0	–	1	1	+1
1	10	1	0.8	+0.1
2	7.2	1	0.75	−0.11111
3	6.5	1	0.73077	−0.18803
4	6.2308	1	0.72222	−0.22085
5	6.1111	1	0.71818	−0.23591
6	6.0545	1	0.71622	−0.24309
7	6.0270	1	0.71525	−0.24659
8	6.0135	1	0.71477	−0.24831
9	6.0067	1	0.71453	−0.24916
10	6.0034	1	0.71441	−0.24958
11	6.0017	1	0.71435	−0.24979
12	6.0008	1	0.71432	−0.24990
13	6.0004	1	0.71430	−0.24995
14	6.0002	1	0.71429	−0.24997
15	6.0001	1	0.71429	−0.24999
16	6.0001	1	0.71429	−0.24999
17	6.0000	1	0.71429	−0.25000

As expected, successive iterates reduce the error by a factor of approximately $|\lambda_2/\lambda_1| = 3/6 = \frac{1}{2}$. The constancy of x_0 results from the normalization $\|\mathbf{x}\|_\infty = 1$.

Since the power method is linear, the Aitken process (*cf* 4.9) can be applied to it, giving quadratic convergence of the eigenvalue with the same linear convergence of the eigenvector, but we shall not pursue that here.

In the case of *symmetric* matrices the convergence may be somewhat better, but can be further improved by taking account of the fact that the exact eigenvectors form an orthonormal set. Let A be symmetric, with exact eigenvectors λ_k corresponding to exact eigenvectors $\mathbf{x}^{(k)}$; and suppose that λ is an approximate eigenvalue and $\mathbf{x}$ the corresponding approximate eigenvector, with $\|\mathbf{x}\|_2 = 1$. We shall impose the stopping condition $\|A\mathbf{x} - \lambda\mathbf{x}\|_2 < \varepsilon$, where ε is some sufficiently small quantity. We already know that the corresponding limit on λ is

$$\text{Min}_k |\lambda_k - \lambda| \le \|A\mathbf{x} - \lambda\mathbf{x}\|_2 < \varepsilon.$$

In the following modification of the `power` routine, this fact is used as a stopping condition on the iteration for the dominant eigenvalue of a symmetric matrix. The function `norm` is the euclidean vector norm, and `scalar` is the scalar product of two vectors; both are declared in the header file `vecmat.h`.

```
state sympow(matrix &A, vector &x, double &p)
{       state s = ITERATING;
        int iter = 0;
        p = 0.0;
        int n = A.getsize();
        for (int i = 0; i < n; ++i)
                x[i] = 1.0/sqrt(n);
        while (s == ITERATING)
        {       vector z = A*x;
                p = scalar(x, z);
                double normz = norm(z);
                if (normz < toosmall)                   // p = 0
                        s = NEARZERO;
                else
                {       z /= normz;
                        double e = norm(x - z);
                        x = z;
                        if (++iter == maxiter)
                                s = WONTSTOP;
                        if (e < tolerance)
                                s = SUCCESS;
                }
        }
        return s;
}
```

Consider the symmetric matrix

$$\begin{pmatrix} 4 & -1 & 1 \\ -1 & 3 & -2 \\ 1 & -2 & 3 \end{pmatrix},$$

the eigenvalues of which are $\lambda_1 = 6$, $\lambda_2 = 3$, $\lambda_3 = 1$, so that the ratio λ_2/λ_1 is the same as before. The non-symmetric routine **power** gives the results in the table below, using tolerance = 5.10^{-6}:

Iteration	λ_1	Eigenvector		
0	–	1	1	1
1	4	1	0	0.5
2	4.5	1	−0.44444	0.55556
3	5.0	1	−0.68889	0.71111
4	5.4	1	−0.83128	0.83539
5	5.6667	1	−0.91140	0.91213
6	5.8235	1	−0.95448	0.95461
7	5.9091	1	−0.97691	0.97693
8	5.9538	1	−0.98837	0.98837
9	5.9767	1	−0.99416	0.99416
10	5.9883	1	−0.99708	0.99708
11	5.9942	1	−0.99854	0.99854
12	5.9971	1	−0.99927	0.99927
13	5.9985	1	−0.99963	0.99963
14	5.9993	1	−0.99982	0.99982
15	5.9996	1	−0.99991	0.99991
16	5.9998	1	−0.99995	0.99995
17	5.9999	1	−0.99998	0.99998
18	6.0000	1	−0.99999	0.99999

The symmetric routine `sympow`, on the other hand, also with `tolerance` $= 5.10^{-6}$, gives:

Iteration	λ_1	Eigenvector		
0	—	0.57735	+0.57735	0.57735
1	2	0.89443	0	0.44721
2	4.6	0.81482	−0.36214	0.45268
3	5.6475	0.71062	−0.48954	0.50533
4	5.9085	0.64700	−0.53783	0.54049
5	5.9767	0.61284	−0.55854	0.55899
6	5.9942	0.59525	−0.56815	0.56823
7	5.9985	0.58634	−0.57280	0.57281
8	5.9996	0.58185	−0.57509	0.57509
9	5.9999	0.57960	−0.57622	0.57622
10	6	0.57848	−0.57679	0.57679
. . . .				
14	6	0.57742	−0.57732	0.57732
. . . .				
18	6	0.57735	−0.57735	0.57735

The convergence of the eigenvalue, although still only linear, is much improved. That of the eigenvector is unchanged.

13.3 Inverse power method

The power method is designed to compute the dominant eigenvalue only, *ie*, the eigenvalue with maximum modulus, and the corresponding eigenvector. It is easy to modify the power method to compute instead the eigenvalue of *minimum* modulus. For if $A\mathbf{x} = \lambda\mathbf{x}$, then $A^{-1}\mathbf{x} = \lambda^{-1}\mathbf{x}$: the eigenvalue of A of minimum modulus is the dominant eigenvalue of the inverse matrix A^{-1}.

The starting point of the inverse power method is therefore to define the auxiliary vector $\mathbf{z} = A^{-1}\mathbf{x}$, instead of $\mathbf{z} = A\mathbf{x}$ as in the power method, and then to proceed exactly as before, except that the resulting eigenvalue will be the reciprocal of the eigenvalue we seek. However, we have no need to compute the inverse matrix A^{-1}. Instead, we compute the vector $\mathbf{z}$ as the solution of the matrix equation

$$A\mathbf{z} = \mathbf{x},$$

at each iteration, using any standard method. We shall use Crout's method.

We may then define a function

```
state invpow(matrix &A, vector &x, double &p);
```

which is identical with `sympow` except that the statement `vector z = A*x;` is replaced by the pair of statements `vector z = x; crout(A, z);`, and the header file `crout.h` must be included in the user program.

The inverse power method, however, may be used more forcefully if approximations to the eigenvalues of a symmetric matrix are known or may be guessed, by Gershgorin's theorem or otherwise. If λ is a sufficient approximation to *any* of the eigenvalues of the matrix A, then we may compute the dominant eigenvalue λ_0 of the matrix $(A - \lambda I)^{-1}$. The eigenvalue which is sought will then be $\lambda + 1/\lambda_0$, and the rate of convergence will be determined by the ratio $(\lambda - \lambda_1)/(\lambda - \lambda_2)$, where λ_1 is the eigenvalue nearest in magnitude to the guessed value λ, and λ_2 is the next nearest. If $|\lambda - \lambda_1| \ll |\lambda - \lambda_2|$, convergence can be quite fast.

We therefore design the function `invpow` to return the "polished" eigenvalue in the variable `p`, which must be initialized with a rough approximation. The auxiliary function `maxmod` is again used, this time to avoid alternating signs in successive iterations. It is also convenient to define a function (*cf* 10.5) which creates the identity matrix, `matrix identity(int n)`. The dominant eigenvalue of $(A - \lambda I)^{-1}$ is stored in the variable `pp`:

```
state invpow(matrix &A, vector &x, double &p)
//      initialize p to approximate eigenvalue of A
//      final value of p is precise eigenvalue
{       state s = ITERATING;
```

```
        int iter = 0;
        double pp;
        int n = A.getsize();
        for (int i = 0; i < n; ++i) x[i] = 1.0/sqrt(n);
        matrix I = identity(n);
        while (s == ITERATING)
        {       vector z = x;
                crout(A - p*I, z);
                pp = scalar(x, z);
                double normz = norm(z);
                if (normz < toosmall)
                        s = NEARZERO;
                else
                {       z /= normz;
                        int q = maxmod(x);
                        if (sign(z[q]) != sign(x[q])) z *= -1.0;
                        double e = norm(x - z);
                        x = z;
                        if (++iter == maxiter)
                                s = WONTSTOP;
                        if (e < tolerance)
                                s = SUCCESS;
                }
        }
        p += 1.0/pp;
        return s;
}
```

Let us apply this routine to the computation of the eigenvalue $\lambda_2 = 3$ of our matrix, assuming that we have an initial approximation $\lambda \approx 3.5$. The results are shown in the table:

Iteration	λ_1	Eigenvector		
0	–	0.57735	0.57735	+0.57735
1	2	0.68825	0.68825	−0.22942
2	2.9412	0.83050	0.43968	−0.34197
3	2.9976	0.81314	0.42127	−0.40167
4	2.9999	0.81715	0.40955	−0.40563
5	3	0.81637	0.40877	−0.40799
6	3	0.81652	0.40830	−0.40815
7	3	0.81649	0.40827	−0.40824

13.4 Jacobi rotation method

The power method, and those derived from it, are designed to compute a single eigenvalue and eigenvector. The remaining methods presented in this chapter compute *all* eigenvalues and eigenvectors simultaneously. The first such method applies successive planar rotations to a symmetric matrix until it has in effect become diagonal.

Consider first a two-dimensional symmetric matrix,

$$A = \begin{pmatrix} \alpha & \gamma \\ \gamma & \beta \end{pmatrix}.$$

The matrix

$$R = \begin{pmatrix} \cos\theta & -\sin\theta \\ \sin\theta & \cos\theta \end{pmatrix}$$

is orthogonal because $RR^T = I$. Applied to a vector in the (x,y)–plane, R rotates it through an angle θ. Let us now calculate the effect on A of the similarity transformation $R^{-1}AR$. Let $c = \cos\theta$, $s = \sin\theta$, so that $c^2 + s^2 = 1$.

Then
$$R^{-1}AR = \begin{pmatrix} \alpha c^2 + 2\gamma cs + \beta s^2 & (\beta-\alpha)cs + \gamma(c^2-s^2) \\ (\beta-\alpha)cs + \gamma(c^2-s^2) & \alpha s^2 - 2\gamma cs + \beta c^2 \end{pmatrix}.$$

The rotation diagonalizes A provided $(\beta - \alpha)cs + \gamma(c^2 - s^2) = 0$. We must therefore choose θ so that

$$\cot 2\theta = d \equiv (\alpha - \beta)/2\gamma.$$

It is convenient to express all the trigonometric functions in terms of the parameter d. Let $t = \tan\theta$; then $c = \cos\theta = 1/\sqrt{(t^2 + 1)}$, and $s = ct$. Now by simple trigonometry $\cot 2\theta = \frac{1}{2}(\cot\theta - \tan\theta) = \frac{1}{2}(t^{-1} - t)$; and therefore,

$$t^2 + 2dt - 1 = 0.$$

For reasons that will appear presently, we choose the *smaller* of the two roots, for which $|\theta| \leq \pi/4$. Then,

$$t = \text{sign}(d)/[\,|d| + \sqrt{(d^2 + 1)}\,],$$

and the rotation matrix which diagonalizes A has been expressed solely in terms of the parameter d.

It remains to express the eigenvalues in terms of t. When we have done this the diagonalized matrix becomes

$$R^{-1}AR = \begin{pmatrix} \alpha + t\gamma & 0 \\ 0 & \beta - t\gamma \end{pmatrix}.$$

The eigenvalues are just those obtained from the characteristic equation

$$(\alpha - \lambda)(\beta - \lambda) - \gamma^2 = 0$$

if we set $(\alpha - \beta) = 2\gamma d$, and $2d = (1 - t^2)/t$.

We must now try to diagonalize a larger matrix using the same technique. The general strategy is to find an off–diagonal element of relatively large modulus, a_{pq} say, and to perform a rotation R_{pq} in the (p,q)-plane so as to reduce this element to zero. That rotation will, however, affect all elements in row p and column q. If we repeat this process with another non-diagonal element, the original a_{pq} may therefore again acquire a non-zero value—but we hope it will be smaller than before. The process is repeated until all the off-diagonal elements are of modulus less than a given small size, when the matrix may be considered to have been diagonalized. If we store the various rotations as we go along, the results will include not only the eigenvalues of A, but also the eigenvectors.

The matrix R_{pq} may be written

$$R_{pq} = \begin{matrix} & & & p & & q & & \\ & \begin{pmatrix} 1 & 0 & 0 & \dots\dots & 0 \\ 0 & 1 & 0 & \dots\dots & 0 \\ & & \dots\dots\dots\dots & & \\ p & 0 & \dots & c & \dots & -s & \dots & 0 \\ & & \dots\dots\dots\dots & & \\ q & 0 & \dots & s & \dots & c & \dots & 0 \\ & & \dots\dots\dots\dots & & \\ 0 & \dots & 0 & \dots\dots & 0 \\ 0 & \dots & 0 & \dots\dots & 1 \end{pmatrix} \end{matrix}$$

It is a unit matrix except in the rows p and q and columns p and q, where the only non-zero elements are, as indicated, determined by $c = \cos\theta_{pq}$, $s = \sin\theta_{pq}$, and

$$\cot 2\theta_{pq} = (a_{pp} - a_{qq})/2a_{pq}.$$

After the similarity transformation $R_{pq}^{-1}AR_{pq}$, the resulting matrix is unchanged except in rows and columns p and q. The affected elements are:

$$a'_{pp} = a_{pp} + ta_{pq},$$

$$a'_{qq} = a_{qq} - ta_{pq},$$

$$a'_{pq} = a'_{qp} = 0,$$

$$a'_{kp} = ca_{kp} + sa_{kq} = a'_{pk}, \quad (k \neq p,q),$$

$$a'_{kq} = -sa_{kp} + ca_{kq} = a'_{qk}, \quad (k \neq p,q).$$

We see now why we chose t to be the smaller of the two possible values. It is because in general we would like the changes in the matrix elements to be relatively small with each iteration, so that the convergence is steady. A large value might give rise to poor stability.

It is now a simple matter to check that the sum of the squares of the diagonal elements of the matrix is increased by every rotation that annihilates an off–diagonal element. It is sufficient to consider the two-dimensional case. If we define $Q = \Sigma_k a_{kk}^2 = \alpha^2 + \beta^2$, then after the rotation we have $Q' = (\alpha + \gamma t)^2 + (\beta - \gamma t)^2 = Q + 2\gamma^2$, and in general,

$$Q' = Q + 2a_{pq}^2.$$

Now the euclidean norm of the matrix A is defined by

$$\|A\|_E = \Sigma_{ij} a_{ij}^2 = \text{trace}(A^\mathsf{T}A).$$

If A is symmetric, then $A^\mathsf{T}A = A^2$, so that $\|A\|_E = \text{trace}(A^2) = \Sigma_k \lambda_k^2$, where the λ_k are the eigenvalues of the matrix A. The euclidean norm is therefore invariant under Jacobi rotations, because the eigenvalues are themselves invariant. Let us now define the sum of the squares of the off-diagonal elements of A, $O = \Sigma_{i \neq j} a_{ij}^2$. By definition, we have $Q + O = \|A\|_E$. Therefore, under a Jacobi rotation,

$$O' = O - 2a_{pq}^2.$$

Each rotation increases the sum of the squares of the diagonal elements, and reduces the sum of the squares of the off-diagonal elements. If in each iteration we choose the off-diagonal candidates for annihilation with care, an iteration consisting of a sequence of Jacobi rotations may be expected to converge to a diagonal matrix. Since a product of orthogonal matrices is itself an orthogonal matrix, the diagonal elements of the diagonalized matrix will be the eigenvalues of the original matrix. Likewise, the columns of the matrix representing the succession of rotations will be the corresponding eigenvectors (cf 13.1).

We must therefore store the rotations so that we can obtain these eigenvectors. If we store them in a matrix $R = (r_{ij})$, then the effect of the rotation R_{pq} is given by the equations:

$$r'_{kp} = cr_{kp} + sr_{kq},$$

$$r'_{kq} = -sr_{kp} + cr_{kq}.$$

We may now write a procedure which carries out the rotation $R_{pq}^{-1}AR_{pq}$, and stores the rotation cumulatively in the matrix R:

```
void   rotate(matrix &A, matrix &R, int p, int q)
//     rotates A thru θ in pq-plane to set A[p][q] = 0
//     rotation stored in R whose columns are eigenvectors of A
{      int n = A.getsize();
       double d = (A[p][p] - A[q][q])/(2.0*A[p][q]);// d = cot 2θ
       double t = sign(d)/(fabs(d) + sqrt(d*d + 1));// t = tan θ
       double c = 1.0/sqrt(t*t + 1);               // c = cos θ
       double s = t*c;                             // s = sin θ
       A[p][p] += t*A[p][q];
       A[q][q] -= t*A[p][q];
       A[p][q] = A[q][p] = 0.0;
       for (int k = 0; k < n; ++k)                  // transform A
       {      if (k != p && k != q)
              {     double Akp =  c*A[k][p] + s*A[k][q];
                    double Akq = -s*A[k][p] + c*A[k][q];
                    A[k][p] = A[p][k] = Akp;
                    A[k][q] = A[q][k] = Akq;
              }
       }
       for (k = 0; k < n; ++k)                       // store rot
       {      double Rkp =  c*R[k][p] + s*R[k][q];
              double Rkq = -s*R[k][p] + c*R[k][q];
              R[k][p] = Rkp;
              R[k][q] = Rkq;
       }
}
```

The function `rotate` will annihilate any element (p,q) that we choose. But we also need a definite strategy to guide our choice of victim systematically. We need an auxiliary function. For the moment, we ask this function to find the indices of the off–diagonal element of greatest modulus, and to return that modulus:

```
double maxoffdiag(matrix &A, int &p, int &q)
{      int n = A.getsize();
       double max = 0.0;
       for (int i = 0; i < n-1; ++i)
       {      for (int j = i+1; j < n; ++j)
              {     double Aij = fabs(A[i][j]);
                    if (Aij > max)
                    {      max = Aij;
                           p = i;
```

```
                        q = j;
                }
        }
    }
    return max;
}
```

Jacobi's iterative function may then be written:

```
state jacrot(matrix &A, matrix &R)
//      final A is diagonal matrix of eigenvalues
//      final R has eigenvectors in rows
{       state s = ITERATING;
        int n = A.getsize();
        for (int i = 0; i < n; ++i) R[i][i] = 1.0;   // unit matrix
            int iter = 0;
            while (s == ITERATING)
            {       int p, q;
                    double max = maxoffdiag(A, p, q);
                    if (max > tolerance)
                    {       rotate(A, R, p, q);
                            if (++iter == maxiter) s = WONTSTOP;
                    }       else
                    {       s = SUCCESS;
                            R = R.transpose();// eigenvectors in rows
                    }
            }
        }
        return s;
}
```

With the matrix studied in the last section, and with $\mathtt{tolerance} = 10^{-10}$, the routine jacrot succeeds in finding all three eigenvalues and eigenvectors, with just *two* iterations. For the larger matrix

$$\begin{pmatrix} 4 & 1 & -2 & 2 \\ 1 & 2 & 0 & 1 \\ -2 & 0 & 3 & -2 \\ 2 & 1 & -2 & -1 \end{pmatrix}$$

eighteen iterations are required with the same tolerance. The resulting eigenvalues and eigenvectors are:

Eigenvalue	Eigenvector			
6.844621	0.7180	0.2212	−0.5574	0.3534
2.268531	0.2017	0.7894	0.5796	0.0103
1.084364	0.6423	−0.5442	0.5202	−0.1440
−2.197517	−0.1767	−0.1781	0.2877	0.9243

Convergence of the Jacobi method is quadratic (provided the eigenvalues are not degenerate), but the choice of strategy represented in the function `maxoffdiag` is not ideal. The individual rotations are clearly $O(n)$, but `maxoffdiag` is $O(n^2)$, so the block within the `while` loop is also $O(n^2)$. This is because `maxoffdiag` searches through all $n(n-1)/2$ non-diagonal elements on every iteration to find the element of maximum modulus. It is true that the search involves comparisons only, but if n is large the quadratic behaviour eventually dominates. Although the elegance of the classical Jacobi method is thereby reduced, alternative strategies have been adopted to remove this expensive search. On each iteration one may, for example, examine each of the $n(n-1)$ off-diagonal elements in strict cyclic fashion, performing the Jacobi rotation only if the modulus exceeds some pre-determined proportion of the sum of the squares of the diagonal elements, the latter sum calculated once for each iteration. This is known as the *cyclic* Jacobi method, and it is implemented in Pascal in the book by Atkinson and Harley. In the *threshold* method, on the other hand, a limited search is performed on each iteration to find the first element whose modulus exceeds some proportion of the sum of the moduli of all the off-diagonal elements. The effective tolerance therefore decreases on each iteration, or on every few iterations. This method is implemented in C in the book by Press *et al.* In any case, the total number of arithmetic operations required in the Jacobi method is $O(n^3)$.

13.5 Householder's method

Householder's method reduces a real symmetric matrix to tridiagonal form by a systematic non-iterative technique, after which very efficient iterative procedures may be used to complete the diagonalization.

This method relies upon the fact that if $\mathbf{w}$ is any real vector with $\|\mathbf{w}\|_2 = 1$, then the matrix

$$P = I - 2\mathbf{w}\mathbf{w}^{\mathrm{T}}$$

is both orthogonal and symmetric. Since $(\mathbf{w}\mathbf{w}^{\mathrm{T}})^{\mathrm{T}} = (\mathbf{w}^{\mathrm{T}})^{\mathrm{T}}\mathbf{w}^{\mathrm{T}} = \mathbf{w}\mathbf{w}^{\mathrm{T}}$, it follows that $P^{\mathrm{T}} = P$, so P is symmetric. Thus $PP^{\mathrm{T}} = (I - 2\mathbf{w}\mathbf{w}^{\mathrm{T}})^2 = I - 4\mathbf{w}\mathbf{w}^{\mathrm{T}} + 4\mathbf{w}\mathbf{w}^{\mathrm{T}}\mathbf{w}\mathbf{w}^{\mathrm{T}} = I$, so P is orthogonal.

Therefore
$$P^{-1} = P^{\mathrm{T}} = P.$$

The utility of such *Householder matrices* is that if $\mathbf{x}$ and $\mathbf{y}$ are any two vectors with $\|\mathbf{x}\|_2 = \|\mathbf{y}\|_2 = 1$, then there exists a Householder matrix P such that $P\mathbf{x} = \mathbf{y}$. For, if

$$\mathbf{y} = P\mathbf{x} = (I - 2\mathbf{w}\mathbf{w}^T)\mathbf{x} = \mathbf{x} - 2\mathbf{w}(\mathbf{w}^T\mathbf{x}) = \mathbf{x} - 2(\mathbf{w}^T\mathbf{x})\mathbf{w},$$

then $\mathbf{w}$ is proportional to $\mathbf{x} - \mathbf{y}$; and since it has to be normalized, $\|\mathbf{w}\|_2 = 1$, we must take

$$\mathbf{w} = (\mathbf{x} - \mathbf{y})/\|\mathbf{x} - \mathbf{y}\|_2.$$

13.5.1 Reduction to tridiagonal form

To reduce a symmetric matrix A to tridiagonal form we therefore construct a sequence of Householder matrices $P^{(k)}$ which successively transform the columns $k = 1, \ldots, n-2$ (and by symmetry the rows) of A so that only the tridiagonal elements survive. It is convenient to keep the leading diagonal element unaffected at each stage.

Let $A^{(1)} = A$. The first column of $A^{(1)}$ is the vector $\mathbf{x} = (a_{11}, a_{21}, \ldots, a_{n1})^T$; we want to transform it into tridiagonal form $\mathbf{y} = (a_{11}, s, 0, \ldots, 0)^T$, where s is to be determined in terms of an appropriate Householder transformation. We have:

$$\mathbf{w} = (\mathbf{x} - \mathbf{y})/\|\mathbf{x} - \mathbf{y}\|_2 = (0, a_{21} - s, a_{31}, \ldots, a_{n1})^T/N,$$

where

$$N^2 = (a_{21} - s)^2 + \sum_{j>2} a_{j1}^2;$$

and in order that $N = 0$ only if $a_{21} = a_{31} = \ldots = a_{n1} = 0$, we choose $s^2 = \sum_{j>1} a_{j1}^2$; while to reduce rounding errors, we choose s to have the opposite sign to a_{21}. Therefore, $s = -(\text{sign}(a_{21}))|s|$, and $N = \sqrt{[2(s^2 - a_{21}s)]}$.

We have thus constructed the first Householder matrix $P^{(1)} = I - 2\mathbf{w}\mathbf{w}^T$, and may now define $A^{(2)} = P^{(1)}A^{(1)}P^{(1)}$, which is of the form:

$$\begin{pmatrix} a^{(2)}_{11} & a^{(2)}_{12} & 0 & 0 & \cdots & 0 \\ a^{(2)}_{21} & a^{(2)}_{22} & a^{(2)}_{23} & a^{(2)}_{24} & \cdots & a^{(2)}_{2n} \\ 0 & a^{(2)}_{32} & a^{(2)}_{33} & a^{(2)}_{34} & \cdots & a^{(2)}_{3n} \\ \cdots & \cdots & \cdots & \cdots & \cdots & \cdots \\ 0 & a^{(2)}_{n2} & a^{(2)}_{n3} & a^{(2)}_{n4} & \cdots & a^{(2)}_{nn} \end{pmatrix}.$$

However, it is not actually necessary to carry out the full similarity transformation. For if we define $\mathbf{v} = A\mathbf{w}$, $r = \mathbf{w}^T.\mathbf{v}$, $\mathbf{u} = \mathbf{v} - r\mathbf{w}$, then $A^{(2)} = A^{(1)} - 2(\mathbf{w}\mathbf{u}^T + \mathbf{u}\mathbf{w}^T)$.

Having obtained $A^{(2)}$, we define a new vector $\mathbf{w} = (0, 0, a^{(2)}_{32} - s, a^{(2)}_{42}, \ldots, a^{(2)}_{n2})/N$, and proceed to calculate a new s and a matrix $A^{(3)}$, and so on until the matrix is tridiagonal. However, in each Householder step, we have to form the product vector $A\mathbf{w}$. This would involve many multiplications by zero which should be removed in the interests of efficiency. The following auxiliary function serves the purpose, omitting contributions from rows and columns less than k to the usual product $\mathbf{m}*\mathbf{v}$ of a matrix with a vector:

```
vector matvec(matrix &m, vector &v, int k)
//      omit rows & cols < k
{       int n = m.getsize();
        vector u(n);
        for (int i = k; i < n; ++i)
        {       double t = 0.0;
                for (int j = k; j < n; ++j)
                        t += m[i][j] * v[j];
                u[i] = t;
        }
        return u;
}
```

We may now write down a procedure which carries out the complete reduction to tridiagonal form. If we store the resulting diagonal as a vector **d**, and the sub-diagonal (or super-diagonal) as another vector **e**, the matrix A may be used thereafter to store the vectors **w** in successive rows for further use, thus reducing memory requirements. It is also convenient at this point to add to A the unit matrix. The resulting procedure is:

```
void    house(matrix &A, vector &d, vector &e))
{       int n = A.getsize();
        int i, j;
        for (j = 0; j < n-2; ++j)               // each col in turn
        {       double s2 = 0.0;                // s squared
                for (i = j+1; i < n; ++i)
                {       double aij = A[i][j];
                        s2 += aij*aij;
                }
                double a = A[j+1][j];
                double s = -sign(a)*sqrt(s2);   // compute off-diag
                double N = sqrt(2.0*(s2 - a*s));
                vector w(n);
                for (i = j+1; i < n; ++i)       // construct w
                        w[i] = A[i][j];
                w[j+1] -= s;
                w /= N;// normalize w
                vector u = matvac(A, w, j+1);   // omit < j+1 in A*w
                double r = scalar(w, v);
                u -= r*w;
                A -= 2.0*(w*u + u*w);            // householder
                                                // transform
                d[j] = A[j][j];                 // store diagonal
                e[j] = A[j][j+1];               // store off-diag
```

```
          A[j] = w;                        // store w in row j
     }
     d[n-2] = A[n-2][n-2];                 // complete d, e, A
     d[n-1] = A[n-1][n-1];
     e[n-2] = A[n-2][n-1];
     for (j = 0; j < n; ++j)
     {      A[n-2][j] = A[n-1][j] = 0.0;
            A[j][j] = 1.0;                  // add unit matrix
     }
}
```

Applied to the matrix studied in the last section, namely,

$$\begin{pmatrix} 4 & 1 & -2 & 2 \\ 1 & 2 & 0 & 1 \\ -2 & 0 & 3 & -2 \\ 2 & 1 & -2 & -1 \end{pmatrix}$$

the Householder procedure produces the tridiagonal matrix

$$\begin{pmatrix} 4 & -3 & 0 & 0 \\ -3 & 3.3333 & -1.6667 & 0 \\ 0 & -1.6667 & -1.3200 & 0.9067 \\ 0 & 0 & 0.9067 & 1.9867 \end{pmatrix}$$

(where we have expressed the result in full matrix form for ease of inspection).

This, however, is not sufficient. We need to contruct the complete transformation matrix as a product of Householder matrices. The vectors $\mathbf{w}$ are stored in the rows of A. From these vectors we have to form the elements of the Householder matrices $P_k = (I - 2\mathbf{w}\mathbf{w}^T)$, where the subscript k here refers to the row of A in which the particular vector $\mathbf{w}$ is stored. Starting with a unit matrix, I, the complete Householder transformation matrix will then be given by $(\prod_{k=0}^{n-3} P_k)I$. At each stage stage we shall need to form the matrix A', where

$$A'_{ij} = \sum_m (\delta_{im} - 2\mathbf{w}_i\mathbf{w}_m)A_{mj}.$$

Again, in order to save memory, it is desirable that the product matrix PA be stored in the matrix A, otherwise there would be an enormous demand for memory in the case of a large matrix. Fortunately this can be done, because the part of A left unoccupied by the vectors $\mathbf{w}$ not yet used is at all stages sufficient, provided A is purged of each vector $\mathbf{w}$ as it is used. We can deal in this fashion with one column of A at a time, and the additional storage requirement is therefore for this column vector only—a very considerable economy.

In the case, for example, of a 4×4 matrix, A initially holds two **w** vectors off-diagonal, and its diagonal elements are all unity. Successive pre-multiplications by the corresponding Householder matrices produces the following sequence of shapes:

$$\begin{pmatrix} 1 & w & w & w \\ 0 & 1 & w & w \\ 0 & 0 & 1 & 0 \\ 0 & 0 & 0 & 1 \end{pmatrix} \rightarrow \begin{pmatrix} 1 & w & w & w \\ 0 & 1 & 0 & 0 \\ 0 & 0 & a & a \\ 0 & 0 & a & a \end{pmatrix} \rightarrow \begin{pmatrix} 1 & 0 & 0 & 0 \\ 0 & a & a & a \\ 0 & a & a & a \\ 0 & a & a & a \end{pmatrix},$$

where the symbols w and a are place markers rather than values. The pattern generalizes in the obvious fashion to matrices of order higher than 4.

A function for carrying out the above procedure may now be written as follows:

```
void  holder(matrix &A)
//    input is output matrix of procedure "house"
//    output is matrix of transformation to tridiagonal form
{     int n = A.getsize();
      int i, j;
      for (int k = n-3; k >= 0; --k)                    // retrieve w
      {     vector w = A[k];
            for (j = k+1; j < n; ++j) A[k][j] = 0.0;// free A[k]
            for (j = k+1; j < n; ++j)
            {     vector aj(n);                          // col j of A
                  for (i = k+1; i < n; ++i) aj[i] = A[i][j];
                  for (i = k+1; i < n; ++i)
            {     double aij = 0.0;
                  for (int m = k+1; m < n; ++m)        // (I - 2ww)A
                        {     double Pim =
                                    (i == m) ? 1.0 : 0.0;
                              Pim -= 2.0 * w[i] * w[m];
                              aij += Pim * aj[m];
                        }
                        A[i][j] = aij;
                  }
            }
      }
}
```

The heart of this function is the concise treatment of matrix multiplication.

Before passing on, it is worth mentioning that if A is a real *non*-symmetric matrix, it is not in general possible to find an orthogonal transformation that reduces it to tridiagonal form. It is always possible, however, to reduce it to *Hessenberg* form, namely an upper triangular matrix with a single sub-diagonal, or equivalently a lower triangular matrix with a single super-diagonal.

13.5.2 Francis diagonalization

The final step is to diagonalize the tridiagonal matrix obtained by Householder transformations. This may be done in a variety of ways, but the most efficient is to use a succession of planar rotations. In the case of a tridiagonal matrix, this is simple to carry out in principle, although complicated in detail. We shall confine ourselves to explaining the essentials of the method, which was proposed by Francis.

Any real symmetric matrix may be factored, $A = QU$, into an upper triangular matrix U premultiplied by an orthogonal matrix Q. We shall not prove this in general, but shall demonstrate it for the case where A is tridiagonal. Alternatively, the matrix A may be factored into a lower triangular matrix pre-multiplied by a different orthogonal matrix. The diagonalization method which results from the first approach is often called the *QR Method*, because when it was first proposed the term "right triangular" matrix was used instead of "upper triangular" which most authors prefer. Since we have already denoted the rotation matrices by R, we shall stick to U for the triangular matrices which occur in the method.

If A is a tridiagonal matrix, we may reduce its sub-diagonal elements to zero by premultiplication by a sequence of planar rotations. In the two-dimensional case, we may set

$$U = R^{T}A = \begin{pmatrix} c & s \\ -s & c \end{pmatrix} \begin{pmatrix} \alpha & \gamma \\ \gamma & \beta \end{pmatrix} = \begin{pmatrix} c\alpha + s\gamma & c\gamma + s\beta \\ 0 & -s\gamma + c\beta \end{pmatrix},$$

where, if $c = \cos\theta$, $s = \sin\theta$, we must take $t = \tan\theta = \gamma/\alpha$.

In general, the sequence of planar rotations R_{12}^{T}, R_{23}^{T}, ..., $R_{n-2,n-1}^{T}$, carried out in that order, will successively annihilate all sub-diagonal elements, but will create a *second* super-diagonal in the matrix U. If we define $Q = R_{12}R_{23}\ldots R_{n-1,n}$, and $A = QU$, then U will have the form:

$$\begin{pmatrix} \times & \times & \times & 0 & 0 & \ldots & 0 & 0 & 0 \\ 0 & \times & \times & \times & 0 & \ldots & 0 & 0 & 0 \\ 0 & 0 & \times & \times & \times & \ldots & 0 & 0 & 0 \\ \multicolumn{9}{c}{\cdots\cdots\cdots\cdots\cdots\cdots\cdots\cdots\cdots} \\ 0 & 0 & 0 & 0 & 0 & \ldots & \times & \times & \times \\ 0 & 0 & 0 & 0 & 0 & \ldots & 0 & \times & \times \\ 0 & 0 & 0 & 0 & 0 & \ldots & 0 & 0 & \times \end{pmatrix}$$

Now that we know both factors Q and U in the factorization $A = QU$, we may define a new matrix $A' = UQ$. It follows that $A' = Q^{-1}AQ$, where $Q^{-1} = Q^T = R_{n-2,n-1}^T \ldots R_{12}^T$, so that A' is similar to A and has the same eigenvalues.

We may also show that A' is symmetric if A is symmetric:

$$A'^T = (Q^{-1}AQ)^T = (Q^TAQ)^T = Q^TA^TQ = Q^{-1}AQ = A'.$$

Moreover, Q will be found to be upper Hessenberg: it is an upper triangular matrix which also possesses a non-zero principal sub-diagonal. The product UQ is therefore tridiagonal, and it is necessary to compute only the diagonal and subdiagonal elements of A'. The subdiagonal elements are on the whole of smaller magnitude than those of A.

Therefore, if we define a sequence $A^{(1)} = A = Q^{(1)}U^{(1)}$, $A^{(2)} = U^{(1)}Q^{(1)}$, and in general, for $k > 1$, $A^{(k)} = Q^{(k)}U^{(k)}$, $A^{(k+1)} = U^{(k)}Q^{(k)}$, we may hope that it converges to a diagonal matrix, which, being similar by induction to the original matrix A, has diagonal elements which are the eigenvalues of A.

The routine QR which follows carries out one step of this procedure for a symmetric tridiagonal matrix defined in terms of its diagonal vector $\mathbf{d}$ and its off-diagonal vector $\mathbf{e}$. It is convenient to define the rotation variables c and s by means of a subroutine in which special care has to be taken with the signs to ensure continuity at $t = -1$. We have chosen to define it in terms of the cotangent rather than the tangent, because we may find that $d (= \alpha) = 0$, whereas the case of $e = (\gamma =) 0$ will later be excluded:

```
void   sincos(double d, double e, double &c, double &s)
{      double t = d/e;                              // t = cot θ
       if (fabs(t) < 1.0)
             c = t*(s = sign(t)/sqrt(t*t + 1.0));
       else                                         // t = tan θ
       {     t = 1.0/t;
             s = t*(c = 1.0/sqrt(t*t + 1.0));
       }
}
```

Then the QR routine is given by:

```
void   QR(vector &d, vector &e)              // d,e from matrix A
{      int n = d.getsize();
       matrix Q(n);
       for (int r = 0; r < n; ++r)           // unit matrix Q
       {     for (int s = 0; s < n; ++s)
                   Q[r][s] = (s == r) ? 1.0 : 0.0;
       }
       vector f(n-2);                        // extra super-diagonal
       vector sube = e;                      // keep sub-diagonal
```

```
for (int p = 0; p < n-1; ++p)          // build up U
{       double c, s;                   // rotation parameters
        sincos(d[p], sube[p], c, s);   // remove orig sub-diag
        double dp = d[p], dp1 = d[p+1];
        double ep1, fp, fp1;
        if (p<n-2) ep1 = e[p+1];
        if (p<n-2) fp  = f[p];
        if (p<n-3) fp1 = f[p+1];
        double ep = e[p];
        d[p] = c*dp + s*sube[p];           // d'ₚ   =  cdₚ + seₚ
        e[p] = c*ep + s*dp1;               // e'ₚ   =  ceₚ + sd_{p+1}
        if (p<n-2) f[p] = c*fp + s*ep1;    // f'ₚ   =  cfₚ + se_{p+1}
        d[p+1] = -s*ep + c*dp1;            // d'_{p+1} = -seₚ + cd_{p+1}
        if (p<n-2) e[p+1] = -s*fp + c*ep1; // e'_{p+1} = -sfₚ + ce_{p+1}
        if (p<n-3) f[p+1] = c*fp1;         // f'_{p+1} =  cf_{p+1}
        for (int k = 0; k < n; ++k)        // build up Q
        {       double qkp = Q[k][p];
                double rkp = Q[k][p+1];
                Q[k][p]   = c*qkp + s*rkp;
                Q[k][p+1] = -s*qkp + c*rkp;
        }
}
for (p = 0; p < n-1; ++p)                  // form UQ = A'
{       double dp = d[p], dp1 = d[p+1], ep = e[p];
        d[p] = dp*Q[p][p] + ep*Q[p+1][p];
        e[p] = dp1*Q[p+1][p];
}
d[n-1] *= Q[n-1][n-1];
}
```

Successive applications of the routine QR produce a sequence like the following:

$$
\begin{pmatrix} 2 & 1 & 0 \\ 1 & 2 & 1 \\ 0 & 1 & 2 \end{pmatrix} \rightarrow \begin{pmatrix} 2.8000 & 0.7483 & 0 \\ 0.7483 & 2.3429 & 0.6389 \\ 0 & 0.6389 & 0.8571 \end{pmatrix} \rightarrow \begin{pmatrix} 3.1429 & 0.5594 & 0 \\ 0.5594 & 2.2484 & 0.1878 \\ 0 & 0.1878 & 0.6089 \end{pmatrix} \rightarrow \cdots
$$

the eventual eigenvalues being 2, $2\pm\sqrt{2}$.

The rate of convergence of this QR sequence is very slow, but it may be greatly speeded up by adoption of a suitable strategy. It is also very wasteful of memory. The strategy we shall adopt contains three elements. In the first place, if at any stage in the computation of a particular eigenvalue, an off-diagonal element should be sufficiently small the matrix splits into two diagonal submatrices, and the diagonalization process for that eigenvalue may thereafter be carried out on only one of the submatrices:

$$
A = \begin{pmatrix} B & . & 0 \\ \cdots\cdots\cdots & & \cdots\cdots\cdots \\ & . & \\ 0 & . & C \end{pmatrix}.
$$

Since, however, we expect the matrix to converge to diagonal form, such a split *must* happen at some stage. It will also happen to the submatrix in which the sought eigenvalue lies, and so on until the eigenvalue is found, or a 2×2 submatrix occurs which may be separately diagonalized—for example, by means of a single Jacobi planar rotation. In the above sequence it is apparent that the element e_1 is tending to zero as the estimated eigenvalue d_2 approaches its exact value $2-\sqrt{2} \approx 0.5858$. The other two eigenvalues may then be obtained directly from the remaining 2×2 block.

Indeed, a 2×2 block may be irreducible by the QR method. To see that this may be the case, consider the following submatrix A, which has eigenvalues ± 1, and the corresponding rotation Q:

$$
A = \begin{pmatrix} 0 & 1 \\ 1 & 0 \end{pmatrix}, \qquad Q = \begin{pmatrix} 0 & -1 \\ 1 & 0 \end{pmatrix}.
$$

It is easy to see that $A' = Q^{-1}AQ = -A$, so that successive QR transformations of A will never cause it to converge to a diagonal matrix.

Second, if it is possible to estimate a first approximation λ to any eigenvalue, then—as in the inverse power method—it is better to diagonalize the "shifted" tridiagonal matrix $A - \lambda I$. If λ_1 is the eigenvalue nearest to λ, and λ_2 is the next nearest, convergence depends on the ratio $|\lambda - \lambda_1|/|\lambda - \lambda_2|$, which—if λ is well chosen—may be quite small and the convergence correspondingly fast.

Third, there is no need for the auxiliary matrix Q in the QR routine, nor for the auxiliary vectors f and sube: all may be taken care of within the matrix A inherited from the Householder reduction to tridiagonal form.

Most of the additional complication arises from the splitting process, which, however, is at the heart of the method. But first we need an auxiliary function which detects an off–diagonal element close to zero, and thus determines where a split may take place:

```
int    split(int j, vector &e)
//     returns index of zero element, n-1 if none
{      int m, n = e.getsize();
       for (m = j; m < n-1; ++m)
             if (fabs(e[m]) < tolerance) break;
return m;
}
```

Using two further functions, `twoblock` and `qrblock`, to be defined, we may now put in place the splitting strategy just described:

```
void francis(vector &d, vector &e, matrix &A)
//    the QR method for tridiagonal matrices
//    on entry, A is output of holder
//    on exit, rows of A store eigenvecs, d the eigenvals, e rubbish
      int n = A.getsize();
      for (int j = 0; j < n-1; ++j)           // eigenvalues in turn
      {      int iter = 0;
             int m = split(j, e);
             while (m != j)
             {     if (m == j+1)               // ie 2x2 block
                   {     twoblock(d, e, A, j++);
                   }     else
                   {     if (++iter == maxiter)
                               error("won't converge!");
                         qrblock(d, e, A, j, m);
                         m = split(j, e);  // find new split
                   }
             }
      }
      A = A.transpose();                       // row eigenvecs for
                                               // output
}
```

The eigenvectors are produced as columns of the vector A; the matrix is transposed so that finally they are to be found in the corresponding rows for ease of output.

Let us consider `twoblock` first. We have to diagonalize a 2×2 submatrix by a single Jacobi rotation, taking account of the effect of that rotation on the other rows and columns of the matrix. It closely resembles the function `rotate` in 13.4, except that the vectors **d** and **e** now take the place of the matrix A, while A now takes the place of the transformation matrix R:

```
void  twoblock(vector &d, vector &e, matrix &A, int j)
{     int n = A.getsize();
      double g = (d[j] - d[j+1])/(2.0*e[j]);            // cot2θ
      double t = sign(g)/(fabs(g) + sqrt(g*g + 1.0)); // tanθ
      double c = 1.0/sqrt(t*t + 1.0);                   // cosθ
      double s = t*c;                                   // sinθ
      d[j]   += t*e[j];
      d[j+1] -= t*e[j];
```

```
        e[j] = 0.0;
        for (int k = 0; k < n; ++k)
        {       double akj =   c*A[k][j] + s*A[k][j+1];
                double bkj = -s*A[k][j] + c*A[k][j+1];
                A[k][j]     = akj;
                A[k][j+1] = bkj;

        }
}
```

It is in the function qrblock that we have to adopt the strategy of shifting the eigenvalue so as to speed up the convergence. The procedure francis finds each eigenvector in turn, indexed by j, and the corresponding eigenvector. If we are seeking the jth eigenvalue, then instead of trying to diagonalize the tridiagonal matrix A, we attempt to diagonalize the matrix $A - s_j I$ instead, where s_j is the shift for that eigenvalue. There are two common ways of choosing the shift, neither foolproof. The first is to set $s_j = d_j$; the second is to diagonalize the submatrix determined by d_j, d_{j+1}, e_j, and to choose the eigenvalue of that submatrix which is closest to d_j. We shall adopt the second strategy in qrblock, which is adapted from the routine **tlqi** in the book by Press *et al.*

```
void   qrblock(vector &d, vector &e, matrix &A, int j, int m)
{       int n = A.getsize();
        double t = tangent(d, e, j);
        double g = d[m] - d[j] + t*e[j];               // shifted
        double s = 1.0, c = 1.0, p = 0.0;
        for (int i = m-1; i >= j; i--)
        {       double f = s * e[i], b = c * e[i];
                double r;
                if (fabs(f) >= fabs(g))
                {       c = g/f;
                        r = sqrt(c*c + 1);
                        e[i+1] = f*r;
                        c *= (s = 1.0/r);
                }       else
                {       s = f/g;
                        r = sqrt(s*s + 1.0);
                        e[i+1] = g*r;
                        s *= (c = 1.0/r);
                }
                g = d[i+1] - p;
                r = (d[i] - g)*s + 2.0*c*b;
                p = s * r;
                d[i+1] = g + p;
                g = c * r - b;
```

```
              // form eigenvector
              for (int k = 0; k < n; ++k)
              {       double aki = A[k][i];
                      double bki = A[k][i+1];
                      A[k][i]    = c*aki - s*bki;
                      A[k][i+1]  = s*aki + c*bki;
              }
      }
      d[j] = d[j] - p;                                // remove shift
      e[j] = g;
      e[m] = 0.0;
}
```

It is a very compressed function, and it works from the bottom up in order to save rounding errors. It is also very fast. Consequently, the function `francis` diagonalizes a tridiagonal matrix to full precision with very few *QR* iterations for each eigenvalue.

Finally, we can bring together the Householder and QR methods to provide for the complete diagonalization of a real symmetric matrix:

```
void  hqr(matrix &A, vector &d)
//      on entry A holds real symmetric matrix
//      on exit eigenvalues in d, eigenvectors as rows of A
{       int n = A.getsize();
        vector e(n);
        house(A, d, e);
        holder(A);
        francis(d, e, A);
}
```

This method is usually preferred over the Jacobi method because it involves many fewer iterations. Many other methods are also available, however, not least for the case of degenerate eigenvalues which we have ignored; for instance, in the EISPACK package developed by the Argonne National Laboratory.

14 Interpolation and data fitting

Given a set of $n + 1$ numbers y_i defined at $n + 1$ *nodes* x_i, where $x_0 < x_1 < \ldots < x_n$, it is required to find a value y which may plausibly be said to approximate the exact value, if it were known, corresponding to the point x. The given values may be from a table, or may result from an expensive piece of computation, or they may be the result of experimental observations. Indeed, we may be interested in the latter case to discover a plausible functional relationship between x and y which would enable us, over some range of the variables concerned, to write $y = f(x)$, to some approximation. If the value of x lies in the range $x_0 < x < x_n$ of known values, we are seeking an *interpolated* value of y; if it lies outside this range, we seek instead an *extrapolated* value. For reasons that will become apparent, it is very much more difficult to extrapolate reliably than it is to interpolate. For the moment it is perhaps sufficient to observe that, even for interpolation, there must be an assumption that any functional relationship between y and x is not only continuous but sufficiently slowly varying, otherwise there is no reason to suppose, given x, that y has any particular value (*cf* 3.11).

There are essentially two approaches. The first is to find a function within some class of functions which fits the known points exactly, that is to say, a function f which satisfies $y_i = f(x_i)$ for each i, and to suppose that the required value of y is given approximately by $y = f(x)$. The second is to find a function, again from some class of functions, which in some sense to be determined fits the known points as well as possible, and then to interpolate using this function. The second method is particularly useful when analysing experimental data, the form of which is already thought to be known in advance (except perhaps for the values of certain parameters), but where some experimental error is superimposed. We study these two approaches in turn.

We are encouraged to try to fit a polynomial to the given points by

Weierstrass's theorem:

> If the function $f(x)$ is defined and continuous in the interval $x \in [a, b]$, and $\varepsilon > 0$ is given, then there exists a polynomial $p(x)$ defined over the same interval, for which $|f(x) - p(x)| < \varepsilon$ for all $x \in [a, b]$.

Suppose, then, that x_i are $n + 1$ distinct nodes at which a set of numbers y_i is given. We seek a polynomial,

$$P_n(x) = a_n x^n + a_{n-1} x^{n-1} + \ldots + a_1 x + a_0,$$

of degree at most n, for which $P_n(x_i) = y_i$, where $i = 0,1, \ldots, n$. If such a polynomial exists it is unique. For suppose there are two distinct polynomials, P and Q, of degree at most n, satisfying the above requirements; and consider their difference, which itself must be a polynomial of degree at most n:

$$D_n(x) = P_n(x) - Q_n(x).$$

Because P and Q are equal at the $n + 1$ nodes x_i, it follows that D must have $n + 1$ zeros, which is contrary to the supposition that it is of degree at most n. The contradiction is resolved if and only if D is identically zero, in which case P and Q are the same polynomial.

We note in passing that the coefficients a_i satisfy the system of equations,

$$\begin{pmatrix} 1 & x_0 & x_0^2 & \cdots & x_0^n \\ 1 & x_1 & x_1^2 & \cdots & x_1^n \\ \cdots\cdots\cdots\cdots\cdots\cdots \\ 1 & x_n & x_n^2 & \cdots & x_n^n \end{pmatrix} \begin{pmatrix} a_0 \\ a_1 \\ \cdot \\ a_n \end{pmatrix} = \begin{pmatrix} y_0 \\ y_1 \\ \cdot \\ y_n \end{pmatrix},$$

where the matrix is known as a *Vandermonde matrix*: it has an inverse that is known analytically, but we shall not use it here. We are interested not so much in the coefficients of the polynomial as in its value at a particular point.

14.1 Lagrangian interpolation

The straight line joining the points (x_0, y_0), (x_1, y_1) in the euclidean plane may be written:

$$(y - y_0)/(y_1 - y_0) = (x - x_0)/(x_1 - x_0).$$

Linear interpolation for a value $x \in I[x_0, x_1]$ (*cf* 3.7 for notation) is obtained by taking the corresponding value of y calculated from this formula. This is the interpolation procedure used every day when rough and ready responses are called for.

The generalization we study first is to fit a polynomial of degree n to the given points, thus utilizing all the available information, before proceeding to find the interpolated value. Note, however, that the straight line above may also be written

$$y = \frac{x - x_1}{x_0 - x_1} y_0 + \frac{x - x_0}{x_1 - x_0} y_1;$$

or

$$y = P(x) \equiv L_0(x)y_0 + L_1(x)y_1,$$

where the elementary polynomials $L_i(x)$ are given by

$$L_0(x) = (x - x_1)/(x_0 - x_1), \text{ and } L_1(x) = (x - x_0)/(x_1 - x_0),$$

and have the values

$$L_0(x_0) = 1, \ L_0(x_1) = 0, \ L_1(x_0) = 0, \ L_1(x_1) = 1.$$

This is Lagrange's form of the linear interpolating polynomial, and it is this that we now seek to generalize.

Let
$$L^{(n)}_k(x) = \prod_{0 \le i \ne k \le n} (x - x_i)/(x_k - x_i),$$

which is a set of $n + 1$ polynomials of degree n having the particular values:

$$L^{(n)}_k(x_j) = 1, \ j = k$$
$$= 0, \ j \ne k.$$

Then the Lagrange interpolating polynomial of degree n is defined to be:

$$P(x) = \sum_k L^{(n)}_k(x) y_k.$$

Clearly, $P(x_i) = y_i$ by construction, and the interpolated estimate of the value of y at the point x is $P(x)$.

14.1.1 Lagrangian algorithm

In the following routine, the values y_i are stored in the vector **yv** and the corresponding values of the nodes x_i in the vector **xv**. The inner loop computes $L^{(n)}_j(x)$, and the outer loop sums the product of this with y_j. The routine returns this sum, being the value of the interpolating polynomial at the point x:

```
double lagrange(double x, vector &xv, vector &yv)
{       double y = 0.0;
      int n = xv.getsize();
      for (int i = 0; i < n; ++i)
      {       double li = 1.0;
            double xi = xv[i];
            for (int j = 0; j < n; ++j)
            if (j != i)
            {       double denom = xi - xv[j];
                  if (fabs(denom) < lagrange_toosmall)
                        error("coincident nodes in lagrange()");
                  li *= (x - xv[j])/denom;
            }
            y += yv[i]*li;
      }
      return y;
}
```

14.1.2 Errors of interpolation

We now have to estimate the error of interpolating a table of values to find an intermediate value. Suppose we are given the values of the function $f(x) = 1/x$ at a number of values of x:

x	$=$	2	3	4	5
$f(x)$	$=$	0.5	0.333333	0.25	0.2

and we wish to estimate its value at $x = 3.5$. We shall do the work by hand. Since $x = 3.5$ lies between $x = 3$ and $x = 4$, linear interpolation between these two values gives $f(3.5) \approx (1/3 + 1/4)/2 = 0.2917$, as compared with the true value $f(3.5) = 0.2857$.

Now let us use a polynomial of degree 2 fitted to the points $x = 2, 3, 4$, and alternatively to the points $x = 3, 4, 5$. In the first case, we have

$$L_0(x) = \tfrac{1}{2}(x-3)(x-4), \quad L_1(x) = -(x-2)(x-4), \quad L_2(x) = \tfrac{1}{2}(x-2)(x-3),$$

so that
$$L_0(3.5) = -0.125, \quad L_1(3.5) = 0.75, \quad L_2(3.5) = 0.375,$$

and
$$f(3.5) \approx -0.125 \times 0.5 + 0.75 \times 0.333333 + 0.375 \times 0.25 = 0.2813.$$

In the second case,

$$L_0(x) = \tfrac{1}{2}(x-4)(x-5), \quad L_1(x) = -(x-3)(x-5), \quad L_2(x) = \tfrac{1}{2}(x-3)(x-4),$$

so that
$$L_0(3.5) = 0.375, \quad L_1(3.5) = 0.75, \quad L_2(3.5) = -0.125,$$

and
$$f(3.5) \approx 0.375 \times 0.333333 + 0.75 \times 0.25 - 0.125 \times 0.2 = 0.2875.$$

We use all the information available to us if we try a polynomial of degree three. In that case we have

$$L_0(x) = -(1/6)(x-3)(x-4)(x-5), \quad L_1(x) = (1/2)(x-2)(x-4)(x-5),$$
$$L_2(x) = -(1/2)(x-2)(x-3)(x-5), \quad L_3(x) = (1/6)(x-2)(x-3)(x-4),$$

so that

$$L_0(3.5) = -0.0625, \quad L_1(3.5) = 0.5625, \quad L_2(3.5) = 0.5625, \quad L_3(3.5) = -0.0625,$$

and $f(3.5) \approx -0.0625 \times 0.5 + 0.5625 \times 0.3333 + 0.5625 \times 0.25 - 0.0625 \times 0.2 = 0.2844$.

The errors of these four interpolations are 0.0060, −0.0045, 0.0018, −0.0013, respectively. These are the actual errors we now have to estimate *a priori*.

Let $P(x)$ represent a Lagrangian interpolating polynomial of degree n, and if $t \in I[x_0, \ldots, x_n]$, define

$$g(t) \equiv f(t) - P(t) - [f(x) - P(x)] \prod_{0 \le i \le n} (t - x_i)/(x - x_i),$$

where $x \in I$. By construction, this function is zero when $t = x_k$ for all $k \in [0, n]$. It is also zero when $t = x$. Therefore $g(t) = 0$ at the $n + 2$ points $x, x_0, \ldots, x_n$ of the interval I. By a simple extension to Rolle's theorem it follows that there exists a number $\xi(x) \in I$ such that $g^{(n+1)}(\xi) = 0$. Thus:

$$0 = g^{(n+1)}(\xi) = f^{(n+1)}(\xi) - P^{(n+1)}(\xi) - [f(x) - P(x)](d^{n+1}/dt^{n+1}) \prod_i (t - x_i)/(x - x_i) \big|_{t=\xi}.$$

But $P^{(n+1)}(\xi) \equiv 0$, because P is a polynomial of degree n, and $(d^{n+1}/dt^{n+1}) \prod(t-x_i) = (n+1)!$.

Therefore,
$$f(x) = P(x) + \frac{f^{(n+1)}(\xi)}{(n+1)!} \prod_{i=0}^{n} (x - x_i).$$

It is interesting to compare this result with the corresponding error term for a Taylor expansion about the single point x_i, for which:

$$f(x) = \sum_{k=0}^{n} \frac{f^{(k)}(x_i)}{k!} \cdot (x - x_i)^k + \frac{f^{(n+1)}(\xi)}{(n+1)!} \cdot (x - x_i)^{n+1}.$$

The error terms are very similar except that the product

$$\Psi_n(x) = \prod_{i=0}^{n} (x - x_i)$$

of all the $(n + 1)$ differences in the Lagrange approximation is replaced in the Taylor approximation by the power $(x - x_i)^{n+1}$ of one of them. The Taylor series may be a good approximation near a single node, but it rapidly gets worse as the difference from that chosen node increases. This is not surprising since it uses only information derived from a single node. The Lagrangian polynomial, on the other hand, remains a reasonable approximation through the central part of the interval $I[x_0, \ldots, x_n]$, where there are several nodes to keep Ψ small. Near the extremities, however, the value of Ψ may be quite large, and the approximation correspondingly poor.

To estimate the magnitude of the errors in the case being studied, where $f(x) = 1/x$, we note that $f^{(k)}(x) = (-1)^k k!/x^{k+1}$, which has its maximum magnitude at the lower end x_0 of the interval spanned by the polynomial. Therefore,

$$\underset{\xi \in I}{\text{Max}} \left| f^{(n+1)}(\xi)/(n + 1)! \right| = 1/x_0^{n+2},$$

and a bound on the magnitude of the error in this case is given by

$$E_{bnd} = (1/x_0^{n+2})\,|\Psi_n(x)|.$$

This immediately explains why the first quadratic interpolation, with $x_0 = 1$, was so much worse than the second with $x_0 = 2$.

The following table compares the actual errors with this error bound. It will be seen that it is not a very good estimate of the actual error $E_{act} = f(x) - P(x)$. Of course, better values could be obtained with intervals between points of, say, 0.2 instead of 1.0.

n	x_i	$(1/3.5)_{est}$	$|E_{act}|$	E_{bnd}
1	3 4	0.2917	0.0060	0.0093
2	2 3 4	0.2813	0.0045	0.0234
2	3 4 5	0.2875	0.0018	0.0046
3	2 3 4 5	0.2844	0.0013	0.0176

It is also interesting to see how the error varies with x. It might be expected that the polynomial fit would be most accurate near the middle of the interval, and this is so. For the third degree polynomial:

$x =$	1.0	1.5	2.0	2.5	3.0	3.5	4.0	4.5	5.0	5.5	6.0
$E =$	0.200	0.037	0	−0.003	0	0.001	0	−0.002	0	0.010	0.033

A polynomial fit is normally hopeless for extrapolation, as we see in this case. It is also clear why the above estimates of the error of interpolation tend to be too large. The "fine structure" of the error is primarily determined by the function $\Psi(x)$.

Lagrangian interpolation is particularly poor when the function being interpolated, though continuous, is insufficiently smooth. Consider interpolating between points in the interval $[-1, 1]$ on which is defined the triangle function:

$$\begin{aligned} T(x) &= 0, & |x| &> 0.5, \\ &= (1 - 2|x|), & |x| &\le 0.5. \end{aligned}$$

It is an even function, so only polynomials of even degree need be considered. Clearly, P_2 will not be very good, being only quadratic; in an attempt to mimic the triangular form better we shall try polynomials of degree 6 and 10 also. We obtain the following values of the polynomials:

x =	0.0	0.1	0.2	0.3	0.4	0.5	0.6	0.7	0.8	0.9	1.0
T =	1.0	0.8	0.6	0.4	0.2	0.0	0.0	0.0	0.0	0.0	0.0
P_2 =	1.00	0.99	0.96	0.91	0.84	0.75	0.64	0.51	0.36	0.19	0.00
P_6 =	1.00	0.92	0.72	0.43	0.15	−0.04	−0.08	0.06	0.30	0.41	0.00
P_{10} =	1.00	0.88	0.60	0.34	0.20	0.14	0.00	−0.19	0.00	0.92	0.00

As the degree of the polynomial increases, so the fit in the central portion of the interval improves; but it becomes very bad elsewhere. We have used equidistant nodes to obtain these interpolating polynomials. The position could be somewhat improved by inserting extra nodes in the extremities where $T(x)$ is zero, but the question arises whether interpolations using polynomials of successively higher degree will converge. It can be shown that there are smooth functions with smooth derivatives, such as $1/(1 + x^2)$, for which the sequence does not converge if the nodes are equidistant, although it is always possible to choose the nodes so that convergence results.

14.1.3 Neville's algorithm

There is an obvious difficulty with Lagrange's algorithm in practice. The degree of the interpolating polynomial required to attain a certain precision is not known until the computations have been done. Each time a new polynomial is tried the whole work has to be repeated, which is very wasteful of effort. This suggests that we should seek a recursive method which terminates when some specified error bound has been satisfied.

Suppose that the value of the function $f(x)$ is known at the points $x_0, x_1, \ldots, x_n$, and let $P_{a,b,\ldots,m}(x)$ be the polynomial that agrees with $f(x)$ at the points $x = x_a, x_b, \ldots, x_m$, respectively. Let $x_i, x_j \in [x_0, \ldots, x_k]$. Then it may easily be shown that

$$P_{r,\ldots,s}(x) = [(x - x_j)P_{r,\ldots,j-1,j+1,\ldots,s}(x) - (x - x_i)P_{r,\ldots,i-1,i+1,\ldots,s}(x)]/(x_i - x_j).$$

Setting $P_i = f(x_i)$, we can then create an array

$$
\begin{array}{c|llll}
x_0 & P_0 \\
x_1 & P_1 & P_{0,1} \\
x_2 & P_2 & P_{1,2} & P_{0,1,2} \\
x_3 & P_3 & P_{2,3} & P_{1,2,3} & P_{0,1,2,3}
\end{array}
$$

in which each successive term may be calculated from those already found. In this way we obtain for the function $f(k) = 1/x$:

$$
\begin{array}{c|llll}
x_0 = 2 & 0.5000 \\
x_1 = 3 & 0.3333 & 0.2500 \\
x_2 = 4 & 0.2500 & 0.2917 & 0.2813 \\
x_3 = 5 & 0.2000 & 0.2750 & 0.2875 & 0.2844
\end{array}
$$

The relationship of these numbers to the ones already obtained is obvious. It is, however, helpful to simplify the notation somewhat. Since we end up with a two-dimensional array, it is sufficient to use merely two indices. Let $Q_{i,j}(x)$, with $i \geq j$, denote the interpolating polynomial of degree j on the $(j + 1)$ points $x_{i-j}, x_{i-j+1}, \ldots, x_{i-1}, x_i$; that is to say:

$$Q_{i,j}(x) \equiv P_{i-j,i-j+1,\ldots,i-1,i}(x).$$

Then we may rewrite the recurrence relation in terms of the Q's to obtain

$$Q_{i,j}(x) = [(x - x_{i-j})Q_{i,j-1}(x) - (x - x_i)Q_{i-1,j-1}(x)]/(x_i - x_{i-j}).$$

The same array now becomes:

$$
\begin{array}{c|cccc}
x_0 & Q_{0,0} \\
x_1 & Q_{1,0} & Q_{1,1} \\
x_2 & Q_{2,0} & Q_{2,1} & Q_{2,2} \\
x_3 & Q_{3,0} & Q_{3,1} & Q_{3,2} & Q_{3,3}
\end{array}
$$

This is called Neville's method, and it is programmed below. The routine begins by finding the point nearest to that at which evaluation is desired, because interpolation is normally more accurate in that case.

```
int    nearest(double x, vector &v)
//     returns k of v[k] nearest x
{      int k = 0; int n = v.getsize();
       double diff = fabs(x - v[0]);
       for (int i = 1; i < n; ++i)
       {      double difi;
              if ((difi = fabs(x - v[i])) < diff)
              {      diff = difi; k = i; }
       }
       return k;
}

double neville(double x, vector &xv, vector &yv, double &err)
{      double y;
       int n = xv.getsize();
       vector c = yv, d = yv;
       int k = nearest(x, xv);
       y = yv[k--];
       double incr;
       // increases degree of polynomial to set value
       for (int m = 1; m < n; ++m)
       {      for (int i = 0; i < n-m; ++i)
              {      double xi  = xv[i] - x;
                     double xim = xv[i+m] - x;
                     double cd = c[i+1] - d[i];
                     double mult = xi - xim;;
                     if (fabs(mult) < neville_toosmall)
                     error("coincident nodes in neville()");
                     mult = cd/mult;
                     c[i] = xi*mult;
```

```
            d[i] = xim*mult;
        }
        incr = (2*k < (n-m-1) ? c[k+1] : d[k-]);
        y += incr;
    }
    err = fabs(incr);          // crude error estimate
    return y;
}
```

Notice that the routine makes an estimate of the error in terms of the last increment; it is usually not a very good estimate.

14.1.4 Divided differences

Suppose that we are fitting a polynomial of degree n to the function $f(x)$ at the $n + 1$ points $x_0, x_1, \ldots, x_n$; and define coefficients $p_0, \ldots, p_n$ by:

$$P(x) = p_0 + p_1(x - x_0) + p_2(x - x_0)(x - x_1) + \ldots + p_n(x - x_0)(x - x_1) \ldots (x - x_n).$$

Then,

$$P(x_0) = f(x_0) = p_0$$
$$P(x_1) = f(x_1) = p_0 + p_1(x_1 - x_0)$$
$$P(x_2) = f(x_2) = p_0 + p_1(x_2 - x_0) + p_2(x_2 - x_0)(x_2 - x_1)$$
$$\cdots\cdots\cdots\cdots\cdots\cdots\cdots\cdots\cdots\cdots\cdots\cdots$$

In terms of the divided differences introduced in 3.7, we have

$$P(x_0) = p_0 = f(x_0) = f[x_0]$$
$$P(x_1) = p_0 + p_1(x_1 - x_0) = f[x_1]$$
$$P(x_2) = p_0 + p_1(x_2 - x_0) + p_2(x_2 - x_0)(x_2 - x_1) = f[x_2]$$
$$\cdots\cdots\cdots\cdots\cdots\cdots\cdots\cdots\cdots\cdots\cdots\cdots$$

so that

$$p_0 = f[x_0]$$
$$p_1 = (f[x_1] - f[x_0])/(x_1 - x_0) = f[x_0, x_1]$$
$$p_2 = (f[x_1, x_2] - f[x_0, x_1])/(x_2 - x_0) = f[x_0, x_1, x_2]$$
$$\cdots\cdots\cdots\cdots\cdots\cdots\cdots\cdots\cdots\cdots\cdots$$

and

$$P(x) = f[x_0] + \sum_{k=1}^{n} f[x_0, x_1, \ldots, x_k](x - x_0)(x - x_1) \ldots (x - x_{k-1}).$$

An algorithm may readily be devised to obtain interpolated values in terms of these divided differences, and the reader may care to do so.

We may also use divided differences to express the error of interpolation. Let $P_1(x)$ be the polynomial which interpolates $f(x)$ at the nodes $x_0, x_1, \ldots, x_n$, and at an additional and distinct node t. Then,

$$P_1(x) = P(x) + (x - x_0)(x - x_1) \ldots (x - x_n) f[x_0, x_1, \ldots, x_n, t].$$

But $P_1(t) = f(t)$; therefore, the distributed error is given by

$$E(t) \equiv f(t) - P(t) = (t - x_0)(t - x_1) \ldots (t - x_n) f[x_0, x_1, \ldots, x_n, t].$$

Comparing this expression for the error with the earlier expression obtained in 12.1.2, we see that $f[x_0, x_1, \ldots, x_n, t] = f^{(n+1)}(\xi)/(n+1)!$, where $\xi \in I[x_0, x_1, \ldots, x_n, t]$.

14.1.5 Inverse interpolation

One often needs to use inverse interpolation. Given a table of values of the function $f(x)$ at certain nodes x_i, what is the value of x for which $f(x)$ has a certain value? The roles of x and y have to be interchanged, defining a set of nodes $y_i = f(x_i)$ corresponding to some of the given values of x, and proceeding as before.

It is somewhat more difficult to estimate the errors of interpolation in advance, however. We are attempting to interpolate values derived from the function

$$x = g(y) \quad \text{where } g(f(x)) = x,$$

which is often abbreviated by writing $g = f^{-1}$.

To estimate or set bounds upon the error, we therefore need the successive differentials of the function g, which may not be readily available. The errors are better estimated in computation, from successive divided differences, or otherwise.

14.2 Cubic spline piecewise interpolation

If interpolation of a function is required only over a relatively narrow range of the independent variable (or, in general, variables) over which the function is sufficiently smooth, Langrangian interpolation may suffice. We have seen, however, that large errors may build up near the extremities of the interval treated, especially if polynomials of high degree are required to match a locally "rough" function. Alternative methods exist where the interval is covered piecemeal. Of these, *cubic spline* interpolation is the most popular. It attempts to fit the interval by a succession of cubic polynomials chosen so as to fit the function exactly at the given nodes, but also to satisfy certain conditions of continuity.

Suppose we wish, as before, to interpolate the function $f(x)$, which is defined at the nodes x_i, where $0 \leq i \leq n$, the corresponding values being $y_i = f(x_i)$. In each interval $[x_i, x_{i+1}]$ let function f be interpolated by the cubic polynomial:

$$S_i(x) \equiv a_i(x - x_i)^3 + b_i(x - x_i)^2 + c_i(x - x_i) + d_i, \ \ 0 \le i < n.$$

Altogether, there are $4n$ coefficients, at present unknown.

Clearly, $S_i(x_i) = d_i = y_i = f(x_i), \ \ 0 \le i < n,$

leaving $3n$ coefficients still to be found.

We shall now demand continuity of the polynomial, and of its first and second derivatives, at the node points. Setting $h_i = x_{i+1} - x_i$, and defining $d_n = S_{n-1}(x_n) = y_n = f(x_n)$, we obtain the three sets of equations

$$d_i = a_{i-1}h_{i-1}^3 + b_{i-1}h_{i-1}^2 + c_{i-1}h_{i-1}h_{i-1} + d_{i-1},$$
$$c_i = 3a_{i-1}h_{i-1}^2 + 2b_{i-1}h_{i-1} + c_{i-1},$$
$$b_i = 3a_{i-1}h_{i-1} + b_{i-1},$$

where, in each case, $1 \le i \le n$, and we have defined $c_n = S_{n-1}'(x_n)$, $b_n = S_{n-1}''(x_n)/2$.

We can immediately solve the last of these for a_{i-1} in terms of b_i and b_{i-1}:

$$a_{i-1} = (b_i - b_{i-1})/3h_{i-1},$$

and this value may be substituted into the equations for c_i and d_i. Eliminating c_i and c_{i-1}, we finally obtain

$$b_{i-1}h_{i-1} + 2b_i(h_i + h_{i-1}) + b_{i+1}h_i = 3(d_{i+1} - d_i)/h_i - 3(d_i - d_{i-1})/h_{i-1},$$

where $1 \le i < n$. and the h_i and d_i are known. We now have a set of $n-1$ equations for the $n + 1$ unknowns b_i, where $0 \le i \le n$. We are therefore at liberty to impose two further conditions.

We choose to set $S_0''(x_0) = S_{n-1}''(x_n) = 0$, so that $b_0 = b_n = 0$, giving us the *natural cubic spline*, as it is called, and a set of tridiagonal equations for the unknown b_i. We could alternatively have set $S_0'(x_0) = f'(x_0)$ and $S_{n-1}'(x_n) = f'(x_n)$, the *clamped spline*, which would also have given us a set of tridiagonal equations, but we cannot always assume knowledge of $f'(x)$. It is also possible to impose cyclic boundary conditions, but then the equations are no longer tridiagonal. In what follows, we shall work with the natural spline. It follows that the fit will continue to be least good near the extremities of the interval.

14.2.1 Cubic spline routine

The main part of the interpolation routine using the natural cubic spline is the routine already developed for computing the solutions of tridiagonal equations (*cf* 10.6). We therefore begin with the two statements:

```
#include <tridiag.h>
double tridiag_toosmall = 1.0E-10;
```

We shall represent the nodes by a vector x and the corresponding function values by a vector y. Since we shall work in terms of the differences $x_{i+1} - x_i$ and $y_{i+1} - y_i$, it is useful to have an auxiliary function for computing intervals:

```
vector intervals(vector &v)
{       int n = v.getsize();
        vector h(n);
        for (int i = 0; i < n-1; ++i)
             h[i] = v[i+1]-v[i];
        h[n-1] = 0.0;       // to complete vector
        return h;
}
```

Note that we are working with n nodes rather than $n + 1$ in order to simplify the notation, although the number of nodes appears explicitly only in the user program. In terms of the function `intervals` we may define new vectors dx and dy. Then we may write a procedure to convert the equations for the b_i above into tridiagonal form:

```
void  nodestoeqns(vector &dx, vector &dy, tridiag &equ)
{       int n = dx.getsize();
        equ[0][L] = 0.0;
        equ[0][D] = 1.0;           // arbitrary non-zero
        equ[0][U] = 0.0;
        equ[0][B] = 0.0;
        for (int i = 1; i < n-1; ++i)
        {       equ[i][L] = dx[i-1];
                equ[i][D] = (dx[i] + dx[i-1])*2.0;
                equ[i][U] = dx[i];
                equ[i][B] = (dy[i]/dx[i] - dy[i-1]/dx[i-1])*3.0;
        }
        equ[n-1][L] = 0.0;
        equ[n-1][D] = 1.0;
        equ[n-1][U] = 0.0;
        equ[n-1][B] = 0.0;
}
```

These equations are solved by the standard procedure `triagonal(equ)` given in 10.6. The spline coefficients have then to be extracted from the solution, which is achieved by the following routine:

```
void  splinecoeffs(tridiag &equ, vector &dx, vector &dy,
                        vector &a, vector &b, vector &c)
{       int n = equ.getnum();
        for (int i = 0; i < n-1; ++i)
```

```
{        b[i] = equ[i][B];
         a[i] = (equ[i+1][B] - equ[i][B])/(dx[i]*3);
         c[i] = dy[i]/dx[i]-
                 (equ[i+1][B]+equ[i][B]*2)*dx[i]/3;
}
}
```

Finally, the interpolated value has to be calculated at the chosen point *x*. A function which returns this value follows:

```
double cubicspline(double x, vector &xv, vector &yv)
{        int n = xv.getsize();
         vector a(n), b(n), c(n);
         vector dx = intervals(xv), dy = intervals(yv);
         tridiag equ(n);
         nodestoeqns(dx, dy, equ);
         tridiagonal(equ);
         splinecoeffs(equ, dx, dy, a, b, c);
         // find appropriate segment
         for (int i = 0; i < n; ++i) if (x < xv[i]) break;
         // evaluate
         double s = x - xv[--i];
         double t = a[i]*s;
         t += b[i]; t *= s;
         t += c[i]; t *= s;
         t += yv[i];
         return t;
}
```

14.2.2 Lagrange polynomial and cubic spline compared

We consider first the "difficult" triangle function $T(x)$, taking nodes at $x = -1.0, 0.5, 0.0, 0.5, 1.0$. The natural cubic spline approximation $S(x)$ results in the table

x =	0.0	0.1	0.2	0.3	0.4	0.5	0.6	0.7	0.8	0.9	1.0
T =	1.0	0.8	0.6	0.4	0.2	0.0	0.0	0.0	0.0	0.0	0.0
S =	1.0	0.92	0.73	0.43	0.21	0.0	−0.12	−0.17	−0.14	−0.08	0.0
P =	1.0	0.88	0.60	0.34	0.20	0.14	0.00	−0.19	0.00	0.92	0.0

where we have also compared it with the approximation by a Lagrange polynomial P of degree 10. The bump near the extremity of the interval is much reduced.

The cubic spline is not necessarily a better approximation than a Lagrange polynomial, especially if the function being interpolated is smooth. The following table compares the natural cubic spline against a polynomial of degree 6 for the function $f(x) = 1/x$, with nodes at $x = 1.0(0.2)2.0$. The tabulation is for the intermediate values of *x*:

x	=	1.1	1.3	1.5	1.7	1.9
S	=	0.91257	0.76826	0.66694	0.58804	0.52679
P	=	0.90918	0.76921	0.66668	0.58822	0.52637
$1/x$	=	0.90909	0.76923	0.66667	0.58824	0.52632

It may be shown that if $h = \text{Max } h_i$, and if the function $f(x)$ is four times differentiable over the interval I over which the spline S is defined, then $|f(x) - S(x)| \leq Ch^4 f^{iv}(\xi)$, where $\xi \in I$ and C is independent of h. Clearly, if the available tabulation permits, the intervals should be chosen small enough to obtain interpolated values of high accuracy.

14.3 Data fitting

We turn now to consider the problem of representing experimental data, which in general will include some errors of measurement, by some formula, usually simple, which is suggested by the theory behind the phenomena being investigated. This is not a textbook on statistics, and we shall content ourselves with a very elementary presentation of the statistical content of the work.

In essence, an experiment returns a value y corresponding to the value x of some independent variable—the distance travelled by a body as a function of time, for example—but the values of y are measured with finite precision, so that there is a certain randomness in the values found. We shall assume that the values of x are known precisely, although in practice this will not usually be the case. In general, there may be several independent variables x, and several simultaneous measurements may be made of different variables y for each value of the variables x; we shall ignore this possible complication.

The experimenter has reason to believe, or wishes to postulate, that there should be a given functional relationship between y and x: $y = f(x; a_0, a_1, \ldots, a_m)$, for example, where the $m + 1$ parameters a_i are to be determined, if possible, and within experimental error, by the measurements. Measurements are performed at n distinct values of x which we shall represent as the components of a vector $\mathbf{x} = (x_0, x_1, \ldots, x_{n-1})$, and the measured values of y are represented as another vector $\mathbf{y} = (y_0, y_1, \ldots, y_{n-1})$. Let $\mathbf{f}$ represent the vector with components $f(x_i; a_0, \ldots, a_m)$. Then it seems natural to attempt to choose the parameters a_k in such a way that the norm $\|\mathbf{y} - \mathbf{f}\|_p$ is a minimum. As usual, the values $p = 1, 2, \infty$ provide the norms of interest, of which the euclidean norm $\|\cdot\|_2$ is the one of immediate concern, leading to the *least squares approximation*. The infinite norm $\|\cdot\|_\infty$ leads to the *minimax approximation* which we shall study more fully in 17.5.

In passing, recall from chapter 8 that if each measured value y_i is normally distributed with standard deviation σ_i (the standard deviation being the square root of the variance of a gaussian distribution), then minimization with respect to the parameters a_k of the quantity

$$\chi^2 = \sum_{i=0}^{m} [(y_i - f(x_i; a_0, a_1, \ldots, a_n))/\sigma_i]^2$$

provides the *maximum likelihood estimate* of the parameters. However, in this section we make the underlying assumption that the experimental errors are small, otherwise the problem becomes one of statistical estimation rather than merely interpolation.

14.3.1 Least squares approximation

We confine ourselves here to the case where the function to be fitted to a range of data can be represented by a polynomial of degree n,

$$P(x) = a_m x^m + a_{m-1} x^{m-1} + \ldots + a_1 x + a_0,$$

and we shall therefore wish to minimize the quantity

$$E = [\sum_i (y_i - \sum_k a_k x_i^k)^2]^{1/2}.$$

The minimum is obtained by setting $\partial E / \partial a_k = 0$, for each parameter a_k. This provides us with a set of $m + 1$ equations for the $m + 1$ parameters a_k, in terms of the n experimental quantities y_i measured at the n points x_i. We assume that $n > m + 1$, otherwise the result would be meaningless. These equations may be written

$$X\mathbf{a} = \mathbf{b},$$

where $\mathbf{a} = (a_0, \ldots, a_m)^T$, $\mathbf{b} = (b_0, \ldots, b_m)^T$, with $b_k = \sum_i y_i x_i^k$, and the matrix X has elements

$$X : x_{jk} = \sum_i x_i^{j+k}.$$

This matrix is closely related to the ill-conditioned Hilbert matrix (*cf* 11.3), so that we may hope to solve these equations for the minimizing values of the parameters a_k only if the polynomial is of low degree. Other methods must be used if this is not so. However, for reasonably low values of m we may use a standard procedure such as `crout` to obtain the coefficients a_k.

To construct a suitable routine we first need to have a function that returns an integer power of a floating point variable:

```
double power(double x, int k)
{       double t = 1.0;
        for (int i = 0; i < k; ++i)
                t *= x;
        return t;
}
```

For large values of the exponent more efficient functions may be devised, but the simple version will suit our present purpose.

We shall also need to evaluate the resulting polynomial in order to find how well it fits the measurements. We may adapt the routine `horner` introduced in 4.10.1:

```
double horner(vector &a, double x)
{       int k = a.getsize();                // k = m + 1
        double p = a[k-1];
        for (int i = k-2; k >= 0; --k)
        {       p *= x;
                p += a[i];
        }
        return p;
}
```

Methods exist for setting confidence limits on the values of the coefficients a_k and on the goodness of fit, but a full exposition would take us too far from our chief objectives. Using the function `horner`, the mean square deviation of the values y_i measured at the points x_i from the polynomial defined by the coefficients a_k, assumed known, is given by the function

```
double msqdev(vector &x, vector &y, vector &a)
{       int n = x.getsize();
        double sqrdev = 0;
        for (int i = 0; i < n; ++i)
        {       double diff = y[i] - horner(a, x[i]);
                sqrdev += diff * diff;
        }
        return sqrdev/n;
}
```

and if that is sufficiently small we shall be content.

We may now write down a function which constructs the matrix X and the vector **b**, finds the vector **a** of polynomial coefficients using the routine `crout`, and calculates the resulting root mean square deviation:

```
double rootmeansquare(vector &x, vector &y, vector &v)
//      on input v = b, on output a = v
{       int s = v.getsize();                // poly degree + 1
        int n = x.getsize();                // number of points
        matrix X(s);// construct matrix X
        for (int j = 0; j < s; ++j)
        {       for (int k = j; k < s; ++k)
                {       double Xjk = 0;
                        for (int i = 0; i < n; ++i)
                                Xjk += power(x[i], j+k);
```

```
                    X[j][k] = Xjk;
                    if (k > j) X[k][j] = Xjk;
            }
    }
    for (k = 0; k < s; ++k)            // construct vector v = b
    {       double bk = 0;
            for (i = 0; i < n; ++i)
                    bk += y[i] * power(x[i], k);
            v[k] = bk;
    }
    crout(X, v);                        // solve for a = v
    double msd = msqdev(x, y, v);
    return sqrt(msd);
}
```

The following data were obtained by adding a random contribution with variance 0.00001 to the exponential function e^x:

−1.0	−0.8	−0.6	−0.4	−0.2	0.0	0.2	0.4	0.6	0.8	1.0
.3636	.4552	.5458	.6698	.8152	.9984	1.2140	1.4894	1.8238	2.2223	2.7203

The results for a polynomial of degree 6 are compared with the Taylor coefficients $1/k!$ below:

$k = 0$	1	2	3	4	5	6
$1/k! = 1$	1	0.5	0.1667	0.0417	0.0033	0.0014
$a_k^{RMS} = 0.9960$	1.0030	0.5091	0.1430	0.0488	0.0319	−0.0119

The overall RMS deviation is 0.0025, corresponding to a variance 0.000006, consistent with the variance 0.00001 of each point. Of course, this is an extremely crude test of the goodness of fit. To do better, the reader should consult a book on statistical estimation.

14.3.2 Other approximations

The least square method of fitting a simple polynomial to a set of data is not always appropriate. As we have observed, the matrix X may not be well behaved, especially if polynomials of high degree are required. In that case improvement may be possible if orthogonal polynomials are used instead.

If $P_j(x)$ forms a set of polynomials of degree j such that at the n data points x_i there are *orthogonality relations*

$$\sum_i P_j(x_i)P_k(x_i) = 0, \quad j \neq k,$$
$$> 0, \quad j = k,$$

then it may be shown that the least squares approximation to the data y_i by a polynomial of degree at most m is defined by

$$p_m(x) = \sum_j c_j P_j(x),$$

where the coefficients c_j are given by $c_j = \sum_i y_i P_j(x_i)/\sum_i P_j^2(x_i)$.

The problem is transformed into the problem of finding polynomials having the required orthogonality properties.

Alternatively, it may be more appropriate to try to represent the data by means of a series of sines and cosines, as in Fourier analysis, or by some other functions suggested by the nature of the data. It is quite often the case that the data behave exponentially, perhaps with a succession of decaying exponentials, in which case it becomes natural to attempt linear fits to the logarithm of the data instead of the data itself.

Throughout this chapter we have been concerned with functions or data given at discrete points, seeking to interpolate between them, or trying to find best fits to discrete data with simple functional forms. In order to examine the errors involved we have had to consider the behaviour of the approximating polynomials as continuous functions. In later chapters we consider the approximation of functions in terms of orthogonal polynomials (chapter 17), and by means of Fourier series (chapter 21).

14.4 Sorting

We have tacitly assumed that the data are presented to us in an ordered fashion: for example, values of a function $f(x)$, or of a set of measurements y_i, obtained at an increasing sequence of given points x_i. Practice may be very different: the sequence of points at which measurements are made may even be random. In that case, before applying the methods of data analysis presented here, it may first be necessary to reorder the sequence of measurements. The resulting process is known as *sorting*, the theory of which forms an important part of computing science (for an authoritative account, see Volume 3 of the book by Knuth). We shall content ourselves here with a simple description of the process, followed by two examples of the more efficient sorting algorithms which may be useful in numerical applications.

Suppose we have an array A of integers, 8 in number,

$$A:\ 51\ \ 12\ \ 83\ \ 37\ \ 19\ \ 96\ \ 24\ \ 45$$

and we wish to rearrange them so that they form an increasing sequence. How do we do this in such a way that the amount of computing effort is not too great, especially if the number of elements is very much larger that 8?

There is a well-known process by which a card player orders a hand of cards. The second card is placed before the first card if that would be its correct relative position; otherwise it remains where it is. The third card is then placed in its correct position with respect to the first two; and so on, until all the cards are in order. The player is successively inserting an additional card into the sub-set of cards already ordered according to a preset convention. In computing science, this process is known as a *linear insertion sort*, or more colloquially as a *picksort*. Let us write it as a procedure for setting in rising order an array *A* of *n* integers:

```
void  picksort(int *A, int n)
{      for (int j = 1; j < n; ++j)       // choose next "card"
       {      int K = A[j];               // examine value
              int i = j - 1;
              while (i >= 0 && A[i] > K)   // make way for it
                     A[i+1] = A[i--];      // in correct position
              A[i+1] = K;                  // and insert it
       }
}
```

Applied to the above array, we obtain as desired:

$$12\ \ 19\ \ 24\ \ 37\ \ 45\ \ 51\ \ 83\ \ 96.$$

However, in practice our data may not simply be integers. In general, we shall have measured a number of variables at given values of some other variables, and we shall wish to order the results according to a regular sequence of one or more of these variables. For example, we might wish to show how the pressure and volume of a given mass of gas are related to a rising sequence of temperatures. In general, therefore, we must suppose that our data will be represented by a *structure* containing members of possibly different types, one of which we shall regard as the independent variable, or *key*, for the purpose of ordering. A great deal of computing time and memory might then be involved in effecting the rearrangements required by the above procedure. For that reason, we shall regard the data array as *fixed*, but define an auxiliary array of pointers, each element of which is in one-to-one correspondence with an element of the data array. Sorting thereafter takes place on the pointers instead of the data. It is convenient for purposes of exposition to continue to work with an array of integers, even though the data array may in practice be much more complicated. (The generalization to data of any type is obvious.) The following simple procedure sets up the desired correspondence for an array of integers,

```
void  makepointers(int *A, int **p, int n)
{      for (int i = 0; i < n; ++i)
              p[i] = &A[i];
}
```

and the data array may then be printed out in terms of the elements pointed to,

```
void  printarray(int **p, int n)
{     for (int i = 0; i < n; ++i)
            cout << *p[i] << tab;
}
```

We may now rewrite `picksort` in terms of pointers,

```
void  picksort(int **p, int n)
{     for (int j = 1; j < n; ++j)
      {     int *k = p[j];
            int i = j - 1;
            while (i >= 0 && *p[i] > *k)
                  p[i+1] = p[i--];
            p[i+1] = k;
      }
}
```

which achieves the same result as before when printed out using `printarray`.

However, the amount of computation required by `picksort` is clearly proportional to n^2, and this might well prove prohibitive for $n \geq 50$, say. It is perfectly adequate for a card player, but not for the analysis of extensive data. A great deal of effort has been devoted to finding processes which behave as some positive power of n less than 2, or as $n \log n$. The merits of so doing are clear from the following table:

n	10^2	10^3	10^4	10^5	10^6
$n \log_2 n$	6.6×10^2	10^4	1.3×10^5	1.7×10^6	2×10^7
n^2	10^4	10^6	10^8	10^{10}	10^{12}

The two procedures which follow are both $O(n \log n)$, at least for an array of "average" initial disorder; we shall come back to the latter point presently.

14.4.1 Quicksort

The fastest sorting method on most machines, for an average array with large n, is Hoare's Quicksort method. The idea is to choose any element of the array, say the first one, and by a sequence of interchanges find its final position in the sorted array by requiring every element before it to have lesser or equal value, and every element after it a higher one. In its final position, the chosen element *partitions* the array into two subarrays that may be separately sorted; and so on recursively.

We shall continue to represent the array by a pointer to an array of pointers int **p. Clearly, we must work with subarrays. A subarray of p will have its first element pointed to by p[low], say, and its last by p[high]. Let us suppose that we have found a suitable partition function which places the first element of the subarray in its final position q in that sub-array,

```
int q = partition(p, low, high);.
```

Then the complete sorting routine for that subarray may readily be written:

```
void qsort(int **p, int low, int high)
{       if (low < high)                 // otherwise nothing to do!
        {       int q = partition(p, low, high);
                qsort(p, low, q-1);             // sort lower subarray
                qsort(p, q+1, high);            // sort higher subarray
        }
}
```

Our problem, therefore, is to devise a suitable partition function. The workhorse will be the swap function

```
inline void swap(int **p, int i, int j)
{ int *tmp = p[i]; p[i] = p[j]; p[j] = tmp; }
```

In the following function the for loop, which is terminated by a break statement (*cf* 2.4.5), seeks the correct final position for the first element of the subarray, after which the two pointers are interchanged.

```
int partition(int **p, int low, int high)
{       int lo = low;
        int hi = high + 1;
        int *k = p[low];
        for (;;)
        {       while (*p[++lo] <= *k) ;                // find new lo
                while (*p[- -hi] > *k) ;                // find new hi
                if (lo < hi) swap(p, lo, hi);
                else break;
        }
        swap(p, low, hi);                               // set low in place
        return hi;
}
```

Finally, we must initialize qsort:

```
inline void quicksort(int **p, int n) { qsort(p, 0, n-1); }
```

To understand just how the quicksort process works we have indicated where internal output may be put in order to obtain intermediate printouts at each level of the recursion. We obtain the following output for the array considered before:

```
initial      51  12  83  37  19  96  24  45

             24  12  45  37  19  51  96  83
             19  12  24  37  45
             12  19
             37  45
             83  96

sorted       12  19  24  37  45  51  83  96
```

where, however, we have marked in italics at each level of recursion the first element of each sub-array which is to be placed in its correct position.

This routine is indeed very fast, of order $n \log n$, for an array initially containing an average amount of disorder. But in the worst case it is of order n^2; and, somewhat perversely, the worst case is the array that is already in order! As it stands, quicksort effectively cannot be used for an array that is already highly ordered. One remedy is to ensure that the array is sufficiently disordered to begin with, which may be done by including a small random number generator (*cf* 7.4) in the initialization of quicksort, which introduces disorder into the initial pointer assignments. Schematically,

```
inline void randquicksort(int **p, int n)
{ randomize(p); qsort(p, 0, n-1); }
```

although we shall not elaborate. A second remedy transfers the n^2 behaviour from the already ordered array to some other array not so likely to occur in practice. This was first implemented by Singleton, and it is of sufficient interest that we present it in full.

In the version of quicksort given above, we first found the correct final position of the *initial* element of the array, put it there, and by a sequence of swaps ensured that all the elements below it had a lesser or equal value, and all those above a larger one. It is the choice of the initial element which leads to the n^2 behaviour of the already ordered array. There is no need to choose the initial element: even a random element could be chosen. The choice made by Singleton, however, was the *median* element. In terms of array indices, instead of setting

```
int k = A[low];
```

he set

```
int q = (low + high)/2;
int k = A[q];
```

With this choice, one merely has to put this element in its final place, scanning through the array as before, except that `lo` skips element `q` instead of element `low`. This is a fine opportunity to make use of the `continue` statement (*cf* 2.4.5).

There is no need to alter the function `qsort`; it is `partition` that must be reworked:

```
int    partition(int **p, int low, int high)
//     Singleton's variant
{      int lo = low - 1; `
       int hi = high + 1;
       int q = (low + high)/2;
       int *k = p[q];
       for (;;)
       {      for (;;)
              {      if (++lo == q && lo < high) continue;
                     if (*p[lo] > *k || lo == high) break;
              }
              while (*p[- -hi] > *k) ;
              if (lo < hi) swap(A, lo, hi);
              else break;
       }
       swap(p, q, (lo + hi)/2);
       return hi;
}
```

14.4.2 Heapsort

A *heap* may be defined as a set of n numbers a_i, $0 \le i < n$, which satisfy the relations:

$$a_j \ge a_{2j+1}, \ a_j \ge a_{2j+2}, \ \text{for } 0 \le j < n/2.$$

If we arrange the numbers in the form of a *binary tree* according to index:

$$0$$

$$1 \qquad 2$$

$$3 \quad 4 \quad 5 \quad 6$$

and so on, then these relations ensure that a_0, the largest of them, is at the top of the tree; and that at a given level in the tree, each node is greater than each of its two "progeny" in the level below.

This is the basis of the sorting algorithm *heapsort* proposed by Williams. An arbitrary array may be efficiently converted into a heap using a sifting function later discovered by Floyd:

```
void   floyd(int **p, int lo, int hi)
{      int i = lo, j;
       while ((j = 2*i+1) <= hi)
       {      if (j < hi && *p[j] < *p[j+1]) j++;
              if (*p[i] < *p[j])
              {      swap(p, i, j);
                     i = j;
              }      else break;
       }
       // include optional internal printout here
}
```

where we have again indicated where internal output may be provided in order to understand how the whole sorting process works. Successive application of this function, beginning with lo = i = (n-1)/2, and reducing *i* by one each time, constructs the heap.

The floyd procedure may also be used successively to extract the maximum element from the heap and to construct a new heap from what remains. This produces the elements in the sorted order.

Using floyd in both contexts, the heapsort algorithm may be written:

```
void   heapsort(int **p, int n)
{      for (int i = (n-1)/2; i > 0; --i)
              floyd(p, i, n-1);                // construct heap
       // provide internal printout if desired
       for (i = n-1; i > 0; --i)
       {      floyd(p, 0, i);                  // extract maximum
              swap(p, 0, i);
       }
}
```

With internal printout included in heapsort as well as floyd, the output for the same array as before is:

```
initial      51  12  83  37  19  96  24  45

construct    45  19  96  24  37
heap         96  45  19  83  24  37
             45  96  37  19  83  24  12
             51  45  96  37  19  83  24  12

heap         96  45  83  37  19  51  24  12
```

```
extract      83   45   51   37   19   12   24
top          51   45   24   37   19   12
             45   37   24   12   19
             37   19   24   12
             24   19   12
             19   12

sorted       12   19   24   37   45   51   83   96
```

Although in this example heapsort may seem slower in execution than `quicksort`, it is in fact only a little slower for large n, and it has the great advantage that its worse case is still only $O(n \log n)$. This is why it is often the preferred sorting method.

The reader may care to write a recursive version of the function `floyd`.

15 Graphics

At this point it is convenient to introduce the elements of graphics, so that functions may be represented graphically and the screen mapped. As every teacher knows, a simple diagram is better than a thousand learned words—and no more or less misleading! Of course, many ready-made graphics packages exist, and it may well be that the reader's preferred C++ implementation operates under systems software which includes such a package. For those who dislike black boxes, however, it is helpful to have some idea how such packages work, at least in the simple context of this book; and for those who do not have convenient access to a graphics package, the present chapter provides some remedy.

However, a word of caution is called for. As we explained in chapter 1, input and output in C++ are not part of the language, and to a greater or lesser extent are implementation dependent. At the time of writing, this is especially so in graphics, where PC implementations are likely to be somewhat idiosyncratic. In this book we have used TurboC++, although features peculiar to that implementation have been suppressed wherever possible. Implementation dependent elements inevitably crept in when we considered streams in 6.3. In this chapter we shall write a simple graphics package which may be freely used, within its limitations; but it cannot be written without reference to the chosen implementation. Therefore, we have drawn upon a number of TurboC++ library functions, especially graphics functions. Whenever we have done so, we have tried to explain exactly what facilities these functions provide. Other implementations may provide similar functions, and the real enthusiast may care to note that the functions we call upon can all, within the limitations of this book, be provided quite simply in assembly language.

Before entering the world of graphics, however, it is instructive first to learn how to draw a rough graph of a function in text mode, for which we already have most of the required apparatus. This will then be refined in the much more sophisticated graphics mode. Later we shall show how to map the screen in various interesting ways. The techniques we shall introduce are useful in many applications; they are almost indispensable for studies in non-linear dynamics.

15.1 Text mode

We shall first develop a procedure for displaying in text mode a rough graph of a function $f(x)$ over some interval $[a,b]$. We have to perform a mapping of the function onto the screen, with a resolution imposed by the graphical limitations of text mode.

There are two distinct approaches. In the first, the screen is represented by a two-dimensional array of characters

```
char screen[rows][cols];
```

where `cols, rows` are the number of characters in a line, and the number of lines, respectively, permitted by the screen. Thus, `screen[j][i]` is the *i*th character in the *j*th line. In the standard PC monitor in text mode it is possible to show 80 characters in a line and 25 lines. Part of the screen, however, should be reserved to label the axes and for other purposes. Some of the elements of the array will contain the characters `HZ = '-'` or `VT = '|'`, representing axes. Values of the function may be represented by the character `FO = 'o'`. All other elements will be blank, represented by the character `SP = ' '`. The problem thus reduces to finding where these four characters appear as elements of the array.

Having determined the array completely it may be displayed using a fast print function which, following TurboC++, we call `putch(ch)`, and which prints the single character `ch` to screen:

```
for (int j = 0; j < rows; ++j)
{       for (int i = 0; i < cols; ++i)
               putch(screen[j][i]);
}
```

Note that if a function such as `putch` is not available, we may replace the function call `putch(ch);` with the output stream statement `cout << ch;`, but there may be some loss of speed due to the administrative overhead of using the full, and here unnecessary, apparatus of streams.

Alternatively, a function `gotoxy(int x, int y)` may be used to position the cursor at whatever point on the screen the next character should go. Note, however, that in TurboC++, which provides this function, the extreme top-left character is at (1,1), not (0,0). This second approach is the simpler to understand, and it is the one we shall adopt.

In passing, note that `putch` and `gotoxy` provide us with all we need to write a function to clear the screen. It is fast enough for many purposes. It overwrites the screen with spaces from top left to bottom right, and then returns the cursor top left:

```
void    myclear()
{       int rows = 25, cols = 80;
        gotoxy(1,1);
        for (int j = 0; j < rows; ++j)
        {       for (int i = 0; i < cols; ++i)
                       putch(' ');
        }
        gotoxy(1,1);
}
```

The function $f(x)$ we wish to graph has to be calculated at values of x corresponding to the discrete positions at which characters may be placed upon the screen, and the values obtained have to be approximated so that they may be represented on the screen. Given the interval of x over which we desire to graph it, we have first to find the corresponding range of values covered by the function $f(x)$.

We represent the interval of x by [xlo, xhi]; and we subdivide it into NX−1 subintervals, each of length dx = (xhi − xlo)/(NX − 1), by an additional NX − 2 points. We shall spread the values of the function over NY lines. Allowing for a border, it is convenient to take NX = 60, NY = 21. The exact values of $f(x)$ at these NX points (including the end points) will be represented by an array:

```
double y[NX].
```

We have to calculate the elements of this array, and in the course of doing so we also compute the maximum and minimum values recorded in the array in order to determine the scale of the graph. (These may not be exactly the maximum and minimum values of the function because we are examining it only at discrete points, but they should be quite close if the function does not vary too rapidly.)

Let the maximum and minimum elements of y be denoted by yhi and ylo, respectively; then dy = (yhi − ylo)/(NY − 1) represents the interval in function values corresponding to the distance between lines on the screen. The x-axis occurs at row jax = 1+int(yhi/dy), and the screen point corresponding to the function value y[i] is the i-th element of row j = 1 + int((yhi − y[i])/dy). Naturally, the values representing x and y on the screen are integers, and are therefore approximate representations only, reflecting the very limited resolution of the screen in text mode.

The following procedure sketches the function $f(x)$ in a box determined by xlo, xhi, ylo, yhi, the values of which are printed at the extremities of the box, and with the x- and y- axes superimposed. The top-left corner of the box is offset at (OX, 0). The header file sketch.h is:

```
#ifndef SKETCH
#define SKETCH

      #include <standard.h>

      const int  OX   = 10;
      const int  NX   = 60;
      const int  NY   = 21;
      const char VT   = '|';
      const char HZ   = '-';
      const char FO   = 'o';

      const long l = ios::right+ios::fixed+ios::showpos+ios::unitbuf;
```

```
const   int p = 3;
const   int w = 8;

typedef double (*func)(double);

void        sketch(func);
```

```
#endif //SKETCH
```

It is a trivial matter to modify these definitions, and the procedure itself, if the function should contain additional parameters.

The procedure `sketch` now follows, based closely on the routine **scrsho** of Press *et al*. It begins with the definition of a constant HUGE, larger than anything likely to occur in the program. It is used to initialize `ylo` and `yhi`. Most of the procedure definition is contained within an unconditional loop denoted by `for(;;)`, indefinite repetition of which can be prevented by a `break` command on deliberately entering equal values of the parameters `xlo`, `xhi`. This loop is convenient because, on first plotting the function, the interval over which it should be considered may require some initial experimentation. As we have seen, there are more elegant ways of presenting such a loop, but this is the most concise.

```
const double HUGE = 1.0E+100;

void  sketch(func f)
{       format lpw(1, p, w);
        for (;;)
        {       double xlo, xhi;
                cout << "Enter xlo xhi (xlo=xhi to stop): ";
                cin >> xlo >> xhi;
                if (xlo == xhi) break;
                clear;
                // compute and store function and find extreme values
                double dx = (xhi-xlo)/(NX-1);
                double iax = 1 - int(xlo/dx); // position of y-axis
                double ylo = HUGE, yhi = -HUGE;
                double y[NX];                  // to store values of function
                int i, j;
                double x = xlo;
                for (i = 0; i < NX; ++i)       // fill y[], and find ylo, yhi
                {       y[i] = f(x);
                        if (y[i] < ylo) ylo = y[i];
                        if (y[i] > yhi) yhi = y[i];
                        x += dx;
                }
                if (yhi == ylo) yhi = ylo + 1.0;      // prevent catastrophe!
```

```
            double dy = (yhi-ylo)/(NY-1);
            // sketch box, and x-axis
            int jhi = 1, jlo = NY;
            int jax = 1 + (int)(yhi/dy);       // position of x-axis
            for (j = jhi; j <=jlo; ++j)        // verticals
            {     gotoxy(OX+1, j);
                  putch(VT);
                  if (iax > 1 && iax < NX)     // sketch y-axis
                  {     gotoxy(OX+iax+1, j);
                        putch(VT);
                  }
                  gotoxy(OX+NX, j);
                  putch(VT);
            }
            for (i = 1; i < NX-1; ++i)         // horizontals
            {     gotoxy(OX+1+i, jhi);
                  putch(HZ);
                  if (jax > 0 && jax < NY-1)    // sketch x-axis
                  {     gotoxy(OX+1+i);
                        putch(HZ);
                  }
                  gotoxy(OX+1+i, jlo);
                  putch(HZ);
            }
            // sketch function
            for (i = 0; i < NX; ++i)
            {     double yi = (yhi - y[i])/dy;
                  j = 1 + int(yi);
                  gotoxy(OX+1+i, j);
                  putch(FO);
            }
            // set scale values
            gotoxy(1, jhi);          cout << lpw << yhi;
            gotoxy(1, jax);          cout << lpw << 0.0;
            gotoxy(1, jlo);          cout << lpw << ylo;
            gotoxy(OX-5, jlo+1);     cout << lpw << xlo;
            if (iax > 1 && iax < NX)
            { gotoxy(OX+iax-5);      cout << lpw << 0.0; }
            gotoxy(X+NX-6, jlo+1);   cout << lpw << xhi;
            gotoxy(1, jlo+3);        // rest cursor
      }
}
```

Notice that scale values are indicated using formatted output as defined in Chapter 6, except that the enumeration there declared in `class ios` now includes an additional term

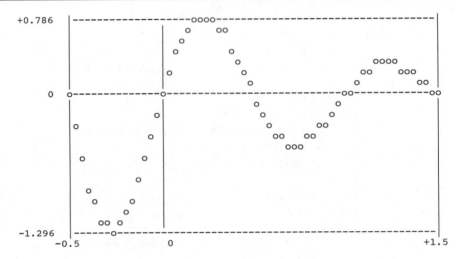

```
Enter xlo xhi (xlo=xhi to stop):
```

Figure 15.1 The function $F(x) = \sin(2\pi x)\exp(-x)$ in text mode.

`unitbuf = 0x2000`. If switched on (*ie*, included in the definition of `lpw`) this has the effect of flushing all streams after insertion has taken place. If this is not included, the normal output operator `<<` conflicts with the cursor placement function `gotoxy`. An alternative, since C++ is mostly compatible with C, is to replace all output operations by the appropriately formatted C library function `printf`.

A plot of the function $f(x) = \sin(2\pi x)\exp(-x)$ over the interval [−0.5, 1.5] as produced by the routine `sketch` is shown in Figure 15.1.

15.2 Graphics mode

This section is in no way a full exposition of the arts and crafts of graphics. It mostly confines itself to seeing how to plot functions with the additional precision available in graphics mode. As far as possible, the procedures we shall offer are straightforward developments of that used in text mode. Again we map a function $f(x)$ onto the screen, but in graphics mode we consider the screen to be composed of an array of *pixels*, the smallest units in a screen display that can be individually programmed. The number of pixels available depends on the monitor used; typically it may have 640×480 pixels, each of which may be individually programmed according to a palette of perhaps 16 colours. A pixel therefore has integer x and y coordinates, and another integer representing colour.

Working in graphics mode thus allows us to display figures and diagrams in colour, both on screen and in print, if appropriate facilities are available. This makes possible displays which are clearer, more attractive, and sometimes more dramatic, than would be the case in black and white. In this chapter, figures are printed in monochrome equivalents for ease of reproduction, but the programs themselves assume that colour is being used.

Let us then define pixel as a class:

```
class pixel {
private:
        int x, y;
        int c;
. . . . . . . . . . . . . . .
};
```

We have made the data members private because there is no need to access them directly in a user program.

A mathemical point (x,y), as we have seen in Chapter 11, may also be defined as a class, although we shall redefine it now. Since it is convenient that all members of the class `point` shall be public, so that they may be directly accessible to the user for mathematical manipulation, we might as well define `point` to be a structure, the members of which are `public` by default:

```
struct point {
        double x, y;
. . . . . . . . . . . . . . .
};
```

Notice that we have used `x` and `y` to identify data members in both classes: they are of type `int` in `pixel` and `double` in `point`. No confusion will result, however, because they can be accessed only with the aid of dot or pointer access operators. Thus, for example:

```
point   p(...);
pixel   s(...);
double  d = p.x;
int     i = s.x;    // in friend funcs only
```

15.2.1 Classes point and pixel

The heart of the problem of representing a mathematical function graphically is to establish two-way relationships between these two classes. This is the equivalent of what we had to do in text mode. We shall require the `point` equivalent of a `pixel` and the `pixel` equivalent of a `point`. Later we shall also wish to represent the complex plane, and we shall therefore wish to convert between `complex` and `point`. The class `point` will thus be declared:

```
struct point {                    // mathematical point
        double x, y;
        point()                   { x = 0; y = 0; }
        point(double a, double b) { x = a; y = b; }
        point(point &p)           { x = p.x; y = p.y; }
        point(complex &z)         { x = real(z); y = imag(z); }
```

```
          // image of pixel in box:
          point(pixel&, point&, point&);
};
```

There are five constructors. Only the last requires any special explanation: it denotes the point which corresponds to a given pixel. A function is displayed relative to a box of values of x and $y = f(x)$. We may define this box by stating the values of the points `tl = (xlo, yhi)` and `br = (xhi, ylo)` which are to correspond to the top-left and bottom-right pixels of the screen, respectively, suitably offset by a border. The point represented by the `pixel` s must therefore be written `point p(s, tl, br)`. Before we can define the corresponding constructor we must evidently define the class `pixel`.

There are also several associated functions for input, output and presentation—which, however, do not have to be declared `friend` because the data members of `point` are `public`. Adequate input and output functions may be written:

```
ostream& operator<<(ostream &s, point &p)
{     s << "(" << p.x << ", " << p.y << ")";
      return s;
}

istream& operator>>(istream &s, point &p)
{     s >> p.x >> p.y;
      return s;
}
```

Let us now proceed to the declaration of the class `pixel`. All non-member functions that are to access the data members directly must in this case be declared `friend` since these members are `private`. The class may be declared as follows:

```
class pixel {                              // pixel image of point
friend point;
private:
      int x, y, c;
public:
      pixel()                              { x = y = c = 0; }
      pixel(int xx, int yy, int cc) { x=xx; y=yy; c=cc; }
      pixel(pixel &p)                      { x=p.x; y=p.y; c=p.c; }
      pixel(point &p, point &tl, point &br, int color);
      pixel& operator=(pixel&);
      void setcol(int col)                 { c = col; }
      int getcol()                         [ return c; }
      int getx()                           { return x; }
      int gety()                           { return y; }
      ..............                       // friend functions
};
```

There are four constructors. The first three are trivial, and are given inline. The assignment operator is similar:

```
inline pixel& pixel::operator(pixel &p)
{     x = p.x;   y = p.y;   c = p.c;
      return *this;
}
```

The fourth constructor constructs the pixel corresponding to the point p in a box defined by the points tl, br. We shall consider it presently. No destructor is required, because no pointers have been created with the operator new. There is a function setcol for changing the colour of a given pixel, and get functions for inspecting the private data members.

The header file stdgraph.h, reproduced in Appendix F, contains these class declarations in full, together with inline definitions, and declarations of self-contained functions for switching on and off graphics mode and defining certain constants, and for presentation of graphs. Included in this header file are instructions to include the header file standard.h of Appendix A, and all the header files in which the various TurboC++ library functions called upon by the package are to be found.

The header file also defines the type func in the simplest way consistent with our objective:

```
typedef double (*func)(double);
```

If this is modified, in order to include extra parameters, for example, it will be necessary to modify all references to objects of type func occurring in the source file stdgraph.cpp and in user programs.

We come now to the graphics source file itself. It begins by including the header file stdgraph.h, which must also be included in all programs which use these graphics techniques. We next define a constant HUGE, as we did in text mode, in order to range the function to be plotted:

```
const double HUGE = 1.0E+100;
```

A number of global variables follow which are the graphics analogues of the text mode variables NX, NY and OX, and in order that there be no possibility of calling them inadvertently from another file containing variables with identical names, these variables are declared static:

```
static int nx;          // max x
static int ny;          // max y
static int ox;          // x offset
static int oy;          // y offset
static int nox;         // nx - 2*ox
static int noy;         // ny - 2*oy
```

The values of these variables have to be assigned. We shall now be dealing with functions in graphics mode, and by convention these are declared as far functions in TurboC++, meaning that they require far pointers (using 4 bytes each (*cf* 1.5)) to call them. We shall abide by that convention here, even though for most purposes with which we are concerned the distinction may be ignored. In the TurboC++ graphics package there are (far) functions `getmaxx()` and `getmaxy()` which give the number of pixels in a line and in a column, respectively, of the connected monitor. Using these we may set `nx` and `ny` with the aid of an inline function which we define to be:

```
inline void far setnxny()    { nx = getmaxx(); ny = getmaxy(); }
```

The choice of offsets requires more direct intervention with a function such as:

```
void   setoffsets()
{      cout << "enter X Y offsets: ";
       cin >> ox >> oy;
}
```

We shall be displaying our functions within a screen box of width `nox` and height `noy` pixels. It is convenient to set these with the aid of the function:

```
void   setnoxnoy()
{ nox = nx-2*ox; noy = ny-2*oy; }
```

Sometimes it is convenient to know the ratio of these lengths:

```
double getaspect()   { return double(nox)/noy; }
```

15.2.2 Block size

Before completing the definition of `point` and `pixel` we need functions defining the block of values which are to be plotted on the screen, because the remaining constructors and other functions depend on these parameters. We define several (overloaded) functions `getblock(...)` which may be required in different circumstances. The first merely asks for the points `tl` and `br` to be entered:

```
void   getblock(point &tl, point &br)
{      cout << "\ntop left at:\n";
       cin >> tl;
       cout << "bottom right at:\n";
       cin >> br;
}
```

The second function assumes that an array of n points $p(x, y)$ has been given, and minimum and maximum values of x and y are to be sought:

```
void   getblock(point p[], int n, point &tl, point &br)
{       tl.x = br.y =   HUGE;
        tl.y = br.x = -HUGE;
        for (int i = 0; i < n; ++i)
        {       double pix = p[i].x, piy = p[i].y;
                if (pix < tl.x) tl.x = pix;
                if (pix > br.x) br.x = pix;
                if (piy < br.y) br.y = piy;
                if (piy > tl.y) tl.y = piy;
        }
}
```

The third assumes that a function $f(x)$ has been defined over the range [xlo, xhi]. The range is divided into nox equidistant points, at which maximum and minimum values of the function are sought:

```
void   getblock(func f, double xlo, double xhi,
                           point &tl, point &br)
{       double ylo = HUGE, yhi = -HUGE;
        double h = (xhi - xlo)/nox;
        double t = xlo;
        for (int i = 0; i < nox; ++i)
        {       double s = f(t);
                if (s < ylo) ylo = s;
                if (s > yhi) yhi = s;
                t += h;
        }
        tl.x = xlo;
        tl.y = yhi;
        br.x = xhi;
        br.y = ylo;
}
```

The last function is the same as the first, but applies to the complex plane:

```
void   getblock(complex &tl, complex &br)
{       cout << "\ntop left at: ";
        cin >> tl;
        cout << "bottom right at: ";
        cin >> br;
}
```

15.2.3 Opening and closing graphics mode

Before any graphics functions can be used graphics mode has to be opened. It is convenient to fix the static variables nx, ny, ox, oy, nox, noy in the course of doing so. The following function makes use of several TurboC++ functions and conventions without further explanation:

```
void  far opengraph()
{     int gdriver = DETECT, gmode;                         // TurboC++
      setoffsets();
      initgraph(&gdriver, &gmode, "\\tcpp\\bgi");   // TurboC++
      int errorcode = graphresult();                       // TurboC++
      if (errorcode != grOk)                               // TurboC++
      printf("error message:%s",grapherrormsg(errorcode));
      setnxny();
      setnoxnoy();
      cleardevice();                                        // TurboC++
}
```

Note that the functions setnxny() and setnoxnoy() have no meaning until initgraph(...) has been called to initiate graphics mode. The function setoffsets, on the other hand, must be called before initiating graphics mode because it is interactive in text mode. The function cleardevice() clears the screen in graphics mode.

Similarly, graphics mode must be shut off before leaving the user program. Again, it is convenient to combine several functions into one, calling for any character to be entered at the keyboard in order to terminate graphics mode:

```
void  far shutgraph(int color)
{     setcolor(color);                                      // TurboC++
      outtextxy(10, 10 , "press any key ...");       // TurboC++
      getch();                                              // TurboC++
      closegraph();                                         // TurboC++
}
```

The TurboC++ function void far outtextxy(int x, int y, char far *str) outputs a given string at screen position x,y in graphics mode. The string is output with a colour set by the TurboC++ function setcolor. The function getch() gets a character from the keyboard without echoing to the screen, and closegraph() finally closes graphics mode.

15.2.4 Presentation

We shall present our output on the screen within a box determined by offsets ox, oy, and with suitable axes superimposed. The axes will show the range of values represented within the box.

For this purpose we need a function which outputs a floating point number in graphics mode. We already have the function `outtextxy` which outputs a given (far) string to a given location on screen. All we need is a procedure which converts the number to a far string. Unfortunately, any such procedure is implementation dependent. We quote without explanation the function we have written to output a floating point number to a given location in TurboC++. We have declared it `static` to give it file scope only:

```
static  void far outnumbxy(int x, int y, double d)
//      all functions are in TurboC++ library
{       char *nbuf = (char*)calloc(20, sizeof(char));
        sprintf(nbuf, "%+8.4f", d);
        char far *fbuf = (char far*)farcalloc(20, sizeof(char));
        movedata(FP_SEG(nbuf), FP_OFF(nbuf),
                    FP_SEG(fbuf), FP_OFF(fbuf), strlen(nbuf));
        free(nbuf);
        outtextxy(x, y, fbuf);
        farfree(fbuf);
}
```

The function `scalegraph` draws a box, erects axes and scales them:

```
void    far scalegraph(point &tl, point &br,
                    int boxcol, int axecol, int numcol)
{       setcolor(boxcol);
        rectangle(ox, oy, nx-ox, ny-oy);       // TurboC++
        setcolor(numcol);                      // TurboC++
        outnumbxy(0, oy, tl.y);
        outnumbxy(0, ny-oy-10, br.y);
        outnumbxy(ox-30, ny-oy+10, tl.x);
        outnumbxy(nx-ox-30, ny-oy+10, br.x);
        point o;                               // origin
        pixel p(o, tl, br, axecol);
        if (p.x > 0 && p.x < nx-2.0*ox)
        {       setcolor(axecol);
                line(p.x, oy, p.x, ny-oy);     // vertical axis
                setcolor(numcol);
                outnumbxy(p.x-30, ny-oy+10, 0.0);
        }
        if (p.y > 0 && p.y < ny-2.0*oy)
        {       setcolor(axecol);
                line(ox, p.y, nx-ox, p.y);     // horizontal axis
                setcolor(numcol);
                outnumbxy(0, p.y, 0.0);
        }
}
```

The TurboC++ library function

```
void far rectangle(int left, int top, int right, int bottom)
```

draws a rectangular box in whatever colour (and line style) has already been set. Four numbers are next output, using `outnumbxy`, to indicate the scale of the box: these numbers are simply the components of the points `tl, br`. Next, we define a point `o` to represent the origin, and enquire whether the box we are about to display contains the origin. This enables us to complete the function `scalegraph` by draw the axes, using the TurboC++ library function:

```
void far line(int x1, int y1, int x2, int y2)
```

15.2.5 Drawing a function

We now consider the conversion constructor which enables us to map points to pixels. We shall need to have available the increments in point space corresponding to neighbouring pixels in screen space:

```
xinc = (br.x - tl.x)/nox;
yinc = (tl.y - br.y)/noy;
```

We define them as local static variables so that they are initialized once only. Provided the point `p` is within the box defined by `tl` and `br`, the corresponding nearest pixel, properly offset, has the constructor:

```
pixel::pixel(point &p, point &tl, point &br, int color)
{       static double xinc = (br.x - tl.x)/nox;
        static double yinc = (tl.y - br.y)/noy;
        x = int(round((p.x-tl.x)/xinc)) + ox;
        y = int(round((tl.y-p.y)/yinc)) + oy;
        c = color;
}
```

We must now consider how to put a pixel to screen in the correct position; *ie*, we need the graphics equivalent of the text mode function `gotoxy`. TurboC++ provides a library function `putpixel(int x, int y, int c)` for this purpose, which illuminates the pixel at screen coordinates `x,y` with the colour `c`. It is now a simple matter to display the point `p` on the screen by illuminating the correct pixel:

```
inline void far plot(point &p, point &tl, point &br, int color)
{       pixel s(p, tl, br, color);
        putpixel(s);
}
```

where

```
inline void far putpixel(pixel &s) { putpixel(s.x, s.y, s.c); }
```

must be declared a friend of `pixel`.

The given data may now be plotted in their entirety, whether presented as an array of points or as a function. In the latter case, a function which displays a given function over a given range of values of its argument is:

```
void  far drawfunc(func f, point &tl, point &br, int color)
{     double xinc = (br.x - tl.x)/nox;
      double x = tl.x;
      for (int i = 0; i < nox; ++i)
      {     point p(x, f(x));
            plot(p, tl, br, color);
            x += xinc;
      }
}
```

A typical user program might then resemble the following:

```
#include <stdgraph.h>
inline double F(double x){ return sin(2.0*PI*x)*exp(-x); }
main()
{     clear;
      double xlo, xhi;
      cout << "xlo xhi = ";
      cin >> xlo >> xhi;
      point tl, br;
      opengraph();
      getblock(F, xlo, xhi, point &tl, point &br);
      scalegraph(tl, br, RED, CYAN, YELLOW);
      drawfunc(F, tl, br, WHITE);
      shutgraph(LIGHTCYAN);
}
```

Notice that this program contains no *explicit* reference to the class `pixel`; it is referred to implicitly, of course, through the invocation of the function `plot` in `drawfunc`. The resulting graph is shown in monochrome in Figure 15.2. It is a considerable improvement on text mode!

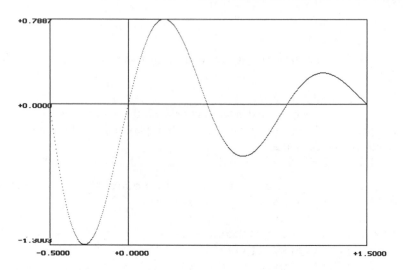

Figure 15.2 The function $F(x) = \sin(2\pi x)\exp(-x)$ in graphics mode.

15.2.6 Scanning the screen

On the other hand, one may desire to scan the whole screen, setting the colour of each pixel according to the result of some computation concerning the corresponding point. In that case, we shall need the conversion constructor:

```
point::point(pixel &s, point &tl, point &br)
{      static double xinc = (br.x - tl.x)/nox;
       static double yinc = (tl.y - br.y)/noy;
       x = tl.x + xinc*(s.x - ox);
       y = tl.y - yinc*(s.y - oy);
}
```

Supposing, for example, that we wish to display on screen the unit circle, filled with white, in a box which is otherwise blue, against a light grey background. Remembering that in screen mode the axes are of different length, if we wish the result to display a circle rather than an ellipse, the aspect ratio must be invoked.

We can most simply define the filled circle as a boolean function determining whether a point p is inside the circle:

```
bool   unitdisk(point &p)
{      static double a = getaspect();
       static double a2 = a * a;
       return (a2*p.x*p.x + p.y*p.y < 1) ? TRUE : FALSE;
}
```

Alternatively, the aspect ratio may be introduced in the selection of the points `tl, br` which determine the range of variables being considered. In either case, a function which displays the circle in the desired manner is:

```
void   drawdisk(point &tl, point &br)
{      int nx = getmaxx();
       int ny = getmaxy();
       for (int j = 0; j < ny; ++j)   // scan screen
       {      for (int i = 0; i < nx; ++i)
              {      pixel s(i, j, LIGHTGRAY);
                     if (inbox(s))
                     {      point p(s, tl, br);
                            if (unitdisk(p)) s.setcol(WHITE);
                            else s.setcol(BLUE);
                     }
                     putpixel(s);
              }
       }
}
```

where `inbox` is a boolean function declared to be a friend of `pixel` and defined as:

```
bool   inbox(pixel &s)
{      return (s.x>=0x && s.x<nx-ox && s.y>=oy && s.y<ny-oy)
              ? TRUE : FALSE;
}
```

The result is shown in monochrome in Figure 15.3.

15.3 The complex plane: quadratic Julia sets

There may be occasions when we wish to display a portion of the complex plane. For example, we may wish to display the modulus of a complex function as contours in the complex plane. In order to make the transition to complex numbers easy we have defined the additional constructor:

```
point::point(complex &z) { x = real(z); y = imag(z); }
```

If we wish to indicate the portion of the complex plane being displayed we must also define the additional function:

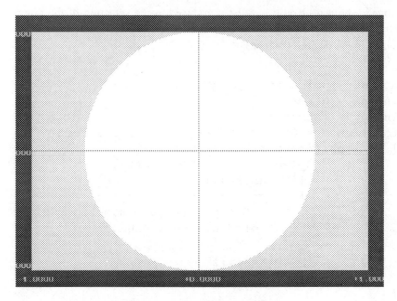

Figure 15.3 The unit disc.

```
void   far scalegraph(complex &tl, complex &br,
            int boxcol, int axecol, int numcol)
{      point ptl(tl), pbr(br);
       scalegraph(ptl, pbr, boxcol, axecol, numcol);
}
```

Let us by way of example consider the graphical representation of the quadratic *Julia sets* which are defined by the iterative scheme

$$\boldsymbol{J}(c): \; z \to f(z) = z^2 + c,$$

where c is a given complex number. The (filled) Julia Set $\boldsymbol{J}(c)$ is defined as the collection of complex numbers z which remain within the set under indefinitely repeated iteration of the function $f(z)$. It may be shown that the set may or may not be connected, depending upon the value of c, but in any case it lies within the circle $|z| = 3$. It is inexact, but sufficient for our purpose, to assume that if

$$|\text{real}(z_n)| < 2.5, \; |\text{imag}(z_n)| < 2.5,$$

for each of 100 iterations, $z_{n+1} = f(z_n)$, the initial value z_0 belongs to the set.

A routine for generating Julia sets may now be written down in analogy with the routine `drawdisk` in the last section. After defining the constants `lim = 2.5` and `maxiter = 100`, and the quadratic iteration function

```
inline complex f(complex &z, complex &c) { return z*z + c; }
```

we need a boolean function which decides whether a particular value of z lies within the allowed limits:

```
inline bool inside(complex &z)
{       return (fabs(real(z)) < lim && fabs(imag(z)) < lim)
                      ? TRUE : FALSE;
}
```

A boolean function may then be written down to decide whether z lies within $J(c)$; this is the equivalent of unitdisk in the last section:

```
bool   julia(complex &z, complex &c)
{       int i = 0;
        bool inset;
        complex w = z;
        while (i++<maxiter && (inset=inside(w)))
                w = f(w, c);
        return inset;
}
```

A routine to decide whether the complex number z represented by the pixel s lies within the Julia set may then be constructed by analogy with the routine drawdisk:

```
void   drawblock(complex &c, complex &tl, complex &br,
                      int bkdcol, int insetcol, int outsetcol)
{       int nx = getmaxx();
        int ny = getmaxy();
        point ptl(tl), pbr(br);
        for (int j = 0; j < ny; ++j)            // scan screen
        {       for (int i = 0; i < nx; ++i)
                {       pixel s(i, j, bkdcol);
                        if (inbox(s))
                        {       point p(s, ptl, pbr);
                                complex z(p.x, p.y);
                                if (julia(z, c)) s.setcol(insetcol);
                                else s.setcol(outsetcol);
                        }
                        putpixel(s);
                }
        }
}
```

The procedure is rather slow, owing to the complex arithmetic required by the iterations, but it serves to illustrate the complexity of the quadratic Julia sets considered here. The following user program nicely demonstrates these properties:

```
main()
{       clear;
        statement("Julia Set of z -> z*z + c\n");
        complex c;
        input("\nenter c: ", c);
        complex tl, br;
        getblock(tl, br);
        opengraph();
        drawblock(c, tl, br, LIGHTGRAY, WHITE, BLUE);
        scalegraph(tl, br, RED, GREEN, MAGENTA);
        shutgraph(LIGHTCYAN);
}
```

Figure 15.4 A Julia set.

Different values of the parameter c lead to radically different patterns. A particularly interesting pattern results if $c = -0.7475 + 0.0963i$, which is shown in monochrome in Figure 15.4 where we have set $tl = -2.0 + 1.3i$, $br = 2.0 - 1.3i$. Apart from its general complexity, its most intriguing feature is the appearance of "sea horses", or "elephant heads", which occur in various sizes and orientations. There are many other patterns, however, obtained with other values of c, and the reader may care to experiment freely.

15.4 Dynamic use of colour: the Mandelbrot set

The celebrated *Mandelbrot set* is obtained if we ask which values of the parameter c give rise to connected quadratic Julia sets $J(c)$. The required routines are very similar to those already described, but the full colour range of the monitor is needed to show the amazing complexity of the Mandelbrot set to its full advantage. This will serve conveniently to illustrate the dynamic use of colour in graphics.

Again we define a "convergence square":

```
const double lim = 2.5;
```

but we do not define `maxiter` as a constant, preferring to consider different values, as we describe presently.

The function $f(z)$ is defined as before, but this time it is more convenient to define a boolean function which decides whether the complex number z lies *outside* the limit box set by the constant `lim`:

```
inline bool outside(complex &z)
{       return (fabs(real(z)) > lim || fabs(imag(z)) > lim)
                    ? TRUE : FALSE; }
```

We now write a generalization of the function `julia` introduced in the last section. Instead of a boolean function, however, we write a function which returns the number of iterations required to send a complex number outside the limit box, up to a given limit `maxiter`:

```
int    mandelbrot(complex &c, int maxiter)
{       int iter = 0;                    // iter gives number of
        complex z = c;                   // iterations before z
        while (++iter < maxiter)         // leaves box, or maxiter,
        {       z = f(z, c);             // whichever is smaller
                if (outside(z)) break;
        }
        return iter;
}
```

If we imagine the Mandelbrot set as an island rising sharply out of the ocean, the number of iterations returned by the function `mandelbrot` is a reciprocal indication of the depth of the water. The larger the number of iterations required before z leaves the limit box, the closer is z to the surface and the edge of the island.

It is natural to try to represent this situation graphically as in a map with different colours separating given contours. We therefore declare a palette of colours and a corresponding series of levels, both represented by arrays of 7 integers, the colours having their integer equivalents as defined in the user's implementation:

```
int color[7];
int level[7];
```

These colours and levels, and their number, are arbitrary; but a suitable set, found by trial and error to represent the complexities of the whole Mandelbrot set most vividly, may be invoked by the two functions:

```
void   palette()
{      color[0]  = WHITE;
       color[1]  = BLACK;
       color[2]  = LIGHTMAGENTA;
       color[3]  = YELLOW;
       color[4]  = LIGHTGREEN;
       color[5]  = LIGHTBLUE;
       color[6]  = BLUE;
}

void   levels()
{      level[0]  =  50;
       level[1]  =  16;
       level[2]  =  11;
       level[3]  =   8;
       level[4]  =   5;
       level[5]  =   3;
       level[6]  =   2;
}
```

The idea is to choose a maximum number of iterations, equal to one of the prescribed levels, and then only those levels equal to or deeper than the maximum will be operative: the shore will be reached more closely the higher the permitted number of iterations. The following routine achieves this objective:

```
void   drawbrot(complex &tl, complex &br, int reslev)
{      int maxiter = level[reslev];
       int nx = getmaxx();
       int ny = getmaxy();
       point ptl(tl), pbr(br);
       for (int j = 0; j < ny; ++j)
       {      for (int i = 0; i < nx; ++i)
              {      pixel s(i, j, LIGHTGRAY);
                     if (inbox(s))
                     {      s.setcol(BLACK);
                            point p(s, ptl, pbr);
                            complex c(p.x, p.y);
```

```
                    int iter = mandelbrot(c, maxiter);
                    for (int k = reslev; k < 14; ++k)
                    {      if (iter >= level[k])
                           {      s.setcol(color[k]);
                                  break;
                           }
                    }
             }
             putpixel(s);
      }
   }
}
```

The routine is slow when the highest numbers of iterations are allowed. The speed is determined essentially by the iterated calls to the function $f(z)$. Some improvement might result by implementing this function in assembly language, but we shall not do that here.

A suitable user program might be:

```
main()
{     clear;
      statement("The Famous Mandelbrot Set\n");
      palette();
      levels();
      bool another = TRUE;
      while (another)
      {     complex tl, br;
            getblock(tl, br);
            int resolution;
            input("\nresolution (0 ... 9) = ", resolution);
            opengraph();
            drawbrot(tl, br, resolution);
            scalegraph(tl, br, RED, GREEN, MAGENTA);
            shutgraph(LIGHTCYAN);
            another = getyes("\nTry another block? - (Y/N): ");
            newline;
      }
      statement("\nGood Riddance, Mandelbrot!\n");
}
```

The function getyes is the useful (text mode) boolean function declared in standard.h and defined in 4.12.

The (filled) Mandelbrot set may be displayed in its entirety by setting tl = (-2.5, 1.5), br = (1.0, -1.5). It is shown in monochrome in Figure 15.5, where, however, we have suppressed the features provided by scalegraph.

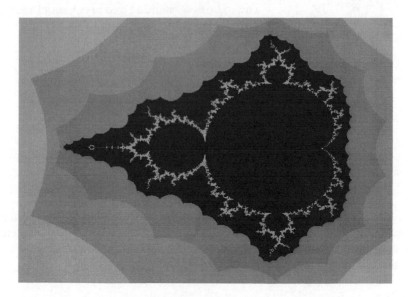

Figure 15.5 The Mandelbrot set.

Other choices of the parameters `tl,br` allow the reader to explore particular "coastal" regions of this fascinating "island" under magnification. In a sense, the Mandelbrot set is a compendium of all connected Julia sets. Indeed, this is how we have constructed it. It is therefore interesting to examine a particular stretch of coastline, that in the immediate neighbourhood of `c = -0.7475 + 0.0963`i, for which we have already computed the Julia set, and displayed it in Figure 15.4. Magnified linearly about 300 times by an appropriate choice of `tl, br`, the results are shown in monochrome in Figure 15.6, with somewhat different colour conventions. The "sea horse" is again the most prominent feature. The greater the magnification the more similar is the Mandelbrot set to its "local" Julia set.

At the magnification shown in Figure 15.6, it is not obvious that the set is connected: if white represents dry land, there appear to be myriad tiny islands in the neighbourhood of the coast. In fact, it can be shown that the Mandelbrot set *is* a single piece, topologically equivalent to a circle (see the book edited by Peitgen and Saupe). The coastline has an infinity of *peninsulae*, but no offshore *insulae*. As the magnification is increased, until it reaches a million or so, the set is found to reproduce itself in its entirety on that reduced scale, and so on in non-ending recursion. The Mandelbrot set is of *infinite* complexity.

15.5 Bezier curves

As a further example of the power of graphics methods, we introduce the *Bezier curves*, frequently used in design (or the preparation of technical books) on account of the ease with

Figure 15.6 The Mandelbrot set—detail of shoreline.

which they represent many different kinds of curve. These are curves which may be defined by recursively dividing by two.

Consider the parametrically defined function

$$B(z_0, z_1; 0 \le t < 1) = (1 - t)z_0 + tz_1,$$

where z_0, z_2 are any two fixed points in the euclidean plane, Then, as t varies from 0 to 1, B traverses the straight line segment from z_0 to z_1.

There is, however, no need to consider B to be a function of the auxiliary variable t. Let $u = (z_0 + z_1)/2$. Then the point $u = B(z_0, z_1; \frac{1}{2})$, and is therefore on the line B. Indeed, $B(z_0, u; t)$, where $0 \le t < 1$, traces out the first portion of the line $B(z_0, z_1; t)$, where $0 \le t < \frac{1}{2}$; and, similarly, $B(u, z_1; t)$ traces out the other portion where $\frac{1}{2} \le t < 1$; and so on, recursively.

We may parody this process in one dimension with the following recursive function, in which the parameter t no longer appears:

```
const double eps = 0.2;
void  B(double x1, double x2)
{     double u = (x1 + x2)/2;
      double d = x1 - x2;
      if (fabs(d) < eps)
            cout << "\n" << x1;
      else
```

```
{       B(x1, u);
        B(u, x2);
}
}
```

If $B(0,1)$ is called, it prints out numbers from 0 to 1 at intervals of 0.125 (except for 1). It is a generalization of this simple process that we shall use to draw more complicated Bezier functions, so a little time spent on understanding how it works will bear fruit.

Returning to the two-dimensional plane, we may now define Bezier functions defined by more than two points. For example, with three points we have

$$B(z_0,z_1,z_2; t) = (1 - t)^2 z_0 + 2t(1 - t)z_1 + t^2 z_2, \ 0 \le t < 1.$$

If we define three additional points

$$z_3 = (z_0 + z_1)/2, \ z_4 = (z_1 + z_2)/2, \ z_5 = (z_3 + z_4)/2,$$

then z_5 lies on the curve $B(z_0,z_1,z_2; t)$ at $t = \frac{1}{2}$, and we may again eliminate the independent variable t by dividing the range by 2 recursively.

In general, with n given points, the Bezier function

$$B(z_0,z_1, \dots, z_{n-1}; t) = (1 - t)^n z_0 + nt(1 - t)^{n-1} z_1 + \dots + nt^{n-1}(1 - t)z_{n-2} + t^n z_{n-1},$$

where the numerical coefficients are binomials, has the same property: if in the same pairwise fashion we define an additional $\frac{1}{2}n(n-1)$ points z_k, $n \le k < \frac{1}{2} n(n+1)$, then the last of these points $z_{1/2 n(n+1)-1}$ lies on the curve at $t = \frac{1}{2}$. Note also that the tangent to the curve $B(z_0, \dots, z_{n-1}; t)$ at the point z_0 (*ie*, at $t = 0$) is the line $B(z_0,z_1; t)$. Similarly, the tangent at z_{n-1} (*ie*, at $t = 1$) is the line $B(z_{n-2},z_{n-1}; t)$.

We wish to write a function which plots out the general n-point Bezier curve in graphics mode. We need two auxiliary functions. The first is very simple; it finds the mid-point of two given points:

```
point midpoint(point &p, point &q)
{       double x = (p.x + q.x)/2;
        double y = (p.y + q.y)/2;
        return point(x, y);
}
```

The second function plays the part in graphics mode of the boolean resolution condition (`fabs(d) < eps`) in our one-dimensional parody. We choose it to give the maximum resolution of which the monitor screen is capable:

```
bool  inres(point &p, point &q, point &tl, point &br)
{       double dx = p.x - q.x;
```

```
        double dy = p.y - q.y;
        double xinc = (br.x - tl.x)/getmaxx();
        double yinc = (tl.y - br.y)/getmaxy();
        if (fabs(dx) < xinc && fabs(dy) < yinc)
              return TRUE;
        return FALSE;
}
```

here we have again used the functions `getmaxx` and `getmaxy` which return the number of pixels in the screen. (Strictly speaking, we should have substituted the constants `nox`, `noy` defined in 15.2.1, but they are not accessible outside the definition of the class `pixel` unless we belatedly make the function `inres` a friend of `pixel`; but this merely results in our choosing a resolution slightly better than is warranted by the display.)

The Bezier function itself is straightforward except for the complication of defining and working with the additional $\frac{1}{2}n(n - 1)$ points. Consider the case of $n = 4$. We begin with an array `point p[4]` of four points which define the curve. Construct an auxiliary array of 10 points, `point t[10]`, of which the first four elements are identical with those of p. The next three are the pairwise midpoints of the first four, the next two the midpoints of those, and the last is the midpoint of the latter pair. As a diagram,

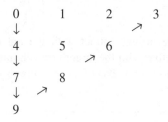

where `4 = midpoint(0,1)`, and so on, up to `9 = midpoint(7,8)`, and we have written the index k to represent the point z_k. The arrows indicate which points are involved in the recursive calls to the sub-curves: thus, $B(0,1,2,3)$ will call $B(0,4,7,9)$ followed by $B(9,8,6,3)$. In general:

```
void  bezier(point p[], int n, point &tl, point &br, int color)
{       if (inres(p[0], p[n-1], tl, br))
        {       pixel s(p[0], tl, br, color);
                putpixel(s);
        }       else
        {       point *t = new point[n*(n+1)/2];
                int i, j, k = 0;
                for (i = 0; i < n; ++i)
                {       for (j = 0; j < n-i; ++j)
                        {       if (i==0) t[k] = p[k];
                                else t[k] =
                                        midpoint(t[k-n+i-1], t[k-n+i]);
```

```
                ++k;
            }
        }
        point *r = new point[n];
        point *s = new point[n];
        k = 0;
        for (i = 0; i <n; ++i)
        {       for (j = 0; j < n-i; ++j)
                {       if (j == 0) r[i] = t[k];
                        if (j == n-i-1) s[j] = t[k];
                        ++k;
                }
        }
        delete t;
        bezier(r, n, tl, br, color);
        delete r;
        bezier(s, n, tl, br, color);
        delete s;
    }
}
```

Note that the recursive calls are brought about with the aid of two auxiliary arrays point r[n], s[n]. These, together with the larger array t, must be deleted as soon as possible, otherwise the call on memory may exceed the machine's capacity. Unfortunately,

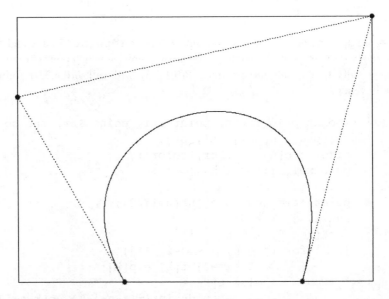

Figure 15.7 Bezier curve defined by four points.

these deletions prevent the function from being strictly tail-recursive, but the damage is not great, much less than the damage of not including them!

One further point is worth making. The routine we have offered proceeds by successive division by two. Since division by two is very fast indeed for integer division, it is tempting to perform all the arithmetic in the integers of graphics mode rather than in floating point. The drawback is that the accumulated rounding errors of integer division result in a gross loss of resolution. For good resolution, the arithmetic must be performed in floating point.

Examples of Bezier curves with $n = 4$ are shown in monochrome in Figures 15.7 and 15.8. Note that it is possible for the curve to cross itself if $n > 3$.

Higher values of n offer the possibility of more complex curves. Parametric Bezier *surfaces* may be defined similarly, if we work instead with three-dimensional points.

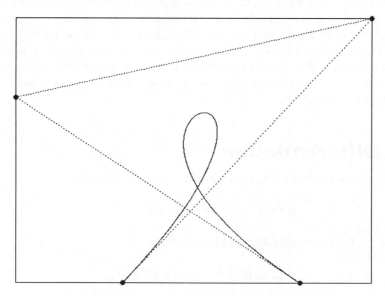

Figure 15.8 The same as Figure 15.7, crossed.

16 Differentiation and integration

The problems discussed in this chapter are very closely related to the problems of interpolation. Data are presented to us, either as a discrete tabulation of some function, or as the result of observation, and it is required that these data be approximately differentiated or integrated. Essentially, the approach is to fit a suitable polynomial to the data, explicitly or implicitly, and to differentiate or integrate the polynomial. The problem is then to estimate the error. We begin with differentiation, and then consider various approximations to integration over a small interval. We consider only the case in which the data are presented in the form of a function. Finally, we develop composite formulae which enable us to integrate a function accurately over a large interval.

16.1 Differentiation

The first order Taylor approximation to a function $f(x + h)$ is given by (cf 3.7)

$$f(x + h) = f(x) + hf'(x) + h^2 f''(\xi))/2!,$$

where $\xi(x) \in (x, x + h)$. Dividing through by h we obtain

$$f'(x) = [f(x + h) - f(x)]/h - \tfrac{1}{2}hf''(\xi),$$

so that, if h is sufficiently small, a first approximation to $f'(x)$ is given by the well-known formula:

$$f'(x) \approx [f(x + h) - f(x)]/h.$$

This is known as the first-order *forward difference formula* if $h > 0$. If $h < 0$, it is known as the first-order *backward difference formula*, which may alternatively be written:

$$f'(x) \approx [f(x) - f(x - h)]/h.$$

In either case, no information is used about the function f except for its values at x and $x + h$ or $x - h$. The corresponding error is relatively large; it is $O(h)$.

In order to obtain more accurate formulae we assume that the data are to be fitted by a Lagrange interpolating polynomial. We know from the last chapter that the polynomial of degree n which interpolates the function $f(x)$ at $n + 1$ values of x is given by

$$f(x) = \sum_{k=0}^{n} f(x_k) L^{(n)}_k(x),$$

where the Lagrange polynomials L are defined by:

$$L^{(n)}_k(x) = \prod_{j \neq k}^{n} [(x - x_j)/(x_k - x_j)].$$

Differentiating, we obtain

$$f'(x) = \sum_k f(x_k) L^{(n)}_k{}'(x) + R^{(n)}_k(x),$$

where the remainder

$$R^{(n)}_k(x) = (d/dx)[f^{(n+1)}(\xi(x)) \prod_j (x - x_j)]/(n + 1)!.$$

Differentiating the product by parts then gives:

$$\prod_j (x - x_j)(d/dx) f^{(n+1)}(\xi(x)) + f^{(n+1)}(\xi(x))(d/dx) \prod_j (x - x_j).$$

It is impossible in general to evaluate the first term because nothing is known in general about the function $\xi(x)$. However, if we choose for x one of the nodes, x_k say, the first term vanishes due to the term $(x - x_k)$ in the product. The second term, on the other hand, is straightforward. At this node, the differential coefficient therefore becomes:

$$f'(x_k) = \sum_{j=0}^{n} f(x_j) L^{(n)}_j{}'(x_k) + \frac{f^{(n+1)}(\xi(x_k))}{(n + 1)!} \prod_{j \neq k}^{n} (x_k - x_j).$$

In practical applications the remainder term is dropped, but we keep it for the present in order to estimate the errors of approximation.

In order to obtain practical formulae we shall for the present assume that the nodes are equally spaced, with spacing h. Consider the first-order approximation obtained by setting $n = 1$. We have

$$L^{(1)}_0(x) = (x - x_1)/(x_0 - x_1), \qquad L^{(1)}_1(x) = (x - x_0)/(x_1 - x_0),$$

so that $\qquad L^{(1)}_0{}'(x) = 1/(x_0 - x_1), \qquad\qquad L^{(1)}_1{}'(x) = 1/(x_1 - x_0).$

There are two possibilities: we may set $x_0 = x$, $x_1 = x + h$, or $x_0 = x - h$, $x_1 = x$. In the first case we retrieve precisely the forward difference formula; in the second, the backward difference formula. The error term is $\pm\frac{1}{2}hf''(\xi)$. Because these formulae require the function f to be evaluated at two points, they are sometimes known as *two-point formulae*.

Consider now the second-order approximation obtained by setting $n = 2$. We have:

$$L^{(2)}{}_0'(x) = \frac{2x - (x_1 + x_2)}{(x - x_1)(x - x_2)},$$

$$L^{(2)}{}_1'(x) = \frac{2x - (x_2 + x_0)}{(x - x_2)(x - x_0)},$$

$$L^{(2)}{}_2'(x) = \frac{2x - (x_0 + x_1)}{(x - x_0)(x - x_1)}.$$

There are three possibilities: we may set $x_0 = x$, $x_1 = x + h$, $x_2 = x + 2h$, or $x_0 = x - h$, $x_1 = x$, $x_2 = x + h$, or $x_0 = x - 2h$, $x_1 = x - h$, $x_2 = x$. In the first case we obtain

$$f'(x) = (1/h)[-(3/2)f(x) + 2f(x + h) - (1/2)f(x + 2h)] + (h^2/3)f^{(3)}(\xi),$$

where $\xi \in (x, x + 2h)$. The error term is quadratic in h. The third case is identical, except that h is replaced by $-h$. These are the forward and backward difference formulae, respectively, in second-order. The second case, however, reduces to:

$$f'(x) = (1/h)[-\tfrac{1}{2}f(x - h) + \tfrac{1}{2}f(x + h)] - (h^2/6)f^{(3)}(\xi).$$

Notice that this *midpoint difference* formula contains only two terms instead of three, and is approximately twice as accurate as the other two second-order formulae. It is *much* more accurate than the two-term, first-order formulae—which is not surprising, because it uses information about the function f drawn from both sides of the chosen point x. By contrast with the two-point, first-order formulae, the second-order formulae are sometimes called *three-point* formulae, even though the midpoint case requires evaluation at only two points.

Still greater accuracy is possible with *five-point* formulae of fourth order, for which the error term is $O(h^4)$. They may be obtained in exactly the same way. The midpoint formula is:

$$f'(x) = (1/12h)[f(x - 2h) - 8f(x - h) + 8f(x + h) - f(x + 2h)] + (h^4/30)f^{(5)}(\xi),$$

where $\xi \in (x - 2h, x + 2h)$. The forward-difference formula in fourth order is

$$f'(x) = (1/12h)[-25f(x) + 48f(x + h) - 36f(x + 2h) + 16f(x + 3h) - 3f(x + 4h)],$$

with a larger error term $(h^4/5)f^{(5)}(\xi)$. This is sometimes used as a left endpoint approximation in the clamped cubic spline method of interpolation (*cf* 14.2). The right endpoint approximation is obtained by changing the sign of h.

Approximations to the higher derivatives may be found in a similar way, but they become laborious to calculate. A useful midpoint approximation to the *second derivative*, however, is easy enough to derive. The third-order Taylor expansions of the function $f(x)$ at the points $x + h$ and $x - h$ are:

$$f(x + h) = f(x) + hf'(x) + \tfrac{1}{2}h^2f''(x) + (1/6)h^3f^{(3)}(x) + (1/24)h^4f^{(4)}(\xi_+),$$

$$f(x - h) = f(x) - hf'(x) + \tfrac{1}{2}h^2f''(x) - (1/6)h^3f^{(3)}(x) + (1/24)h^4f^{(4)}(\xi_-),$$

Adding them together, we get

$$f(x + h) + f(x - h) = 2f(x) + h^2f''(x) + (h^4/12)f^{(4)}(\xi),$$

where by the intermediate value theorem (cf 3.7) $\xi \in (\xi_+, \xi_-)$.

Thus,
$$f''(x) = (1/h^2)[f(x - h) - 2f(x) + f(x + h)] - (h^2/12)f^{(4)}(\xi),$$

where $x - h \leq \xi \leq x + h$.

It may be wondered why it is necessary to endure the additional complications of the higher order formulae when the value of h can be reduced until the truncation error involved lies within acceptable limits. Up to a point this may be true, but rounding errors will eventually predominate over truncation errors. Let us compare the second- and fourth-order midpoint formulae for computing the first derivative of $\exp(x)$ at $x = 0$, where all derivatives are exactly one, and may be taken to be approximately one in the immediate neighbourhood of $x = 0$. The second-order formula requires two evaluations of the function f, each with a rounding error bounded in magnitude by the small number ε, say. The total error involved in the second-order formula is therefore, schematically,

$$E_2(h) = (\varepsilon/h) + (h^2/6),$$

which has a minimum value at $h_2 = (3\varepsilon)^{1/3}$. Any attempt to decrease h below this point will simply increase the rounding errors. If we are using a tabulation of the exponential function accurate to 5 decimal figures it is reasonable to assume $\varepsilon \approx 5.10^{-6}$. Then

$$h_2 \approx (15.10^{-6})^{1/3} \approx 0.025.$$

The fourth-order formula, on the other hand, has a total error of the form

$$E_4(h) = (\varepsilon/3h) + (h^4/30),$$

which has a minimum value at $h_4 = (5\varepsilon/2)^{1/5}$. Using the same value of ε we obtain:

$$h_4 \approx (25.10^{-6}/2)^{1/5} \approx 0.023.$$

The values of h giving minimum total error are not much different in the two cases. The minimum errors, however, are very different:

$$E(h_2) = (3\varepsilon)^{2/3}/2 \approx 0.0003, \qquad E(h_4) = (5\varepsilon/2)^{4/5}/6 \approx 0.00002.$$

The fourth-order approximation is much better.

The main value of these approximations is not so much in estimating the derivative of a function as their application to the solution of differential equations, as we shall see in chapter 18. For that reason we present no routines for computing differential coefficients as such; they are in any case trivial to write.

16.2 Richardson extrapolation

The development of higher-order approximations may be approached from another point of view. Quite generally, suppose that we wish to compute the value of a quantity the exact value of which is Q, and we have a second-order formula $Q_1(h)$ for doing this, where h is a small parameter, so that

$$Q = Q_1(h) + O(h^2).$$

Suppose further that Q may be expressed in terms of even powers of h only; the midpoint formulae just developed for the derivative of a function are examples of this. Then,

$$Q = Q_1(h) + C_1 h^2 + C_2 h^4 + \ldots$$

where $C_1, C_2, \ldots$, are independent of h.

We shall now show how to eliminate these terms systematically. Since this formula for Q is true for all sufficiently small h, we may also write:

$$Q = Q_1(h/2) + C_1(h/2)^2 + C_2(h/2)^4 + \ldots$$

By multiplying this second formula throughout by 4, subtracting the first, and dividing the result throughout by 3 we eliminate C_1:

$$Q = Q_2(h) - (1/4)C_2 h^4 + O(h^6) = Q_2(h) + O(h^4),$$

where
$$Q_2(h) = [4Q_1(h/2) - Q_1(h)]/3.$$

Similarly,
$$Q = Q_4(h) + O(h^6),$$

where
$$Q_4(h) = [16Q_2(h/2) - Q_2(h)]/15,$$

and so on with higher-order terms if desired.

Let us now apply this technique to the computation of the first derivative at $x = 0$ of the exponential function $\exp(x)$. The starting approximation may be taken to be:

$$Q_1(h) = [(f(x + h) - f(x - h)]/2h.$$

It is not difficult to show that $Q_2(h)$, when written out in terms of the values of the function f, is the same as the fourth-order midpoint approximation to f', except that h is replaced by $\frac{1}{2}h$, namely:

$$Q_2(h) = [f(x - h) - 8f(x - \tfrac{1}{2}h) + 8f(x + \tfrac{1}{2}h) - f(x + h)]/6h.$$

With $h = 0.2$, we obtain for the exponential function $\exp(x)$ at $x = 0$:

$$Q_1 = 1.00668, \qquad Q_2 = 0.999997, \qquad Q_4 = 1.000000.$$

16.3 Integration

We now consider the inverse process, namely the numerical integration of a function over some given finite interval. We shall again approximate the function $f(x)$ by a Lagrange interpolating polynomial:

$$f(x) = \sum_{k=0}^{n} f(x_k) L^{(n)}{}_k(x) + \prod_k (x - x_k) f[x_0, \ldots, x_n, x].$$

The required integral may then be written

$$I(f) \equiv \int_a^b f(x)\ dx = I_n(f) + E_n,$$

where the approximate value of the integral is

$$I_n = \sum_k f(x_k) \int_a^b L^{(n)}{}_k(x)\ dx,$$

and the corresponding truncation error is

$$E_n = \int_a^b \prod_k (x - x_k)\ f[x_0, \ldots, x_n, x]$$

$$= f[x_0, \ldots, x_n, \eta] \int_a^b \prod_k (x - x_k), \quad \eta \in [x_0, x_n].$$

Remembering that $f[x_0, \ldots, x_n, \eta] = f^{(n+1)}(\xi)/(n + 1)!$, where $\xi \in [a, b]$,

$$E_n = \frac{1}{(n + 1)!}\ f^{(n+1)}(\xi) \int_a^b (x - x_0)(x - x_1) \ldots (x - x_n)\ dx,$$

which is a more convenient expression for the error. The approximate value of the integral may be written

$$I_n(f) = \sum_k c_k f(x_k),$$

where
$$c_k = \int_a^b L^{(n)}{}_k(x)\, dx.$$

These formulae are correct no matter where the $n+1$ nodes x_k are chosen within the interval $[a, b]$, but the value of the error term may be difficult to estimate in general. We shall from now on assume that the nodes are equally spaced, $x_0 = a$, $x_1 = a + h, \ldots,$ $x_n = b$, where $h = (b - a)/n$, although we shall later see that this does not minimize the error term.

Let us first examine the cases $n = 1$ and $n = 2$. If $n = 1$:

$$L^{(1)}{}_0(x) = (x - x_1)/(x_0 - x_1), \qquad L^{(1)}{}_1(x) = (x - x_0)/(x_1 - x_0),$$

where $x_0 = a$, $x_1 = b$; so that $c_0 = \int_a^b (x - b)/(a - b)\, dx = \tfrac{1}{2}(b - a) = c_1$.

The error term is

$$E_1 = \tfrac{1}{2} \int_a^b (x - x_0)(x - x_1) f[x_0, x_1, x]\, dx = f[x_0, x_1, \eta] \int_a^b (x - x_0)(x - x_1)\, dx.$$

for some $\eta \in [a, b]$. This last step is permitted by the weighted mean value theorem (see 3.7) since the product $(x - x_0)(x - x_1)$ does not change sign in $[a, b]$. Evaluating the integral is elementary, and the error estimate is therefore

$$E_1 = -(h^3/12) f''(\xi),$$

where $h = b - a$, for some $\xi \in [a, b]$.

Thus,
$$\int_a^b f(x)\, dx = \tfrac{1}{2} h[f(a) + f(b)] - (h^3/12) f''(\xi).$$

Dropping the error term gives us the the well known *trapezoidal rule*. We shall see later, however, that a better two-point result is obtained if we set the nodes at:

$$\alpha_\pm = (1/2)(a + b) \pm (1/2\sqrt{3}) h,$$

which is the result of gaussian quadrature (see chapter 17). For the moment it will suffice to note that

$$\int_a^b f(x)\, dx = h[f(\alpha_+) + f(\alpha_-)] + (h^5/135) f^{(4)}(\xi).$$

where now $h = (b - a)/2$. We shall call this the *Gauss rule*.

We may proceed in a similar way if $n = 2$ to obtain *Simpson's rule*:

$$\int_a^b f(x)\, dx \approx (h/3)[f(x_0) + 4f(x_1) + f(x_2)].$$

The error term, however, is a little more complicated to estimate. It is given by:

$$E_2 = \int_a^b (x - x_0)(x - x_1)(x - x_2) f[x_0, x_1, x_2, x]\, dx.$$

The product $(x - x_0)(x - x_1)(x - x_2)$ changes sign in the interval $[a, b]$, so the weighted mean value theorem cannot be used. Instead, define a weighting function by

$$w(x) = \int_a^x (x - x_0)(x - x_1)(x - x_2) \, dx,$$

so that

$$E_2 = \int_a^b w'(x) f[x_0, x_1, x_2, x] \, dx.$$

It is easy to show that $w(x) = \frac{1}{2}(x - a)^2(x - b)^2$, so that $w(a) = w(b) = 0$, and $w(x) > 0$, $x \in (a, b)$. Integrating by parts we find:

$$E = w(x) f[x_0, x_1, x_2, x] \Big|_a^b - \int_a^b w(x) f'[x_0, x_1, x_2, x] \, dx$$

$$= -\int_a^b w(x) f[x_0, x_1, x_2, x, x] \, dx$$

$$= -f[x_0, x_1, x_2, \eta, \eta] \int_a^b w(x) \, dx$$

$$= -(1/4!) f^{(4)}(\xi) \int_a^b w(x) \, dx,$$

where $\xi, \eta \in [a, b]$. The coefficient is the integral

$$\int_a^b w(x) \, dx = \frac{1}{4} \int_a^b (x - a)^2 (x - b)^2 \, dx.$$

Set $t = (x - a)/h$. Then this coefficient becomes:

$$\frac{1}{4} h^5 \int_0^2 t^2 (t - 2)^2 \, dt = (4/15) h^5.$$

Combining terms, the error estimate is finally:

$$E_2 = -(1/90) h^5 f^{(4)}(\xi).$$

Somewhat surprisingly, the error term in Simpson's rule is $O(h^5)$ compared with $O(h^3)$ for the trapezoidal rule. The additional computation is well worth it!

These two formulae are particular cases of the closed *Newton–Cotes* formulae obtained when $n \geq 1$. When $n = 3$ we have the "three-eights rule",

$$\int_a^b f(x) \, dx = \frac{3h}{8} [f(a) + 3f(a + h) + 3f(b - h) + f(b)] - \frac{3h^5}{80} f^{(4)}(\xi);$$

and when $n = 4$ we have "Boole's rule",

$$\int_a^b f(x) \, dx = \frac{2h}{45} [7f(a) + 32f(a + h) + 12f((a + b)/2) + 32f(b - h) + 7f(b)] - \frac{8h^7}{945} f^{(6)}(\xi).$$

Notice that the error term in the three-eights rule is of the same order as in Simpson's rule, although slightly smaller in magnitude. The additional computation involved may not therefore seem justified. The fourth-order formula, Boole's rule, however, is $O(h^7)$ and if h is sufficiently small it should be very accurate.

This behaviour of the error term results from the evaluation by parts of the integral:

$$\int_a^b w'(x) f[x_0, \ldots, x_n, x]\, dx = w(x) f[x_0, \ldots, x_n, x] \Big|_a^b - \int_a^b w(x) f[x_0, \ldots, x_n, x, x]\, dx.$$

where
$$w(x) = \int_a^x (t - x_0) \ldots (t - x_n)\, dt.$$

Clearly, $w(a) = 0$ for all values $n > 0$. However, $w(b) = 0$ only if n is even. Therefore, if n is even, the first term vanishes, and the leading error term is $O(h^{n+3} f^{(n+2)}(\xi))$; but if n is odd, the first term does *not* vanish and the leading error term is only $O(h^{n+2} f^{(n+1)}(\xi))$.

There are also *open* Newton–Cotes formulae in which one or both of the endpoints are deleted from the list of nodes. The most well known is the *midpoint formula*:

$$\int_a^b f(x)\, dx = h f((a + b)/2) + (1/24) h^3 f''(\xi)$$

where $h = (b - a)$, for some $\xi \in [a, b]$. We leave this to the reader to derive.

To program the various integration formulae is trivial, so we merely quote some straightforward results. In particular, we estimate the integral $\int_0^2 f(x)\, dx$ for some elementary functions:

$f(x)$:	x	x^2	x^3	x^4	$1/(1+x)$	$\sqrt{(1+x^2)}$	$\sin x$	$\exp(x)$
trap:	2	4	8	16	1.3333	3.2361	0.9093	8.3891
midp:	2	2	2	2	1.0000	2.8284	1.6829	5.4366
gaus:	2	2.667	4	6.2222	1.0909	2.9533	1.4102	6.3681
simp:	2	2.667	4	6.6667	1.1111	2.9643	1.4251	6.4207
3h/8:	2	2.667	4	6.5185	1.1048	2.9604	1.4201	6.4033
bool:	2	2.667	4	6.4000	1.0993	2.9575	1.4161	6.3892
true:	2	2.667	4	6.4000	1.0986	2.9579	1.4161	6.3891

All approximations give the exact answer for the linear function, as expected; the midpoint method is somewhat better than the trapezoidal; all except trapezoidal and midpoint are exact for quadratic and cubic functions; while the simple example of gaussian quadrature given above has a result somewhat better than Simpson's formula—clearly it is $O(h^5 f^{(4)}(\xi))$. The three-eighths rule is also somewhat better than Simpson, but Boole's rule is very accurate for the cases studied, being exact for quartic functions. It should be borne in mind that these results were obtained for the integration interval [0, 2] which can hardly be described as small.

16.4 Composite formulae

The sequence of approximations obtained with $n = 2, 4, \ldots, \infty$ does not necessarily converge, especially if the interval of integration is wide. For example, a divergent sequence is obtained with the integral:

$$\int_{-4}^{+4} (1 + x^2)^{-1} \mathrm{d}x = 2 \tan^{-1}(4).$$

Since the sequence seems likely to converge when h is small, however, it is natural to consider breaking up the interval into a number of small sub-intervals, to each of which one of the formulae derived above is applied. This is much more straightforward than relying upon formulae of higher and higher order.

Consider first the composite trapezoidal rule, dividing the interval of integration into N equal sub-intervals of width $h = (b - a)/N$, to each of which the trapezoidal rule is applied. Let $x_i = a + ih$, where $i \in [0, N]$. Then

$$\int_{x_i}^{x_{i+1}} f(x) \, \mathrm{d}x = \tfrac{1}{2}h[f(x_i) + f(x_{i+1})] - (h^3/12)f''(\xi_i),$$

for some $\xi_i \in [x_i, x_{i+1}]$. Summing the contributions from the N sub-intervals we obtain

$$\int_a^b f(x) \, \mathrm{d}x = \tfrac{1}{2}h[f(a) + f(b) + 2\sum_{i=0}^{n-1} f(x_i)] - (b - a)(h^2/12)f''(\xi),$$

for some $\xi \in [a, b]$.

Similarly, the composite midpoint rule, with N sub-intervals each of width $h = (b - a)/N$, is

$$\int_a^b f(x) \, \mathrm{d}x = h\sum_{i=0}^{n-1} f(x_i + h/2) + (1/24)(b - a)h^2 f''(\xi).$$

The composite Simpson's rule, with $2N$ sub-intervals each of width $h = (b - a)/2N$, is

$$\int_a^b f(x) \, \mathrm{d}x = (h/3)[f(a) + 4\sum_{i=1}^{n} f(x_{2i-1}) + 2\sum_{i=1}^{n-1} f(x_{2i}) + f(b)] - (1/180)(b - a)h^4 f^{(4)}(\xi).$$

Since h can be made as small as we like (subject to rounding errors) simply by subdividing the interval into sufficiently many sub-intervals, we would not expect there to be convergence difficulties if the function f is reasonably smooth. Below, we offer routines for computing the composite trapezoidal, midpoint, Gauss and Simpson's rules:

```
typedef double (*func)(double);

double trapezoidal(func f, double a, double b, int n)
{       double h = (b - a)/n;
        double t = (f(a) + f(b)) * 0.5;
        double x = a;
        for (int i = 1; i < n; ++i)
        { x += h;   t += f(x); }
        return t * h;
}

double midpoint(func f, double a, double b, int n)
{       double h = (b - a)/n;
        double x = a + 0.5*h;
        double t = f(x);
        for (int i = 1; i < n; ++i)
        { x += h;   t += f(x); }
        return t*h;
}


double gauss(func f, double a, double b, int n)
{       static double r3 = sqrt(3.0);
        double h = (b - a)/n;
        double v = h/(2.0*r3);
        double u = a + h/2.0;
        double t = f(u+v) + f(u-v);
        for (int i = 1; i < n; ++i)
        {       u += h;
                t += f(u+v) + f(u-v);
        }
        return (h/2.0)*t;
}
```

```
double simpson(func f, double a, double b, int n)
{       double h = (b - a)/n;
        double s = f(a + 0.5*h);
        double t = 0.5*(f(a) + f(b));
        double x = a, y = a + 0.5*h;
        for (int i = 1; i < n; ++i)
        {       x += h;
                y += h;
                s += f(y);
                t += f(x);
        }
        t += 2.0 * s;
        return (t * h)/3.0;
}
```

Below, we present the results of applying these composite rules to the integral

$$\int_0^{2\pi} \exp{(-x)} \cos{(x)} \, dx \approx 0.499066278634.$$

N	trap	midpoint	gauss	simpson
1	3.147459	−0.271521	0.205692	0.868139
2	1.437969	0.000000	0.510059	0.479323
4	0.718985	0.383810	0.501389	0.495535
8	0.551398	0.472520	0.499235	0.498823
16	0.511959	0.492596	0.499077	0.499050
32	0.502277	0.497459	0.499067	0.499065
64	0.499868	0.498665	0.499066	0.499066
128	0.499267	0.498966	0.499066	0.499066
256	0.499116	0.499041	0.499066	0.499066
512	0.499079	0.499060	0.499066	0.499066
1024	0.499069	0.499065	0.499066	0.499066
2048	0.499067	0.499066	0.499066	0.499066
4096	0.499066	0.499066	0.499066	0.499066

It may be of interest to see the program which produced this tabulation, in which formatted output is employed:

```
const double twopi = 2.0*PI;

main()
{       clear;
        double a = 0.0, b = twopi;
        long fl = ios::left;
```

```
      long f2 = ios::left + ios::showpos + ios::fixed;
      format fpw1(f1, 4, 10), fpw2(f2, 6, 20), reset(0, 0, 0);
      cout << reset << fpw1 << "n"
           << reset << fpw2 << "trapezoidal"
                    << fpw2 << "midpoint"
                    << fpw2 << "gauss"
                    << fpw2 << "simpson" << endl;
   cout << "_____"
        << "_____\n";
   for (int i = 0; i < 13; ++i)
   {      int n = (int) pow(2, i);
          cout << reset << fpw1 << n
             << reset << fpw2 << trapezoidal(F, a, b, n)
                      << fpw2 << midpoint(F, a, b, n)
                      << fpw2 << gauss(F, a, b, n)
                      << fpw2 << simpson(F, a, b, n) << endl;

   }
   cout << "_____"
        << "_____\n";
}
```

16.5 Asymptotic errors

Let us look at the convergence properties of the sequence of approximations as N, the number of intervals, becomes large. For the composite trapezoidal rule we have seen that the error within a single interval $I_i = [x_{i-1}, x_i]$ is given by $E_1 = -(h^3/12)f''(\xi_i)$, where $\xi_i \in I_i$. Summing over all intervals, the error is therefore:

$$E_1^{(N)} = \sum_{i=0}^{N-1}(-h^3/12)f''(\xi_i)$$

$$= -(h^2/12)\sum_{i=0}^{N-1}hf''(\xi_i).$$

Since $a = x_0 \leq \xi_0 \leq x_1 \leq \xi_1 \leq \ldots \leq \xi_{n-1} \leq x_n = b$, however, the sum, in the limit $N \to \infty$, may be replaced by an integral in the Riemann sense, so that:

$$E^{\mathrm{trap}} \approx -(h^2/12)\int_a^b f''(x)\mathrm{d}x$$

$$\approx -(h^2/12)[f'(b) - f'(a)].$$

This is the asymptotic error formula for the trapezoidal rule. It can be shown, but it is not straightforward, that it may be expanded in even powers of h only, beginning:

$$E^{\mathrm{trap}} = -(h^2/12)[f'(b) - f'(a)] + [(b - a)/720]h^4 f^{(4)}(\xi).$$

Thus, for large N, the trapezoidal error should decrease by a factor 4 if the value of h is halved, *ie*, if the value of N is doubled.

Similarly, the asymptotic error of the composite Simpson's rule is:

$$E^{\text{simp}} = -(h^4/180)[f^{(3)}(b) - f^{(3)}(a)] + (b - a)O(h^5 f^{(5)}(\xi)).$$

The error should decrease by a factor 16 if the value of h is halved, or the value of N is doubled. Note, however, that unlike the trapezoidal case the asymptotic error of the composite Simpson's formula, and of higher order formulae more generally, cannot be expressed solely in terms of even powers of h.

The following table illustrates these asymptotic properties. The asymptotic ratio is asserted after the first few doublings, but large values of N may lead to instabilities due to rounding errors, as we see in the table. Normally, the computation is terminated when the desired accuracy has been reached.

N	trapezoidal	ratio	simpson	ratio
1	3.147459		0.868139	
2	1.437969	2.8	0.479323	−18.7
4	0.718985	4.3	0.495535	5.6
8	0.551398	4.2	0.498813	13.9
16	0.511959	4.1	0.499050	15.5
32	0.502277	4.0	0.499065	15.9
64	0.499868	4.0	0.499066	16.0
128	0.499267	4.0	0.499066	16.0
256	0.499116	4.0	0.499066	16.0
512	0.499079	4.0	0.499066	16.0
1024	0.499069	4.0	0.499066	15.9
2048	0.499067	4.0	0.499066	16.3
4096	0.499066	4.0	0.499066	45.4

16.6 Adaptive integration

Many integrands will vary at different rates in different parts of the range. It is wasteful to apply the same sub-interval width throughout the range. If the variation of the function is sufficiently well known, the necessary adjustments can be made by dividing up the range into a number of sub-ranges, in each of which a different width parameter is used. It is obviously advantageous if this can be done automatically. We present a recursive function which does this.

Let us, for example, choose as the basic integration formula the simple gaussian quadrature function of 16.3:

```
double gauss(func f, double a, double b)
{       static double r = 1.0/sqrt(3);
        double m = (b+a)/2;
        double h = (b-a)/2;
        double rh = r*h;
        double t = f(m + rh) + f(m - rh);
        return h*t;
}
```

The idea is as follows. First run the function gauss over the full interval $[a, b]$. Let the approximate value of the integral so obtained be Z = gauss(f, a, b). Divide the interval into two equal parts, separated at the point c = (a + b)/2. The equivalent approximation to the integral over the left-hand sub-interval is X = gauss(f, a, c), and over the right sub-interval Y = gauss(f, c, b). Let tolerance be a very small number. Then, if $|X + Y - Z|$ < tolerance, Z is a sufficiently accurate estimate of the value of the integral; otherwise, it will be necessary to divide the two sub-intervals in the same way. The following recursive procedure achieves the required end:

```
const int max = 100;
const double tolerance = 1.0E-7;
void  adapt(func f, double &sum, double x, double y, int depth)
{       if (depth > max) error("function varies too rapidly!");
        double z = (x+y)/2;
        double X = gauss(f, x, z);     // left sub-interval
        double Y = gauss(f, z, y);     // right sub-interval
        double Z = gauss(f, x, y);     // whole interval
        if (fabs(X + Y - Z) < tolerance) sum += Z;
        else
        {       adapt(f, sum, x, z, ++depth);
                adapt(f, sum, z, y, depth);
        }
}
```

Here, the integer variable depth refers to the number of sub-divisions that have taken place. It should therefore be initialized to zero, as should sum also. This may be done automatically by a simple non-recursive function:

```
double adaptint(func f, double a, double b)
{       int depth = 0;
        double sum = 0.0;
        adapt(f, sum, a, b, depth);
        return sum;
}
```

A simple, if somewhat artificial, function to test the adaptive properties of this routine is

$$F(x) = (100/x^2)\sin(10/x),$$

integrated from $a = 1$ to $b = 3$. The user program might then be:

```
main()
{       clear;
        double a = 1.0, b = 3.0;
        double sum = adaptint(F, a, b);
        output("\nintegral = ", sum);
}
```

With the function offerred, and the given value of `tolerance`, the program divides the interval $[a, b]$ into 83 sub-interval, distributed over 8 levels. To the precision stated the value of the integral is -1.426025. It is a simple matter to modify the function `adapt` so that it prints out the contribution from each sub-interval. One merely has to include in the function `adapt` a variable defined as `static int slice = 1;` and replace the second `if` statement by, for example:

```
if (fabs(X+Y-Z) < tolerance)
{       sum += Z;
        cout << "slice = " << slice++
                << "     depth = " << depth
                << "    z = " << z
                << "    Z = " << Z << endl;
}
```

The program then prints out the contribution of each sub-interval, and the depth at which the sub-interval occurs.

16.7 Romberg integration

Higher precision may also be achieved by using the Richardson device introduced in 16.2. Fully developed, this is the Romberg method. We shall begin by sketching the first Richardson steps; then we shall present the full Romberg treatment. The object is to reduce the number of function calls, which is a major determinant of the speed of computation in integration.

We have seen that the composite trapezoidal rule, with its error terms, may be written

$$I \equiv \int_a^b f(x)\,\mathrm{d}x = T_1(h) + K_1 h^2 + K_2 h^4 + \dots,$$

where $T_1(h) = \frac{1}{2}h[f(a) + f(b) + 2\Sigma_i f(x_i)]$, and the remaining terms represent the asymptotic error, with $K_1 = -(1/12)[f'(b) - f'(a)]$, and so on. This is precisely the form required by the Richardson method, and we can immediately find a better approximation for the integral, namely,

$$I = T_2(h) - \frac{1}{2}K_2h^4,$$

where
$$T_2(h) = [4T_1(h/2) - T_1(h)]/3.$$

The h^4 term may be eliminated just as in 16.2, by setting:

$$T_4(h) = [16T_2(h/2) - T_2(h)]/15,$$

and so on, thus obtaining a sequence which is much more rapidly convergent than the original. We show the results below for the integral considered in 16.4 and 16.5:

N	Trapezoidal	First diff	Second diff	Third diff
1	3.147459			
2	1.437969	0.868139		
4	0.718985	0.479323	0.453402	
8	0.551398	0.495535	0.496616	0.497302
16	0.511959	0.498813	0.499031	0.499069
32	0.502277	0.499050	0.499066	0.499066
64	0.499868	0.499065	0.499066	0.499066

The performance of the first differences is the exact equivalent of Simpson's rule; the second is of order $O(h^6)$, and the third $O(h^8)$.

In the Romberg method, full advantage is taken of the fact that the error of the trapezoidal approximation may be expressed as a series asymptotic in h^2. This series may be terminated at some point; the result is therefore a polynomial in h^2. This polynomial may be extrapolated to $h = 0$ by means of the `neville` algorithm introduced in 14.1.3, to obtain a value for the integral together with an error estimate. The process may then be repeated with a smaller starting value of h until the error is within the required tolerance.

Since the objective is to minimize the number of function calls, we first design a function to reduce by two the number of calls to the trapezoidal routine. Normally we shall not know how many calls to *trapezoidal* we need to make; the number will be determined by some tolerance condition. To obtain significant improvement using this routine we need to halve the interval parameter h. This, however, means that we have to repeat all the function calls already made at internal points. This waste of effort may be successfully averted (*cf* 2.7) with the use of static variables local to the function which remember the results already obtained with larger values of h, or equivalently smaller values of the number n of sub-intervals:

```
double seqtrap(func f, double a, double b, int n)
// called with n=0 gives single interval trapezoidal approximation
// subsequent calls with n=1,2,3,..., are for 2ⁿ sub-intervals
// non-sequential calls terminate with error message
{       static double s;        // return variable
        static int next;        // add next points on next call
        static int good = 0;    // correct value of n
        if (n != good++) error("non-sequential call to  seqtrap(...)");
        if (n == 0)
        {       next = 1;
                return (s = 0.5*(b-a)*(f(a) + f(b)));
        }       else
        {       int numb = next;
                double h = (b-a)/numb;  // length of current intervals
                double x = a + 0.5*h;
                double sum = 0.0;         // sum over added internal points
                for (int i = 0; i < numb; ++i)
                {       sum += f(x);
                        x += h;
                }
                next *= 2;                // double number of intervals
                return (s = 0.5*(s + (b-a)*sum/numb));
        }
}
```

The process may be illustrated by a simple diagram. If `seqtrap` is called with $n = 0$ the function returns the simple trapezoidal approximation with function calls $f(x)$ at the two ends only of the single interval. If the function `seqtrap` is *subsequently* called with $n = 1$ an additional function call at the midpoint of the interval, and the estimate of the integral is adjusted to take this into account. Later subsequent calls add the midpoints of further sub-intervals. The static variables look after the house-keeping. Calling `seqtrap` in a non-sequential manner earns an error message.

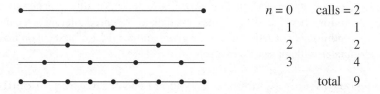

	$n = 0$	calls $= 2$
	1	1
	2	2
	3	4
	total	9

We may simplify the former routine `neville` slightly to allow for the fact that we intend only to extrapolate to zero, and there is no need to check that nodes do not coincide, these nodes being the successive even powers of h:

```cpp
double neville2(vector &xv, vector &yv, double &err)
{       double y;                       // return variable
        int n = xv.getsize();           // degree of polynomial + 1
        vector c = yv;
        vector d = yv;
        int k = n-1;                    // nearest node is last
        y = yv[k--];
        double incr;
        // raise degree of polynomial to set value
        for (int m = 1; m < n; ++m)
        {       for (int i = 0; i < n-m; ++i)
                {       double xi  = xv[i];
                        double xim = xv[i+m];
                        double cd = c[i+1] - d[i];
                        double mult = xi - xim;;
                        mult = cd / mult;
                        c[i] = xi * mult;
                        d[i] = xim * mult;
                }
                incr = (2*k < (n-m-1) ? c[k+1] : d[k--]);
                y += incr;
        }
        err = fabs(incr);               // crude error estimate =
        return y;
}
```

We shall include in the signature of the function romberg the desired degree of the polynomial that is to be used for interpolation, and the allowable error tol. The routine evaluates a succession of trapezoidal approximations using the routine seqtrap, the value of h being reduced by *four* on successive calls. This is to reflect the fact that the error of the trapezoidal approximation is expressable as a series in h^2. When it has computed the first deg+1 terms, vectors t and h, of size deg+1 are constructed to hold these terms and feed them to the function neville. For each subsequent term, these vectors have to be reassembled, and the routine neville run again. When the error estimate (normally a generous estimate) computed by the routine neville2 becomes less than the required tolerance, the process terminates. If this condition is not satisified after a given number lim of calls upon neville2, the process terminates with an error message:

```cpp
double romberg(func f, double a, double b, int deg, double tol)
{       const int lim = 20;
        double s, err;                  // value and error estimate
        vector t(lim), h(lim);          // store values and intervals
        h[0] = 1.0;
        for (int i = 0; i < lim; ++i)
```

```
{       t[i] = seqtrap(f, a, b, i);
        if (i > deg)            // extrapolate poly to h=0
        {       vector hv(deg+1), tv(deg+1);
                for (int j = 0; j <= deg; ++j)
                {       hv[j] = h[i-deg-1+j];
                        tv[j] = t[i-deg-1+j];
                }
                s = neville2(0.0, hv, tv, err);
                if (fabs(err) < tol*fabs(s)) return s;
        }
        h[i+1] = 0.25*h[i];     // polynomial in h*h
}
error("too many steps in romberg(...)");
}
```

This routine, adapted from the routine **qromb** in Press *et al.*, seems rather complicated, and it is. (We have wherever possible tried to avoid complication.) The reason is that we have been trying to reduce the number of function calls, the prime determinant of computation time. Let us run `romberg`, and compare it with the simple routine `trapezoidal`, on the function:

```
int c = 0;
double F(double x)
{       ++c;
        return cos(x)*exp(-x);
}
```

The integral will be taken between the limits $a = 0$ and $b = 2\pi$. Here, the variable c is an integer defined globally, and its job is to count the number of function calls, which may be printed out in the user program if desired.

The value of this integral to 8 significant figures is 0.49906628. To achieve this precision with the simple function `trapezoidal` requires over 64 000 calls to the function F. Using `seqtrap` instead we can halve that number; but with the routine `romberg`, using a polynomial of degree 5 (which appears to be the optimum), the number of calls to F is only 129, a truly remarkable improvement in performance.

Of course, the routine `trapezoidal`, being of error $O(h^2)$, is rather inefficient. The `adaptint` routine discussed in the last section requires 1530 calls to the function to which we applied it there in order to achieve 8 figure accuracy. Applying `romberg` instead to that function requires 513 calls, an important improvement, but not nearly so dramatic. The routine `adaptint` relies upon the two-point gauss formula, much more accurate than `trapezoidal`.

16.8 Improper integrals

An integrable integral is called improper if its integrand, or one or more of the first few derivatives of the integrand, have discontinuities within the interval of integration, or if the integrand cannot be evaluated at one or other of the limits of the integral, or if one or both of these limits is infinite. Some integrals can be defined by a limiting process appropriate to the problem in which they occur. In practice, one may also regard an integral as improper if the computation of it is very slowly convergent, as may well happen with infinite limits. For all of these cases there are ways of computing the value of the integral. But there are also integrals which cannot be integrated at all, because there is no way to define their values; no amount of computation will remove that problem! It is a large and complex subject, so we shall merely sketch some of the more common ways of dealing with it.

The simplest case is that of an integrand with a single step discontinuity at a known position within the interval of integration; for example,

$$\int_a^b f(x)\,dx,$$

where $f(x) = f_1(x)$, $x \in [a, c]$, $f(x) = f_2(x)$, $x \in (c, b)$, $c \in (a, b)$, and the two functions f_1, f_2 are continuous, but $f_1(c) \neq f_2(c)$. The integral may simply be defined as the sum of two integrals,

$$\int_a^b f(x)\,dx = \int_a^c f_1(x)\,dx + \int_c^b f_2(x)\,dx,$$

each to be evaluated separately by standard methods.

Less trivially, the integrand $(\sin x)/x$, is undefined at $x = 0$, but becomes integrable if the value there is defined by the limiting process $\lim_{x \to 0} (\sin x)/x = 1$. Similarly, the integrand $f(x)/\sqrt{x}$ is discontinuous at $x = 0$; while $f(x)\sqrt{(1 - x^2)}$ is discontinuous at $x = \pm 1$, which may cause difficulty if one of the limits of integration is unity.

It is sometimes possible in these and other cases to perform a change of variable to remove the difficulty before computing the value of the integral. For example:

$$\int_0^b f(x)\,dx\,/\sqrt{x} = 2\int_0^{\sqrt{b}} f(u^2)\,du.$$

Similarly,

$$\int_0^1 f(x)\sqrt{(1 - x^2)}\,dx = 2\int_0^1 u^2\sqrt{(2 - u^2)}f(1 - u^2)\,du.$$

The problem caused by infinite limits of integration may also sometimes be removed by a change of variable which replaces infinite limits by finite ones. The behaviour of the function so transformed will, however, need to be examined at the new limits. Suppose that $f(x)$ is a smooth function tending to a constant limit $c \neq 0$ as $x \to \infty$ such that its asymptotic behaviour is $f(x) \approx c + c_1/x + c_2/x^2 + \ldots$ Then

$$\int_1^\infty \frac{f(x)}{x^{3/2}} \, dx = 6 \int_0^1 u^2 f(1/u^6) \, du,$$

and the integrand is smooth at $u = 0$ because $u^2 f(1/u^6) = cu^2 + c_1 u^8 + \ldots$

It is also possible to avoid the problem of an integrand that is not well defined at the limits by avoiding the limits altogether. The composite midpoint method does this; moreover, it shares with the trapezoidal method the property that the asymptotic error may be expressed as a series in even powers of h, so that a routine can be written for it that is easily derived from romberg. We shall not do that here.

Many integrals may be expressed in the form $\int_a^b w(x) f(x) \, dx$, where $w(x)$ is an integrable weight function which may be singular (or near-singular), and $f(x)$ is smooth. In suitable cases, these may be dealt with by approximating $f(x)$ by Lagrange interpolation before evaluating the integral. Another approach is to use the weight function w to define an orthogonal set of functions over the interval of the integral, and to expand $f(x)$ as a finite series of these functions. We shall develop this process of *gaussian quadrature* in the next chapter, noting here only that we have already used one simple version of it in the "Gauss rule" of 16.3.

16.9 Multiple integrals

Multiple integrals occur frequently in the physical sciences when, for instance, one is required to calculate the area of a surface or the volume of a solid body. We shall confine ourselves to the simple case of two dimensions: the extension to more than two dimensions should be fairly obvious; only the computer power needs to be greater.

We consider, then a double integral:

$$I = \iint_A f(x,y) \, dx \, dy,$$

where the integration is over some area A of the (x,y) plane. We shall consider the y integration first, assuming that the circumference of A may be defined by the curves $y = c(x)$ and $y = d(x)$, and the values a and b of x. Accordingly, we define the function

$$g(x) = \int_{y=c(x)}^{y=d(x)} f(x,y) \, dy,$$

in terms of which the x integration may in turn be written:

$$I = \int_a^b g(x)\mathrm{d}x.$$

This is the process we now have to express in C++. We shall make use of the composite formulae of 16.4, rewriting them slightly to fit the present purpose. Note first that we have two types of function, the integrand $f(x, y)$, and the limits $c(x)$, $d(x)$; accordingly, we write:

```
typedef     double (*onefunc)(double);
typedef     double (*twofunc)(double, double);
```

Then a typical composite formula, for given x, may be expressed as a function

```
double oneintegrate(twofunc f, onefunc c, onefunc d,
                const double x, int m);
```

for the y integration, and another function

```
double      twointegrate(twofunc f, onefunc c, onefunc d,
            double a, double b, int m, int n);
```

which repeatedly calls `oneintegrate` for the final x integration. If the parameters are all properly defined, the value of the integral is given by:

```
double I = twointegral(f, c, d, a, b, m, n);
```

As an example we have computed the integral o the function $f(x, y) = e^{-(x+y)}$, over the area enclosed by the curves $y = x^2$ and $y = \sqrt{x}$, between $x = 0$ and $x = 1$, using the composite gaussian formula of 16.4. The integral converges to 6 decimals with $m = n = 512$.

17 Orthogonal polynomials

In this chapter we shall consider how to construct and use orthogonal systems of polynomials. They occur frequently in the physical sciences, and are very useful in computation. Having introduced the general concepts and the instances most commonly found in numerical work, we apply the technique first to the least square approximation of functions by polynomials (comparing the results with the minimax approximation); and second, to the important integration technique known as gaussian quadrature.

17.1 The Gram–Schmidt process

Let $w(x)$ be a non-negative function on the interval $[a, b]$, such that for any integer $n \geq 0$ $\int_a^b |x|^n w(x)\,dx$ is integrable. We shall call it a *weight function*, because it may be used to give more emphasis to one part of the interval than another. Furthermore, we shall assume that $w(x)$ is such that if $g(x)$ is a non-negative continuous function on the interval $[a, b]$, then $\int_a^b w(x)\,g(x)\,dx = 0$ implies $g(x) \equiv 0$. Commonly used weight functions are shown below, with the corresponding intervals of integration, labelled by the conventional names of the polynomials they generate:

$$
\begin{array}{lll}
w(x) \equiv 1 & a \leq x \leq b & \text{(Legendre)} \\
w(x) = 1/\sqrt{(1 - x^2)} & -1 \leq x \leq 1 & \text{(Chebyshev)} \\
w(x) = \exp(-x) & 0 \leq x \leq \infty & \text{(Laguerre)} \\
w(x) = \exp(-x^2) & -\infty \leq x \leq \infty & \text{(Hermite)}
\end{array}
$$

Define the *inner product* of two continuous functions with respect to w by

$$(f, g) = \int_a^b w(x)\,f(x)\,g(x)\,dx.$$

Then the corresponding euclidean norm

$$\|f\|_2 = \{\int_a^b w(x)\,[f(x)]^2\,dx\}^{1/2} = \sqrt{(f, f)}$$

satisfies all the requirements for a norm (see 11.1). In particular, the Cauchy–Schwartz inequality

$$|(f, g)| \leq \|f\|_2 \|g\|_2$$

guarantees the triangle inequality:

$$\|f + g\|_2 \leq \|f\|_2 + \|g\|_2.$$

Suppose now that we have a sequence of polynomials $\{\phi_k(x)\}$, where the index $k \geq 0$ represents the degree of each polynomial, such that

$$(\phi_m, \phi_n) \equiv \int_a^b w(x) \phi_m(x) \phi_n(x) \, dx = \gamma_m \delta_{mn},$$

where the γ_m are constants and δ is Kronecker's delta. Then the polynomials ϕ are said to form an *orthogonal system* with respect to the weight function w. If, in addition, they are chosen so that the coefficients γ are all unity, we say that we have an *orthonormal system*. This may always be done, since if the polynomials ϕ_k are orthogonal, the renormalized polynomials $\phi'_k = \gamma_k^{-1/2} \phi_k$ are orthonormal.

Let us first see that a system of orthogonal polynomials may in principle be constructed by recursion. We may arbitrarily set the coefficient of x^k in ϕ_k equal to unity. Then $\phi_0(x) = 1$, so that $\gamma_0 = \int_a^b w(x) \, dx$. Further, we may take $\phi_1(x) = x + a^{(1)}_0 \phi_0(x)$, where if ϕ_1 is to be orthogonal to ϕ_0, we must set

$$(\phi_1, \phi_0) = (x, \phi_0) + a^{(1)}_0 (\phi_0, \phi_0) = 0,$$

and therefore $$a^{(1)}_0 = -(x, \phi_0)/\gamma_0.$$

In general, we similarly set

$$\phi_n(x) = x^n + a^{(n)}_{n-1} \phi_{n-1}(x) + a^{(n)}_{n-2} \phi(x) + \ldots + a^{(n)}_0 \phi_0(x),$$

and choose the n constants so that $(\phi_n, \phi_k) = 0$, for all k such that $0 \leq k < n$.

Thus, $$a^{(n)}_k = -(x^n, \phi_k)/\gamma_k,$$

where the polynomials ϕ_k and the normalization constants γ_k are known from earlier stages of the calculation. This recursion is known as the *Gram–Schmidt* process. Given that the coefficient of x^k in $\phi_k(x)$ is unity for all k, the polynomials so generated are manifestly unique. And since the polynomials ϕ_k are of different degree, they are linearly independent.

They are also *complete* in the following sense. Let $f(x)$ be an arbitrary polynomial of degree m. In the Gram–Schmidt process, the monomials of degree k may be written

$$x^k = \phi_k(x) - a^{(k)}_{k-1} \phi_{k-1}(x) - \ldots - a^{(k)}_0 \phi_0(x),$$

which is a linear combination of orthogonal polynomials of degree no higher than k. The polynomial f, which is of degree m, may therefore be written as a linear combination of such expressions:

$$f(x) = b_m \phi_m(x) + \ldots + b_0 \phi_0(x),$$

where

$$b_j = (f, \phi_j)/\gamma_j.$$

Therefore the orthogonal system of polynomials $\phi_k(x)$ is complete in the sense that *any* polynomial of given degree may be expressed as a linear combination of the ϕ_k of no greater degree. It is also clear that if f is of degree m and $k > m$, then ϕ_k is orthogonal to f.

What is not immediately obvious, however, is that the polynomials ϕ_k satisfy a *three-point recursion relation*. Consider the polynomial $C(x) = \phi_{n+1}(x) - x\phi_n(x)$, where the polynomials $\phi_k(x)$ have been generated by the above Gram–Schmidt process. $C(x)$ is of degree n, because the coefficient of x^{n+1} vanishes. Therefore:

$$C(x) = c_n \phi_n(x) + \ldots + c_0 \phi_0(x),$$

where the coefficients $c_j = (C,\phi_j)/(\phi_j,\phi_j) = [(\phi_{n+1}, \phi_j) - (x\phi_n, \phi_j)]/\gamma_j$. But $(\phi_{n+1}, \phi_j) = 0$ if $j < n + 1$, and $(x\phi_n, \phi_j) = 0$ if $j < n - 1$. Therefore $C(x) = c_n \phi_n(x) + c_{n-1}\phi_{n-1}(x)$, and

$$\phi_{n+1}(x) = (x + c_n)\phi_n(x) + c_{n-1}\phi_{n-1}(x),$$

where $c_n = -(x\phi_n, \phi_n)/\gamma_n$ and $c_{n-1} = -(x\phi_n, \phi_{n-1})/\gamma_{n-1}$ are already known.

This is a three-point recursion relation from which in principle as many polynomials may be calculated as desired. There are the usual limitations of rounding errors which occur with three-point recursion, but with the weighting factors given above the coefficients, or at any rate their ratios, are rational numbers and may therefore be calculated without error up to some degree at which overflow takes place.

Thus a sequence of orthogonal polynomials may be constructed once a suitable weight function has been chosen. The normalization of the polynomials, however, is a matter of convention and may be changed without affecting the orthogonality properties.

17.2 Zeros of orthogonal polynomials

We prove the following

Theorem.

> If the polynomials $\phi_k(x)$ are an orthogonal system on the interval $[a, b]$ with weight function $w(x)$, then $\phi_n(x)$ has precisely n distinct zeros on the open interval (a, b).

For suppose the zeros of ϕ_n which lie on the interval (a, b) are m in number. Since ϕ_n is of degree n, $m \leq n$. Suppose $m < n$, and represent the zeros by $x_1, \ldots, x_m$. Since they are distinct zeros, $\phi_n(x)$ changes sign at each x_i.

Let $$\Psi(x) = (x - x_1) \ldots (x - x_m);$$

then $F(x) \equiv \Psi(x)\phi_n(x)$ does not change sign on (a, b). Therefore $(\Psi, \phi_n) \neq 0$. But Ψ is of degree $m < n$. Therefore $(\Psi, \phi_n) = 0$, which is a contradiction. Therefore $m = n$.

It may also be shown that zeros of orthogonal polynomials of successive degrees interleave. If the zeros of ϕ_n are $x^{(n)}_i$, and those of ϕ_{n+1} are $x^{(n+1)}_j$:

$$a < x^{(n+1)}_0 < x^{(n)}_0 < x^{(n+1)}_1 < x^{(n)}_1 < \ldots < x^{(n)}_n < x^{(n+1)}_{n+1} < b.$$

These are most important results, as we shall see when we come to Gaussian quadrature.

17.3 Particular systems

For the more familiar choices of weight function instanced at the beginning of 17.1, orthogonal systems of polynomials can be derived without direct reference to the Gram–Schmidt process. They are known as solutions to certain second-order linear differential equations which occur in physical problems. Having obtained these solutions the weight functions can be deduced from the orthogonality relations. They are all useful for the approximation of functions and for the numerical integration of certain types of function. Other weight functions are possible: for example, $w(x) = \log(1/x)$ on the interval $[0, 1]$, for which, however, the polynomial coefficients become very large.

17.3.1 Legendre polynomials

The Legendre polynomials are conventionally denoted by $P_n(x)$. They satisfy the three-point recursion relation:

$$(n + 1)P_{n+1}(x) = (2n + 1)xP_n(x) - nP_{n-1}(x).$$

The first few polynomials are

$$P_0(x) = 1$$
$$P_1(x) = x$$
$$P_2(x) = (3x^2 - 1)/2$$
$$P_3(x) = (5x^3 - 3x)/2$$
$$P_4(x) = (35x^4 - 30x^2 + 3)/8$$

and they are shown graphically in monochrome in Figure 17.1. Their normalization is given by

$$\int_{-1}^{+1} [P_n(x)]^2 dx = 2/(2n + 1).$$

Notice that $P_n(1) = 1$. The weight function is this case is $w(x) = 1$, and the interval over which they are defined is $[a, b] = [-1, +1]$.

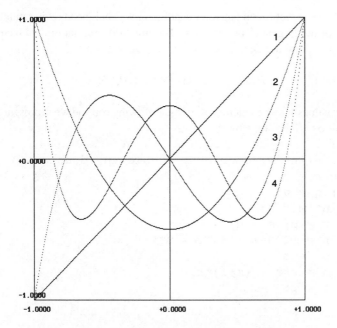

Figure 17.1 Legendre polynomials.

Apart from the normalization convention, the Legendre polynomials may equally well be derived from the Gram–Schmidt process. The orthonormal system is

$$\phi_n(x) = [(2n + 1)/2]^{1/2} P_n(x).$$

Legendre polynomials may be defined for an arbitrary finite interval $[a, b]$ by a simple change of variable, $x = [(b + a) + (b - a)t]/2$ where $t \in [-1, +1]$.

It is tempting to write a doubly recursive function to evaluate a Legendre polynomial, directly translating the three-point recursion relation. We resist this temptation for the reasons given in 3.11. Instead, here is an iterative function that returns the value of $P_n(x)$:

```
double legendre(int n, double x)
{       double r = 0, s = 1;
        for (int m = 0; m < n; ++m)
        {       double t = r;   r = s;
                s = (2*m+1)*x*r - m*t;
                s /= m+1;
        }
        return s;
}
```

In this function, the value of the polynomial $P_{m+1}(x)$ is developed in the variable s, while r holds the value of $P_m(x)$ and t the value of $P_{m-1}(x)$.

However, in the next section on gaussian quadrature we shall need to compute the slope as well as the value of the Legendre polynomials. By direct differentiation, we have $P_0'(x) = 0$, $P_1'(x) = 1$, $P_2'(x) = 3x$, and in general there is a recursion relation

$$P_{n+1}'(x) = xP_n'(x) + (n+1)P_n(x),$$

from which other values may be calculated. In the following routine we represent the value of the polynomial by p and the slope by q:

```
void  legendre(int n, double x, double &p, double &q)
{       double r = 0, s = 0, t = 0;
        p = 1; q = 0;
        for (int m = 0; m < n; ++m)
        {       r = p; s = q;
                p = (2.0*m+1)*x*r - m*t;
                p /= m+1;
                q = x*s + (m+1)*r;
                t = r;
        }
}
```

The initialization of t must be chosen to produce the correct value for the polynomial of degree one; in this case it is initialized to zero. During the loop, r carries the value of the polynomial of one less degree, t of two less, while s carries the differential coefficient corresponding to r. Finally, t is updated.

17.3.2 Chebyshev polynomials

The Chebyshev polynomials are also defined on the interval $[-1, +1]$, but the weight function is $w(x) = 1/(1 - x^2)^{1/2}$. They are conventionally denoted by $T_n(x)$, and they satisfy the three-point recursion relation:

$$T_{n+1}(x) = 2xT_n(x) - T_{n-1}(x).$$

The first few Chebyshev polynomials are shown in monochrome in Figure 17.2; they are given by

$$T_0(x) = 1$$
$$T_1(x) = x$$
$$T_2(x) = 2x^2 - 1$$
$$T_3(x) = 4x^3 - 3x$$
$$T_4(x) = 8x^4 - 8x^2 + 1$$

and their normalization is

$$\int_{-1}^{+1}(1 - x^2)^{-1/2}[T_n(x)]^2 dx = \pi, \qquad n = 0$$
$$= \pi/2, \qquad n > 0.$$

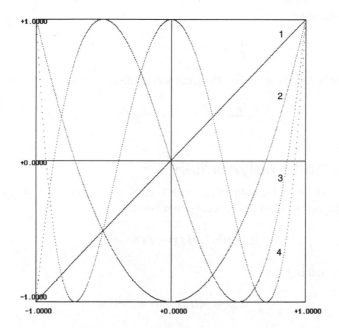

Figure 17.2 Chebyshev polynomials.

The differential coefficients obey the recursion relation

$$T_{n+1}'(x) = [(n + 1)x/n]T_n'(x) + (n + 1)T_n(x).$$

Note that $T_n(x) = \cos[n \cos^{-1}x]$, in which form the recursion relation is still satisfied. In this form it is particularly easy to see that if $n > 0$ the polynomials T_n vary between +1 and −1, with n simple zeros at $\cos[(2k − 1)\pi/(2n)]$, where $k = 1, 2, \ldots, n$.

We leave it to the reader to write a suitable routine to evaluate these polynomials. If the function `legendre` is used as a model, it will be found necessary to initialize the local variable t to the value x to give the correct value for the polynomial of degree one.

17.3.3 Laguerre polynomials

The Laguerre polynomials are denoted by $L_n(x)$; the weight function is $\exp(−x)$, and the interval of integration $[0, \infty]$. They satisfy the recursion relation:

$$(n + 1)L_{n+1}(x) = (2n + 1 − x)L_n(x) − nL_{n-1}(x).$$

The first few are

$$L_0(x) = 1$$
$$L_1(x) = 1 − x$$
$$L_2(x) = 2 − 4x + x^2$$
$$L_3(x) = 6 − 18x + 9x^2 − x^3$$
$$L_4(x) = 24 − 96x + 72x^2 − 16x^3 + x^4$$

and the normalization is

$$\int_0^\infty \exp(-x)[L_n(x)]^2 dx = 1.$$

The differential coefficients satisfy the recursion relation

$$L_{n+1}'(x) = L_n'(x) - L_n(x).$$

17.3.4 Hermite polynomials

The weight function for the Hermite polynomials is $\exp(-x^2)$, and the interval of integration is $[-\infty, +\infty]$. The polynomials satisfy the recursion relation

$$H_{n+1}(x) = 2xH_n(x) - 2nH_{n-1}(x),$$

and the normalization is given by:

$$\int_{-\infty}^{+\infty} \exp(-x^2)[H_n(x)]^2 dx = 2^n n! \sqrt{\pi}.$$

The first few polynomials are:

$$H_0(x) = 1$$
$$H_1(x) = 2x$$
$$H_2(x) = 4x^2 - 2$$
$$H_3(x) = 8x^3 - 12x$$
$$H_4(x) = 16x^4 - 48x^2 + 12$$

and the differential coefficients satisfy the simple relations $H_{n+1}'(x) = 2(n + 1)H_n(x)$.

17.4 Approximation of functions

Suppose that $f(x)$ is defined and continuous on the interval $[a, b]$. We wish to approximate f by a polynomial $p_n(x)$ of degree at most n, in such a way that the total squared error S, given by

$$S = \int_a^b (f(x) - p_n(x))^2 dx,$$

is as small as possible. The local error $e(x) = f(x) - p_n(x)$ is not in general distributed uniformly throughout the interval, and we shall therefore be interested in trying to choose families of polynomials which distribute the error as uniformly as possible.

To begin with, let $p_n(x) = a_n x^n + a_{n-1} x^{n-1} + \ldots + a_1 x + a_0 = \Sigma_k a_k x^k$.

Then the total squared error will be a function of the unknown coefficients a_k,

$$S(a_0, a_1, \ldots, a_n) = \int_a^b (f(x) - \sum_{k=0}^n a_k x^k)^2 dx,$$

and this must be minimized with respect to each coefficient separately. Since

$$\partial S / \partial a_j = -2 \int_a^b x^j f(x) \, dx + 2 \sum_k a_k \int_a^b x^{j+k} \, dx$$

we have $n + 1$ linear *normal equations* for the $n + 1$ coefficients a_k:

$$\sum_{k=0}^n a_k \int_a^b x^{j+k} \, dx = \int_a^b x^j f(x) \, dx.$$

Notice the appearance of the ill-conditioned Hilbert matrix on the left-hand side (see 11.3). The equations may be solved by any reasonable procedure, but errors will seriously impair accuracy if the polynomial is of too high degree.

We have approximated the function f by a single polynomial expressed as a linear combination of monomials with unknown coefficients; if we approximate the function by a linear combination of orthogonal polynomials instead, the corresponding matrix will be diagonal and much labour will be saved. Let us therefore try to approximate the function $f(x)$ by a linear combination of orthogonal polynomials ϕ_k with $k \leq n$, say. The weighted total square error is given by:

$$S = \| f(x) - \sum_{k=0}^n a_k \phi_k(x) \|_2^2.$$

If we now set $\partial S / \partial a_j = 0, j = 0, 1, \ldots, n$, we obtain the normal equations

$$\int_a^b w(x) f(x) \phi_j(x) \, dx = \sum_k a_k \int_a^b w(x) \phi_j(x) \phi_k(x) \, dx,$$

which are just what we had before if we replace the monomials x^k by the polynomials $\phi_k(x)$ and include the weighting function $w(x)$.

However, the polynomials ϕ_k are orthogonal, unlike the monomials; therefore

$$(\phi_j, \phi_k) = \int_a^b w(x) \phi_j(x) \phi_k(x) \, dx = \gamma_j \delta_{jk},$$

and we can immediately solve for the coefficients a_k which minimize the square error:

$$a_k = (f, \phi_k) / \gamma_k$$
$$= \int_a^b w(x) f(x) \phi_k(x) \, dx / \int_a^b w(x) [\phi_k(x)]^2 \, dx.$$

In what follows we shall often take as an example the exponential function $f(x) = e^x$ considered over the interval $[-1, 1]$, because this will simplify algebraic detail, but the reader should consider other functions also. Any finite interval, as we have seen in 17.3.1, may be reduced to $[-1, 1]$ if desired by a simple change of variable.

17.4.1 Taylor approximation

Consider first, however, as a standard of comparison, the Taylor polynomial approximation to the exponential function, expanding e^x about $x = 0$. If we limit ourselves to the polynomial of degree 3, we have

$$P_3^{(T)}(x) = 1 + x + x^2/2! + x^3/3! \,.$$

The distribution of the error of approximation is given by

$$e(x) = e^x - 1 - x - x^2/2 - x^3/6,$$

some values of which are given in the following table. Of course, $e(0) = 0$. Remembering that the remainder term in the Taylor expansion in this case is $R_3(x) = x^4 e^{\xi(x)}/4!$, we may for interest derive the required value of ξ as a function of x:

x	$e(x)$	$\xi(x)$	x	$e(x)$	$\xi(x)$
−1.0	.0345	−.187	0.0	0.0	0.0
−0.9	.0231	−.170	0.1	4.2E−6	.020
−0.8	.0147	−.152	0.2	6.9E−5	.041
−0.7	.0088	−.134	0.3	.0004	.061
−0.6	.0048	−.115	0.4	.0012	.082
−0.5	.0024	−.097	0.5	.0029	.103
−0.4	.0010	−.078	0.6	.0061	.125
−0.3	.0003	−.059	0.7	.0116	.147
−0.2	6.4E−5	−.039	0.8	.0202	.169
−0.1	4.1E−6	−.020	0.9	.0331	.191
			1.0	.0516	.214

As expected, the error is smallest in the vicinity of $x = 0$, the point about which the Taylor expansion has taken place. It is largest at the extremities of the interval. Notice that in this case $\xi(x) \approx 0.2x$.

17.4.2 Legendre approximation

The simplest generalization of the least square approximation is obtained by setting $w(x) = 1$ on the interval $[-1, 1]$, so that all parts of the interval are weighted alike. The resulting polynomials are therefore the Legendre polynomials.

We shall now compute the least square approximation to the exponential function e^x using Legendre polynomials of degree up to three, in order to compare the distribution of error with that given by a Taylor polynomial of degree three. We shall set:

$$p(x) = a_0 P_0(x) + a_1 P_1(x) + a_2 P_2(x) + a_3 P_3(x)$$

where in order to minimize the total square error $\int_{-1}^{+1}[\exp(x) - p(x)]^2 dx$ the coefficients a_k are taken to be $[(2k + 1)/2] \int_{-1}^{+1} \exp(x) P_k(x) dx$. The distribution of error is then given by

$$e(x) = \exp(x) - p(x).$$

It would be natural to compute the integrals involved from first principles, but all that is involved in this simple case is the calculation of the integral

$$I_m = \int_{-1}^{+1} x^m \exp(x) dx,$$

where m is a non-negative integer. Partial integration then gives us the recursion relation:

$$I_m = e - (-1)^m e^{-1} - m I_{m-1}.$$

Given that $I_0 = e - e^{-1}$, it is a simple matter to represent I_m by a recursive function:

```
const double IE = EXP - 1/EXP;
const double IO = EXP + 1/EXP;
double I(int m)
{       if (m == 0) return IE;
        double t = m % 2 ? IO : IE;
        return t - m * I(m-1);
}
```

The values thus obtained are given in the table:

m	I_m	m	I_m
0	2.350402	4	0.552373
1	0.735759	5	0.324297
2	0.878885	6	0.404618
3	0.449507	7	0.253834

The coefficients a_k may now be calculated in terms of these integrals, expanding the polynomial in powers of x in each case. Since we are working only to degree 3 this is very little hard work. We find $a_0 = 1.17520$, $a_1 = 1.10364$, $a_2 = 0.35782$, $a_3 = 0.07045$. The following error distribution is then obtained:

x	$e(x)$	x	$e(x)$
−1.0	0.0089	0.1	0.0035
−0.9	0.0021	0.2	0.0026
−0.8	−0.0019	0.3	0.0011
−0.7	−0.0037	0.4	−0.0008
−0.6	−0.0039	0.5	−0.0027
−0.5	−0.0029	0.6	−0.0042
−0.4	−0.0014	0.7	−0.0045
−0.3	0.0004	0.8	−0.0028
−0.2	0.0020	0.9	0.0020
−0.1	0.0031	1.0	0.0112
0.0	0.0037		

The error is much more uniformly distributed than in the case of the Taylor polynomial, and the maximum error, which occurs in both cases at $x = 1$, is reduced by a factor of nearly 5. The error near $x = 0$ is much larger, but still not so large as at the extremities.

Expanded in powers of x, the "best" Legendre approximation of degree three may be written:

$$p(x) = (a_0 - a_2/2) + (a_1 - 3a_3/2)x + 3a_2x^2/2 + 5a_3x^3/2$$
$$= 0.99629 + 0.99796x + 1.07345x^2/2! + 1.05677x^3/3!.$$

We have written it in this way to facilitate comparison with the Taylor polynomial, where the numerical coefficients would all be unity.

17.4.3 Chebyshev approximation

The weighting function $w(x) = 1/\sqrt{(1 - x^2)}$ gives greater weight to the extremities of the interval $[-1, 1]$ than to the neighbourhood of $x = 0$. The Chebyshev polynomials, which have this weighting function, therefore offer the possibility that the error of approximation will be more uniformly distributed over the interval than in the Legendre case.

This time the coefficients are given by

$$a_k = \int_{-1}^{+1} \exp(x) T_k(x)/\sqrt{(1 - x^2)} . dx/N_k,$$

where $N_0 = \pi$, $N_k = \pi/2$, $k > 0$, and in general they have to be computed by quadrature. The results for the exponential function are $a_0 = 1.26607$, $a_1 = 1.13032$, $a_2 = 0.27150$, $a_3 = 0.04434$, and the distributed error is given below:

x	$e(x)$	x	$e(x)$
−1.0	+0.0050	0.1	+0.0053
−0.9	−0.0010	0.2	+0.0042
−0.8	−0.0041	0.3	+0.0024
−0.7	−0.0051	0.4	+0.0001
−0.6	−0.0045	0.5	−0.0024
−0.5	−0.0030	0.6	−0.0046
−0.4	−0.0009	0.7	−0.0058
−0.3	+0.0014	0.8	−0.0052
−0.2	+0.0033	0.9	−0.0017
−0.1	+0.0047	1.0	+0.0061
0.0	+0.0054		

Notice that there is a marked improvement over the Legendre case, the maximum error being reduced by a further factor of about 2. The central error is a little worse, however, reflecting the more uniform distribution of error.

Expanded in powers of x, the "best" Chebyshev approximation of degree three is:

$$P(x) = (a_0 - a_2) + (a_1 - a_3)x + 2a_2x^2 + 4a_3x^3$$

$$= 0.99457 + 1.08598x + 1.08598x^2/2! + 1.06408x^3/3!.$$

Chebyshev polynomials give almost the best possible polynomial approximation of functions. Notice that the magnitude of the distributed error at its maxima or minima is almost a constant, about 0.005 to 0.006. The reason is very simple. If we were to represent the function $f(x)$ by an *infinite* series, $f(x) = \sum_{k=0}^{\infty} a_k T_k(x)$, and if the series is rapidly convergent—as it is in the case of $f(x) = e^x$—then the error of approximating $f(x)$ by the truncated series $\sum_{k=0}^{n} a_k T_k(x)$ is approximately $a_{n+1} T_{n+1}(x)$. Thus we would expect the error of the third-degree Chebyshev approximation of e^x to vary approximately like $T_4(x)$, which it does. Further calculation shows that $a_4 \approx 0.0055$, which is the approximate magnitude of the maximum and minimum errors shown in the table.

17.5 Minimax approximation

The minimax approximation to a function $f(x)$ over an interval $[a,b]$ by a polynomial $Q(x; a_0, a_1, \ldots, a_n)$ is one which minimizes the infinite norm $\|f - Q\|_\infty$. Thus, we seek to minimize

$$\underset{x \in [a, b]}{\text{Max}} |f(x) - Q(x)|$$

with respect to the polynomial coefficients a_k, and hence the name "minimax".

In general it is not easy to obtain the true minimax approximation. Our present objective is a more limited one: it is to show how to obtain the minimax polynomial of degree three for the function $f(x) = e^x$ over the interval $[-1, 1]$, in order to illustrate the nature of the problem. Of particular interest will be the comparison of the minimax approximation with the Chebyshev approximation.

We are guided by the following two pieces of intuition, both of which may be expressed as formal theorems. First, the minimax polynomial if it exists is unique. Second, the successive maxima and minima of the distributed error are of equal magnitude, but alternate in sign. We shall appeal to the fact that we successfully constructed the Chebyshev approximation and that it very nearly possessed the property of alternating error. In short, we *assume* that the distributed error of the minimax polynomial of degree n which approximates the function $f(x)$ is roughly proportional to $T_{n+1}(x)$, and we therefore write

$$f(x) - Q_n(x; a_k) \approx \varepsilon T_{n+1}(x),$$

where ε is a constant representing the nearly equal magnitudes of the maximum and minimum errors. We have to choose the coefficients a_k of the polynomial Q so that this becomes as nearly true as we are pleased to prescribe. We shall take as starting approximation to the minimax process the Chebyshev approximation of degree n, and we shall attempt to iterate on the coefficients in order to approach the minimax approximation more nearly. It is sufficient for our purpose to assume that the maxima and minima of the error lie near to the maxima and minima of the Chebyshev polynomial of degree $n + 1$. Remembering that $T_n(x) = \cos(n \cos^{-1} x)$, these occur at the $n + 2$ nodes given by

$$x_j = \cos(j\pi/(n + 1)).$$

We therefore start by finding the coefficients a_k of the polynomial Q, for which

$$f(x_j) - Q(x_j; a_0, \dots, a_n) = (-1)^j \varepsilon.$$

This gives us $n + 2$ linear equations (corresponding to the values $j = 0, 1, \dots, n + 1$) for the $n + 1$ coefficients a_k and the additional quantity ε, which we may solve by any standard method provided the degree desired is not too large.

We then calculate the distributed error $e(x) \equiv f(x) - Q(x)$ of the approximation. Its maxima and minima will not be exactly at the nodes x_k, but we may hope they will be close by. They should occur at points x'_k which are the roots of the equation

$$e'(x) = f'(x) - Q'(x) = 0,$$

which may be found by Newton's method, provided that the second derivative $f''(x)$ is known (it certainly is if $f(x) = e^x$), and provided the nodes x_k are sufficiently close to the desired x'_k. The process may now be repeated until the ratio of the optimum of largest magnitude to that of smallest magnitude differs from unity by less than some small quantity such as 2%. The result will approximate more closely the minimax limit than the Chebyshev approximation with which we started.

The following procedure, which was devised by Remes, achieves the desired objective:

```
void  remes(func f, func f1, func f2, vector &coeff, double &err)
//    approximate f with polynomial of degree n-2
{     int n = 1 + coeff.getsize();
      vector node(n);
      chebynodes(node);
      matrix m(n);
      double range = maxrange * 2;
      while (range > maxrange)
      {     for (int i = 0; i < n; ++i)
            {     double t = 1.0;
                  for (int j = 0; j < n-1; ++j)
                  {     m[i][j] = t;
                        t *= node[i];
                  }
                  m[i][n-1] = -1+2*(i%2); // forced error term
            }
            vector b(n);      // because coeff.size != node.size
            for (int k = 0; k < n; ++k) b[k] = f(node[k]);
            crout(m, b);
            for (k = 0; k < n-1; ++k) coeff[k] = b[k];
            newnodes(f1, f2, coeff, node);
            range = maxtomin(f, coeff, node);
            err = fabs(b[n-1]);               // estimate of error
      }
}
```

The function $f(x)$ and its first two derivatives are supplied by the user as

```
double func f(double);   double f1(double);   double f2(double);
```

respectively. The successive approximations to the nodes are stored in the vector node, and the routine chebyshev computes the starting approximation to these nodes. The polynomial coefficients, and the value of the "maximum" error ε are calculated using the standard crout routine, using an auxiliary matrix m and vector b. The routine newnodes straightforwardly uses Newton's algorithm to compute new nodes, and the function maxtomin calculates the ratio of the maximum magnitude optimum to the minimum magnitude optimum. If this ratio exceeds a preset value maxrange the while loop is repeated. Successive estimates of the maximum error are stored in the variable err.

With the function $f(x) = e^x$ over the interval $[-1, 1]$, and maxrange = 1.02, the minimax polynomial of degree three gives a distributed error:

x	$e(x)$	x	$e(x)$
-1.0	$+0.0055$	0.1	$+0.0055$
-0.9	-0.0009	0.2	$+0.0046$
-0.8	-0.0043	0.3	$+0.0029$
-0.7	-0.0055	0.4	$+0.0006$
-0.6	-0.0050	0.5	-0.0019
-0.5	-0.0035	0.6	-0.0041
-0.4	-0.0013	0.7	-0.0054
-0.3	$+0.0010$	0.8	-0.0050
-0.2	$+0.0031$	0.9	-0.0018
-0.1	$+0.0046$	1.0	$+0.0055$
0.0	$+0.0055$		

The maxima and minima are almost equal in magnitude. The minimax polynomial is

$$p(x) = 0.99453 + 0.99568x + 1.08616x^2/2! + 1.07711x^3/3!.$$

The Chebyshev approximation is seen to be a close approximation to the minimax approximation in this case.

17.6 Gaussian quadrature

Orthogonal polynomials can be applied most powerfully to the evaluation of integrals. The work of this section may be regarded as a generalization and extension of the methods of integration presented in the last chapter.

We consider the integral $\qquad I(f) = \displaystyle\int_a^b w(x) f(x) \, dx$

where $w(x)$ is a suitable weight function as described in 17.1. Using Lagrangian interpolation as in 14.1, we may express the function $f(x)$ in terms of $n + 1$ nodes x_i, chosen in a manner yet to be decided:

$$f(x) = \sum_{i=0}^{n} L^{(n)}_i(x) f(x_i) + E_n,$$

where the interpolation polynomials of degree n are given by

$$L^{(n)}_i(x) = \prod_{j \neq i}^{n} (x - x_j)/(x_i - x_j)$$

and the error term is $\qquad E_n = \prod_i (x - x_i) \cdot f^{(n+1)}(\xi)/(n + 1)!.$

Recall that Lagrangian function approximation is exact if f is a polynomial of degree no greater than n, as this formula for the error makes plain. The value of the integral is given approximately by

$$I(f) \approx \sum_{i=0}^{n} w_i f(x_i),$$

where

$$w_i = \int_a^b w(x) L^{(n)}{}_i(x) \, dx$$

is independent of the function f and may be calculated once the nodes x_i have been determined.

The object of gaussian quadrature is to choose the nodes x_i in such a way that this approximation to the integral is *exact* for a polynomial of as high a degree as possible; this degree is called the *precision* of the approximation. Clearly, it has precision at least n. It will be recalled that the trapezoidal approximation has precision 1, and Simpson's approximation has precision 3.

Suppose we have a system of orthogonal polynomials such that $\int_a^b w(x)\phi_j(x)\phi_k(x)\,dx = 0$, when $j \neq k$. Let $q(x)$ and $r(x)$ be polynomials of degree no greater than m, say. Then consider the polynomial:

$$p(x) = q(x)\phi_{n+1}(x) + r(x).$$

It is of degree no greater than $m + n + 1$. Since $q(x)$ is a polynomial of degree no greater than m, it may be expressed as a linear combination of orthogonal polynomials of degree no greater than m: $q(x) = \sum_0^m q_k \phi_k(x)$, say.

Then, $$I(p) \equiv \int_a^b w(x)p(x)\,dx = \sum_0^m q_k \int_a^b w(x)\phi_k(x)\phi_{n+1}(x)\,dx + \int_a^b w(x) r(x)\,dx.$$

If $m < n + 1$ the first term on the right-hand side vanishes owing to the orthogonality of the functions ϕ_k. The highest value of m for which it vanishes is $m = n$, and henceforth we give it that value. Then the polynomial $r(x)$ is of degree no greater than n, and the second term, by Lagrangian interpolation, is therefore $\sum_0^n w_k r(x_k)$, where the x_k are the nodes still to be determined.

To determine the nodes we return to the definition of the polynomial p, considering its value at any one of the nodes, x_i, say:

$$p(x_i) = q(x_i)\phi_{n+1}(x_i) + r(x_i).$$

If we now choose the $n+1$ nodes to be the roots of $\phi_{n+1}(x) = 0$, then $p(x_i) = r(x_i)$. We already know that these roots all lie within the interval of integration. It follows that

$$I(p) \equiv \int_a^b w(x)p(x)\,dx = \sum_{i=0}^{n} w_i p(x_i)$$

where the polynomial $p(x)$ now has degree $2n + 1$, which is by definition the precision of the approximation when we substitute $f(x)$ for $p(x)$.

The calculation of the error is a little more complicated, but it is clearly proportional to $f^{(2n+2)}(\xi)/(2n+1)!$. The final formula for Gaussian quadrature is therefore

$$I(f) = \int_a^b w(x)f(x)\,dx = \sum_{i=0}^{n} w_i f(x_i) + O(f^{(2n+2)}(\xi)/(2n+2)!),$$

where the x_i are the roots of $\phi_{n+1}(x) = 0$, and the weights w_i are determined in terms of the weighting function $w(x)$, the Lagrange polynomials $L^{(n)}_i(x)$, and the roots x_i, and are independent of f.

Extensive tabulations of the nodes and the corresponding weights are available for a wide range of weight functions. Nevertheless, we indicate in the following sections how they may be computed directly.

17.6.1 Computation of nodes

We may compute the roots of $\phi_k(x) = 0$ by Newton's method if we know the first differential coefficients of the orthogonal polynomial ϕ_k. In 17.3.1 we presented a routine for calculating these for the Legendre polynomials. Similar routines may be written for other systems of orthogonal polynomials. We now describe a routine for computing the roots which is a modification of the function `polyroot` presented in 4.10.1, which uses Horner's algorithm.

First define the function type `polynomial`:

```
typedef void (*polynomial)(int, double, double&, double&);
```

An instance of this type is the function name `legendre`, and we assume that similar functions have been written for Chebyshev, Laguerre and Hermite polynomials. The definition of this function type allows us to present a routine for computing roots that applies to all these systems of polynomials. The routine is:

```
state polyroot(polynomial P, int n, double &x)
{       int iter = 0;
        state s = ITERATING;
        while (s == ITERATING)
        {       double p, q;        // value p, slope q
                P(n, x, p, q);
                if (fabs(p) < tolerance)
                        s = SUCCESS;
                else if (fabs(q) < toosmall)
                        s = NEARZERO;
                else
                {       x -= p/q;    // Newton's algorithm
```

```
            if (++iter == maxiter)
                s = WONTSTOP;
        }
    }
    return s;
}
```

where the user has to provide suitable values for the constants `tolerance`, `toosmall` and `maxiter`, and an approximation close enough to the root desired that it will not be confused with any other root. All the roots $x^{(n)}_i$ of a given polynomial ϕ_n may conveniently be stored in a vector of dimension n. In finding starting values it is helpful to recall from 17.2 that roots of successive degrees interleave.

17.6.2 Computation of weights

Once the nodes are known, the weights may be calculated from the formula we obtained in 17.5, namely:

$$w_i = \int_a^b w(x) L^{(n)}_i(x)\,dx.$$

This, however, is not a very convenient formula to use for computation. Instead, we shall go back to the definition of error:

$$E_n(f) = \int_a^b w(x) f(x)\,dx - \sum_{k=0}^{n} w_i f(x_i).$$

We require the weights w_i to satisfy these equations, with zero error for all polynomials of degree not greater than the precision. It is, of course, sufficient that they satisfy the equations for monomials x^j, where $j = 0, 1, \ldots, n$. This gives us $n + 1$ equations for the $n + 1$ weights:

$$\sum_{i=0}^{n} w_i x_i^j = q_j,$$

where

$$q_j = \int_a^b w(x) x^j\,dx, \qquad j = 0, 1, \ldots, n,$$

are the moments of the weight function.

These equations may also be written in matrix form:

$$
\begin{pmatrix}
1 & 1 & \ldots & 1 \\
x_0 & x_1 & \ldots & x_n \\
\ldots\ldots\ldots\ldots\ldots \\
x_0^{n-1} & x_1^{n-1} & \ldots & x_n^{n-1} \\
x_0^n & x_1^n & \ldots & x_n^n
\end{pmatrix}
\begin{pmatrix}
w_0 \\
w_1 \\
. \\
w_{n-1} \\
w_n
\end{pmatrix}
=
\begin{pmatrix}
q_0 \\
q_1 \\
. \\
q_{n-1} \\
q_n
\end{pmatrix}.
$$

In this form the weights may be found by normal matrix methods; and it is easy to check, when $n = 0, 1, 2$, that the solutions are exactly those given in terms of the Lagrange polynomials. The matrix on the left, however, is a Vandermonde matrix, similar to that introduced in chapter 14.

There exists a very concise algorithm for extracting the weights from these equations, due to Rybicki (see the routine **vander** in the book by Press *et al.*):

```
void  rybicki(vector &x, vector &w, vector &q)
{       int n = x.getsize();
        if (n != w.getsize() || n != q.getsize())
            error("inconsistent vectors in rybicki(...)");
        if (n == 1) w = q;
        else
        {     int i, j;
              vector v(n);
              v[n-1] = -x[0];
              for (i = 1; i < n; ++i)
              {     double xi = -x[i];
                    for (j = n-1-i; j < n-1; ++j)
                          v[j] += xi*v[j+1];
                    v[n-1] += xi;
              }
              for (i = 0; i < n; ++i)
              {     double xi = x[i];
                    double t = 1.0, b = 1.0, s = q[n-1];
                    int k = n-1;
                    for (j = 1; j < n; ++j)
                    {     b = v[k--] + xi*b;
                          s += q[k]*b;
                          t = xi*t + b;
                    }
                    w = s/t;
              }
        }
}
```

17.6.3 Nodes and weights combined

It is convenient to have a procedure for calculating nodes and weights of polynomials of a given degree at the same time. First we need to calculate the moments of the weight function. In the case of the Legendre polynomials, where $w(x) = 1$, and $q_i = 2/(i + 1)$ if i is even, and zero otherwise, we may write a very simple function:

```
double legemom(int n)
{       if (n%2) return 0.0;
        return 2.0/(n+1);
}
```

Similarly:

```
double chebmom(int n)
{       if (n==0) return PI;
        if (n%2) return 0.0;
        double t = 1.0;
        for (int i = 1; i < n; i+=2)
                t *= double(i)/(i+1);
        return t * PI;
}

double lagumom(int n)
{       if (n==0) return 1.0;
        double t = 1.0;
        for (int i = 1; i <= n; ++i)
                t *= i;                     // n!
        return t;
}

double hermmom(int n)
{       static double rtpi = sqrt(PI);
        if (n==0) return rtpi;
        if (n%2) return 0.0;
        double t = 1.0;
        for (int i = 1; i < n; i+=2)
                t *= double(i)/2;
        return t * rtpi;
}
```

If we now define the function type

```
typedef double (*moment)(int);
```

we may refer to any of the moment functions without revealing our choice until the final implementation.

Let us suppose that we wish to compute all the zeros and weights of degree no greater than some integer N defined globally. When computing the zeros of degree n, we shall need to employ a starting vector x0 containing n approximate values. In estimating these we shall use the fact that they interleave the actual zeros of degree $n - 1$. We shall also need a vector m which stores the moments of the weight function, and it is convenient to compute these once and for all up to degree N. Therefore we shall need two vectors of degree N, which it is convenient to declare globally:

```
vector x0(N), m(N);
```

Then for any particular polynomial system we may compute the moment vector:

```
void   makemoments(moment mf)
{      for (int i = 0; i < N; ++i)
           m[i] = mf(i);
}
```

Let us suppose that the zeros of degree $n - 1$ are known. First approximations to the zeros of degree n may be taken to lie halfway between the zeros of degree $n - 1$, and at the extremities halfway between the first zero and the lower limit of integration a, and halfway between the last zero and the upper limit of integration b. This works well in all cases studied, except for the Laguerre polynomials where it is better to take as starting values the actual zeros of degree $n - 1$, supplemented by a last approximate zero some distance further on; otherwise there is a danger of missing zeros.

We shall wish to store the zeros and weights on file for future use. We therefore define data files with names such as:

```
char   *legefile = "LEGEDATA.DTA";
char   *chebfile = "CHEBDATA.DTA";
char   *lagufile = "LAGUDATA.DTA";
char   *hermfile = "HERMDATA.DTA";
```

and an output file

```
ofstream ofile;
```

(in the manner of 6.4) which will eventually be opened for writing.

The following function calculates the zeros of degree n and the corresponding weights given the zeros of degree $n - 1$ stored in the global vector x0 and the moments stored in the global vector m:

```
void   quadrate(polynomial P, int n, double a, double b)
{      vector x(n);                    // zeros of degree n
       if (n == 1) x0[0] = 0.0;
       int i;
       for (i = 0; i < n; ++i)                      // set starting values
       {      if (P == laguerre)
              {      if (i == 0) x[i] = x0[0];
                     else if (i == n-1) x[i] = x0[n-2] + 5.0;
```

```
                      else x[i] = x0[i];
            }     else
            {     if (i == 0) x[i] = (a + x[0])/2;
                  else if (i == n-1) x[i] = (x0[n-2] + b)/2;
                  else x[i] = (x0[i-1] + x0[i])/2;
            }
            polyroot(P, n, x[i]);        // compute x[i]
      }
      vector q(n), w(n);
      for (i = 0; i < n; ++i)
      {     q[i] = m[i];                 // read moments into q
            x0[i] = x[i];                // update x0 from x
      }
      rybicki(x, w, q);                  // compute weights
      for (i = 0; i < n; ++i)
      {     ofile.write((char*)&x[i], sizeof(double));
            ofile.write((char*)&w[i], sizeof(double));
            cout << fpw << x[i] << fpw << w[i] << reset << endl;
      }
}
```

where we have chosen also to echo the results to screen, with formatting given by:

```
long  L = ios::left + ios::showpos + ios::fixed;
format fpw(L, 10, 16), reset(0, 0, 0);
```

Of course, the starting values are not known unless the zeros of degree $n - 1$ have already been computed. Therefore we must call the function quadrate for each value of n up to the largest required:

```
void  alldegs(polynomial P, moment mf, double a, double b)
{     makemoments(mf);
      for (int n = 0; n <= N; ++n)
      {     cout << "n = " << n << endl;
            quadrate(P, n, a, b);
            newline;
      }
}
```

Finally, we need to call this function in an executable file after opening the output file properly. In the case of the Legendre polynomials:

```
main()
{     clear;
      ofile.open(legefile, ios::out | ios::binary);
```

```
    alldegs(legendre, legemom, -1, 1);
    ofile.close();
}
```

In the cases of the Laguerre and Hermite polynomials the infinite limits of integration should be replaced by a suitable large value such as const double HUGE = 100;. We have computed all the zeros and weights for the four systems of polynomials given above, for all degrees not exceeding $N = 16$.

The following sample nodes and weights are for Gauss–Legendre quadrature:

n	x_i	w_i
0	0.0	2.0
1	±0.57735027	1.0
2	±0.77459667	0.55555556
	0.0	0.88888889
3	±0.86113631	0.34785485
	±0.33998104	0.65214515

The case of $n = 0$ is the midpoint method, while the case of $n = 1$ was referred to as the "Gauss rule" in 16.3.

For the Chebyshev polynomials the results are already known: the zeros are at $x^{(n)}_i = \cos((2i - 1)\pi/2n)$, and the corresponding weights are $w^{(n)}_i = \pi/n$. It is useful, however, to run quadrate in this case in order to get some idea of the accuracy with which zeros and weights can be calculated. Setting tolerance = 10^{-12} we obtain the Chebyshev weights correct to about $0.5.10^{-10}$ in the case of $n = 16$.

17.6.4 An example

If we wish to compute an integral by gaussian quadrature, it is first necessary to extract the nodes and weights from the appropriate file. As in 6.4, we therefore declare a global input file:

```
ifstream ifile;
```

where ifile will be initialized to the appropriate data file when it is opened. The data were written to file as pairs of floating point numbers of double precision, each node followed by the corresponding weight, beginning with $n = 1$ and ending with $n = 16$ (although we could have pushed the computation to higher degrees had we wished). In order to extract nodes and weights of a given degree n we have to position the file reading pointer to the right place in the file; this we can do using the ostream member function seekg introduced in 6.4.2. The data referring to degree n begin at byte number $\frac{1}{2}n(n - 1) \times 2 \times$ sizeof(double), so we write a function:

```
void   seekdeg(int n)
{      long offset = n*(n-1)*sizeof(double);
       ifile.seekg(offset, ios::beg);
}
```

to position the file pointer correctly.

Having positioned the file, we may now read out pairs of doubles, x and w, and construct the value of the integral:

```
typedef double (*func)(double);
double gaussint(func f, int n)
//     limits of integration implicit in polynomials used
{      seekdeg(n);
       double I = 0.0;
       for (int i = 0; i < n; ++i)
       {      double x, w;
              ifile.read((char*)&x, sizeof(double));
              ifile.read((char*)&w, sizeof(double));
              I += w * f(x);
       }
       return I;
}
```

Given a particular function $F(x)$, and data file `datafile`, the following program displays the values of the integral for successive numbers of nodes n. The constants `npw` and `ipw` are appropriate objects of the class `format` introduced in 6.3.4:

```
main()
{      clear;
       int N = 16;
       ifile.open(datafile, ios::in | ios::binary);
       cout << npw << "n" << ipw << " integral" << rst << "\n\n";
       for (int n = 1; n <= N; ++n)
       {      double I = gaussint(F, n);
              cout << npw << n << ipw << I << rst << "\n";
       }
       ifile.close();
}
```

For the integral

$$I(2\pi) = \int_{0}^{2\pi} \exp(-x)\cos(x)\,dx \approx 0.499066278634$$

studied in the last chapter, Legendre polynomials (after the appropriate change of variable to make the range of integration $[-1, +1]$) give the following printout:

```
n        integral
0        -0.27152106
1        +0.20569247
2        +0.53735632
3        +0.50436942
4        +0.49893518
5        +0.49905461
6        +0.49906638
7        +0.49906629
8        +0.49906628
9        +0.49906628
. . . . . . . . . . . . . . . . .
```

8-figure accuracy is achieved with just 9 nodes: that means just 9 function calls!

If we change the upper limit of integration to infinity, then we may use Laguerre polynomials to compute the value of the integral

$$I(\infty) = \int_0^\infty \exp(-x)\cos(x)\,dx = 1/2$$

This time we may expect rather more nodes to be required because of the oscillatory behaviour of the integrand $\cos x$ (the exponential is now the weighting function). To achieve 8 figure accuracy 13 nodes are needed, but this is still a very good performance:

```
n        integral
0        +0.54030231
1        +0.57020877
2        +0.47652084
3        +0.50249371
4        +0.50053849
5        +0.49973750
6        +0.50004249
7        +0.50000121
8        +0.49999777
9        +0.50000051
10       +0.49999997
11       +0.49999999
12       +0.50000000
13       +0.50000000
. . . . . . . . . . . . . . . . .
```

18 Differential equations

We are concerned in the first instance with the computation of the solution of a first-order differential equation with one independent variable and a given initial condition, sometimes known as the initial value problem:

$$dy/dx = f(x, y), \quad y(x_0) = y_0.$$

The general idea, given that we know the value of y at $x = x_0$, is to compute its value at $x = x_0 + h$, where h is sufficiently small. In first approximation, also known as *Euler's approximation*, this value may be given by the first terms of the Taylor expansion of $y(x)$ about x_0 and is:

$$y(x_0 + h) \approx y_0 + hf(x_0, y_0).$$

Once this value is known, the approximate value at $x = x_0 + 2h$ may be computed, and so on. We shall be interested in whether the values of y so calculated increasingly depart from their true values the larger is $|x - x_0|$. We shall also need to use higher order terms in the Taylor expansion in order to obtain greater accuracy. Clearly, we must distinguish between the local, or truncation, error of such a formula, and the global, or accumulated, error after many steps.

First, however, we need to know in what circumstances a differential equation has a solution at all, and if it does whether it is unique. We also have to examine its stability. In the next chapter we shall extend the theory to systems of differential equations with several dependent variables, and to differential equations of higher order. We shall in general follow the analytical treatment of K. E. Atkinson.

18.1 Existence and uniqueness

It is obvious that there are initial value problems for which unique solutions exist. For example, if $y' = \lambda y$, $y(a) = \eta$, so that $f(x,y) = \lambda y$, then

$$y(x) = Ce^{\lambda x}, \quad \text{where } y(a) = Ce^{\lambda a} = \eta, \quad \text{so that } C = \eta e^{-\lambda a}, \text{ and } y(x) = \eta e^{\lambda(x-a)}.$$

For the general case, we state without proof the following

Theorem.

Let the function f be defined and continuous for all points (x, y) in a domain D defined by $a \leq x \leq b$ with a and b finite, and $-\infty < y < +\infty$; and let there exist a finite constant $L \geq 0$ such that

$$\left| f(x, y_1) - f(x, y_2) \right| \leq L \left| y_1 - y_2 \right|$$

for all points (x, y_1), (x, y_2) in D: then there exists a unique solution $y(x)$ which is continuous and differentiable at all points in D. The inequality is known as a *Lipschitz condition*, and the constant L is called a Lipschitz constant.

Note that if $\partial f(x, y)/\partial y$ exists and is bounded on D, then by the mean value theorem

$$f(x, y_1) - f(x, y_2) = [\partial f(x, y)/\partial y]_{y=\eta} (y_1 - y_2), \text{ for some } \eta \in [y_1, y_2].$$

We may therefore take $L = \mathrm{Max}_{(x,y) \in D} \left| \partial f(x,y)/\partial y \right|$, and the conditions of the theorem are satisfied. Note that the theorem provides a *sufficient* condition for the existence of a unique solution; it is not a necessary condition.

To show the Lipschitz condition at work we may convert the initial value problem to an integral equation:

$$y(x) = y_0 + \int_{x_0}^{x} f(t, y(t)) \, dt.$$

Define a sequence of functions,

$$y_0(x) = y_0, \quad y_{n+1}(x) = y_0 + \int_{x_0}^{x} f(t, y_n(t)) \, dt, \quad n > 0.$$

Then the difference,

$$y(x) - y_{n+1}(x) = \int_{x_0}^{x} [f(x, y(t)) - f(x, y_n(x))] dt,$$

may be estimated for an interval $I = [x_0 - \alpha, x_0 + \alpha]$ about x_0, where $\alpha > 0$ is some number, not necessarily small:

$$\left| y(x) - y_{n+1}(x) \right| \leq L \int_{x_0}^{x} \left| y(t) - y_n(t) \right| dt$$

$$\leq \alpha L \left\| y - y_n \right\|_{\infty}.$$

Here, the right-hand side is independent of x, so on the left-hand side we may take the maximum value and obtain:

$$\left\| y - y_{n+1} \right\|_{\infty} \leq \alpha L \left\| y - y_n \right\|_{\infty}.$$

The sequence of functions $y_n(x)$ therefore converges linearly to the solution $y(x)$, provided α is chosen sufficiently small that $\alpha L < 1$. It may similarly be shown that if $y^{(1)}$ and $y^{(2)}$ are solutions of the same initial value problem, then $\left\| y^{(1)} - y^{(2)} \right\|_\infty \le \alpha L \left\| y^{(1)} - y^{(2)} \right\|_\infty$, so that the solutions are identically equal, which is to say that the solution is unique.

Although hardly to be recommended, this process—known as *Picard iteration*—may actually be used to obtain a solution. Consider for example the initial value problem $y' = -2xy$, $y(0) = 1$, the solution of which is $y(x) = \exp(-x^2)$. The successive iterations,

$$Y_0(x) = 1, \; Y_1(x) = 1 - x^2, \; Y_2(x) = 1 - x^2 + x^4/2!,$$

$$Y_3(x) = 1 - x_2 + x^4/2! - x^6/3!, \ldots\ldots\ldots\ldots,$$

obviously converge to the Taylor expansion in x^2 of the exponential function $\exp(-x^2)$.

Consider next the problem $y' = y \cos x$, $0 \le x \le 1$, $y(0) = 1$. We have $f(x, y) = y \cos x$, so that $\partial f/\partial y = \cos x$, and f satisfies a Lipschitz condition on the domain D given by $0 \le x \le 1$, $-\infty < y < +\infty$, with $L = 1$. The function is continuous on D, so there is a unique solution which happens to be $y(x) = e^{\sin x}$. On the other hand, consider the problem $y' = y^{1/2}$, $y(0) = 0$, for which $\partial f/\partial y = \frac{1}{2}y^{-1/2}$, which is discontinuous at $y = 0$. On any interval containing the origin, the value of L is infinite; thus the conditions of the theorem are not satisfied, and in this case there are *two* solutions, $y(x) = 0$, and $y(x) = x^2/4$. Clearly, it is important to study the application of a Lipschitz condition before embarking upon a possibly expensive program of computation.

18.2 Stability and conditioning

In seeking a computed solution to an initial value problem we obviously need to know whether or not we are computing an approximation to a unique solution, and we have just dealt with that problem. Given that computation is of limited accuracy, however, we need to press the matter further. In particular, we need to know whether a small change in the statement of the initial value problem induces only a correspondingly small change in the computed solution. If it does not, the result is likely to be meaningless.

Consider the perturbed problem, $y' = f(x, y) + \delta(x)$, $y(x_0) = y_0 + \varepsilon$,

where it is assumed that the perturbing function $\delta(x)$ is continuous. We assume also that $|\varepsilon| < \varepsilon_0$, and $\|\delta\|_\infty < \varepsilon_0$. Denote the perturbed solution by $y(x; \delta, \varepsilon)$; it satisfies:

$$y(x; \delta, \varepsilon) = y_0 + \varepsilon + \int_{x_0}^x f(t, y(t; \delta, \varepsilon))\, dt.$$

Subtracting the unperturbed equation, we have

$$y(x; \delta, \varepsilon) - y(x) = \varepsilon + \int_{x_0}^x [f(t, y(t; \delta, \varepsilon)) - f(t, y(x))]\, dt + \int_{x_0}^x \delta(t)\, dt.$$

Then, proceeding as before, we can show that on an interval $I = [x_0 - \alpha, x_0 + \alpha]$ such that $\alpha L < 1$,

$$\| y(x; \delta, \varepsilon) - y \|_\infty \leq k(1 + \alpha)\varepsilon,$$

where $k = 1/(1 - \alpha L)$. Thus, the solution y depends continuously on the data of the problem and the initial value problem is said to be *stable*. Notice, however, that the constant k could in principle be quite large.

Even a stable problem may be ill-conditioned for numerical work. Consider as an example the somewhat artificially stated problem:

$$y' = 100y - 101e^{-x}, \qquad y(0) = 1,$$

the solution of which is $y(x) = e^{-x}$. Suppose we perturb just the initial value, so that $y(0) = 1 + \varepsilon$. Then the solution becomes

$$y(x; \varepsilon) = e^{-x} + \varepsilon e^{100x}$$

as may readily be seen by direct substitution. For any perturbation $\varepsilon \neq 0$, the solution will diverge from the unperturbed solution very rapidly. Notice that $\partial f(x, y)/\partial y = 100 > 0$ in this case. In well-conditioned problems this quantity would be negative. If we state the same problem differently, namely

$$y' = -y, \qquad y(0) = 1,$$

then $\partial f(x,y)/\partial y = -1 < 0$. The exact solution of the perturbed problem is now

$$y(x; \varepsilon) = (1 + \varepsilon)e^{-x}.$$

To understand this behaviour in a more general case, let the perturbed initial condition produce a change Δy in the exact solution:

$$\Delta y(x) \equiv y(x; \varepsilon) - y(x) = \varepsilon + \int [f(t, y(t; \varepsilon)) - f(x, y(t))]\, dt$$

$$\approx \varepsilon + \int [\partial f(t, y(t))]\Delta y(t; \varepsilon)\, dt,$$

where the approximation involves the mean value theorem and is valid so long as the perturbation does not become too large. This approximate equation is linear in Δy and the corresponding approximate solution is readily seen to be:

$$\Delta y \approx \varepsilon \exp[\int (\partial f/\partial y)\, dt]$$

which grows exponentially if $\partial y/\partial y > 0$.

18.3 Euler's method

We have already given the basis of Euler's method, which may be regarded as the first approximation to all other methods, and which therefore deserves special study even though it is not very accurate. We state it afresh. We seek an approximate solution to the initial value problem:

$$y' = f(x, y), \quad a = x_0 \le x \le x_n = b, \quad y(x_0) = y_0.$$

Using Taylor's theorem, we may set

$$y(x + h) = y(x) + hy'(x) + \tfrac{1}{2}h^2 y''(\xi), \quad \text{where } x < \xi < x + h.$$

Dropping the truncation term, setting $y' = f$, and taking $h = (b - a)/n$, we may replace the differential equation by the system of difference equations relating the increment in y to the increment in x at each step:

$$x_{i+1} = x_i + h, \quad y_{i+1} = y_i + hK_i$$

where $K_i = f(x_i, y_i)$, and where in each case y_i is the approximate value of y for $x = x_i \equiv a + ih$, as determined from the previous step. Of course, y_0 is given as the initial condition. The local error at each step, as we have seen, is $O(h^2)$, but the accumulated error as x approaches b might still become quite large.

In order that the routines we develop may be applicable to different functions we first define a function type to represent the function $f(x, y)$:

```
typedef double (*func)(double, double);
```

The Euler increment is then given by the function:

```
double euler(func f, double x, double y, double h)
{       double K = f(x,y);
        return h*K;
}
```

The reasons for this somewhat elaborate nomenclature will appear when we generalize the Euler method to higher orders of approximation.

For the sake of definiteness we shall study the problem

$$y' = x - y + 1, \quad y(0) = 1.$$

We have $f(x, y) = x - y + 1$, so that $\partial f/\partial y = -1$ and $L = 1$. Moreover, f is continuous, so there exists a unique solution:

$$y(x) = x + \exp(-x).$$

In what follows, we shall refer to this as "our usual problem". If we set $h = 0.2$, then over the range $0 \leq x \leq 2$ we obtain the comparison with the exact solution shown in the table:

x	Exact	Euler	Error	hD
0.0	1.0	1.0	0.0	0.0
0.2	1.018731	1.0	0.0187	0.0164
0.4	1.070320	1.04	0.0303	0.0268
0.6	1.148812	1.112	0.0368	0.0329
0.8	1.249329	1.2096	0.0397	0.0359
1.0	1.367879	1.32768	0.0402	0.0368
1.2	1.501194	1.462144	0.0391	0.0361
1.4	1.646597	1.609715	0.0369	0.0345
1.6	1.801897	1.767772	0.0341	0.0323
1.8	1.967299	1.934218	0.0311	0.0298
2.0	2.135335	2.107374	0.0280	0.0271

The column labelled hD we shall come to presently. However, it is interesting to see, at least in this case, that the errors do *not* greatly build up as the integration proceeds; indeed, they pass through a maximum at about $x = 1$. Moreover, if we examine the dependence of the error at $x = 1$ upon the step-size h, we see that it is roughly proportional to h:

h	0.2	0.1	0.05	0.025	0.01
Error	0.0402	0.0192	0.0094	0.0046	0.0018

In order to understand this behaviour we need to study how the truncation error at each step accumulates. Including the truncation term, the *exact* solution satisfies

$$y(x_{i+1}) = y(x_i) + hf(x_i, y(x_i)) + \tfrac{1}{2}h^2 y''(\xi_i),$$

for each i, whereas the *approximate* values y_i are given in the Euler method by

$$y_{i+1} = y_i + hf(x_i, y_i).$$

Denote by E_i the accumulated error at each step, namely $E_i = y(x_i) - y_i$. Then:

$$E_{i+1} \equiv y(x_{i+1}) - y_{i+1} = y(x_i) - y_i + h[f(x_i, y(x_i)) - f(x_i, y_i)] + \tfrac{1}{2}h^2 y''(\xi_i),$$

Assume y'' is bounded, so that there exists a number $M = \|y''\|_\infty$. Using the Lipschitz condition with constant L we can then set bounds on the errors:

$$|E_{i+1}| \leq (1 + hL)|E_i| + \tfrac{1}{2}h^2 M.$$

Let $h = (b - a)/n$ and $x_i = a + ih$. The magnitude of the accumulated error at any step is thus limited by the inequality:

$$|E_{i+1}| \leq (1 + hL)|E_i| + \tfrac{1}{2}h^2M$$

$$\leq (1 + hL)^{i+1}|E_0| + [1 + (1 + hL) + (1 + hL)^2 + \ldots + (1+hL)^i] \cdot \tfrac{1}{2}h^2M.$$

But $E_0 = 0$, and the remaining term is a geometric series. Therefore,

$$|E_i| \leq [((1 + hL)^i - 1)/L] \cdot \tfrac{1}{2}h^2M.$$

Now, if $t > -1$, and $m > 0$, then the fact that $0 \leq 1 + t \leq e^t$ implies $0 \leq (1 + t)^m \leq e^{mt}$.

Thus, $$|E_i| \leq (hM/2L)[\exp(L(x_i\text{-}x_0)) - 1]$$

for each x_i, where $i = 0,1, \ldots ,n$. In particular,

$$|E_n| \leq (hM/2L)[\exp(L(b - a)) - 1].$$

The most significant feature of this result is that the error is expected to be proportional to h as we found in the particular case above. However, this error bound allows the error to grow exponentially as the interval of integration increases. This may well be pessimistic, especially since in general a rough estimate is necessary for $M = \|y''\|_\infty$, which in any case reflects the greatest curvature of the true solution over the entire interval of integration. In our usual problem, we have $M = L = 1$, $b - a = 2$, so the bound we have calculated, namely $|E_n| \leq \tfrac{1}{2}h(e^2 - 1) \approx 0.6$, is much larger than the true error. That at least is to err on the right side. But, in any case, we have proved that the Euler method converges, because as $h \to 0$, $E_i \to 0$, and $y_i \to y(x_i)$.

Note, however, that in the problem $y' = y^{1/2}$, $y(0) = 0$, for which $L = \infty$, the solution generated by the Euler method is $y_{i+1} = y_i + hy_i^{1/2} \equiv 0$, since $y_0 = 0$, yielding the trivial solution $y(x) = 0$. The non-trivial solution $y(x) = x^2/4$ does not appear.

It is instructive to carry the analysis of the Euler method one step further in order to understand why the accumulated error behaves as it does. Proceeding to one further term in the Taylor expansion, we obtain

$$E_{i+1} = E_i + h[f(x_i, y(x_i)) - f(x_i ,y_i)] + \tfrac{1}{2}h^2y''(x_i) + O(h^3),$$

where $$f(x_i, y(x_i)) - f(x_i, y_i) = E_if_y(x_i, y(x_i)) + O(h^2),$$

and f_y is the usual abbreviation for $\partial f/\partial y$.

Therefore $$E_{i+1} = [1 + hf_y(x_i, y(x_i))]E_i + \tfrac{1}{2}h^2y''(x_i) + O(h^3).$$

We have shown that E_i is proportional to h. We may therefore set $E_i = hD_i$, say. Dropping the higher order terms we then have:

$$D_{i+1} = D_i + h[f_y(x_i, y(x_i))D_i + \tfrac{1}{2}y''(x_i)].$$

If we define $F(x, D) \equiv f_y(x, y(x))D + \tfrac{1}{2}y''(x)$, then $D_{i+1} = D_i + hF(x_i, D_i)$, which is the Euler method for the initial value problem:

$$D' = F(x, D), \quad D(0) = 0.$$

If we apply this analysis to our usual initial value problem with $y' = x - y + 1$, $y(0) = 1$, we have $\partial f/\partial y = -1$, $y'' = e^{-x}$, $F(x, D) = -D + \tfrac{1}{2}e^{-x}$, so that

$$D'(x) + D(x) = \tfrac{1}{2}e^{-x}, \quad D(0) = 0,$$

the solution to which is $D(x) = \tfrac{1}{2}x\exp(-x)$. Thus we expect the accumulated error to be

$$E_i = hD_i + O(h^2).$$

The values of hD are given in the above table; it will be seen that the actual errors follow hD rather closely, in particular showing a maximum at $x = 1$.

It is also worth noting that the effect of rounding errors can be included in the above analysis, simply by modifying the difference equations slightly:

$$y_{i+1} = y_i + hf(x_i, y_i) + \tfrac{1}{2}h^2y''(\xi_i) + \delta_i,$$

where δ_i represents the typical rounding error of about $\delta \approx 0.5.10^{-d}$, if one is working to d decimal digits. In the error bound which results, the factor $hM/2L$ is replaced by $[(hM/2) + (\delta/h)]/L$. Any attempt to improve accuracy by reducing the step size below about $h_0 \approx \sqrt{(2\delta/M)}$ will *increase* the accumulated error. If we work to double floating point precision we are not likely to meet this difficulty unless extreme accuracy is desired, and in that case Euler's method would certainly not be used.

Finally, we can examine the stability of the Euler method just as we examined the stability of the initial value problem itself. Confining ourselves to variations in the initial value, $y_0 = y(x_0) + \varepsilon$, we find

$$\text{Max}\,|y_i(\varepsilon) - y_i| \leq e^{L(b-a)}|\varepsilon|,$$

so that Euler's method offers a stable solution to a stable initial value problem.

18.4 Higher order methods

Since we derived Euler's method from a Taylor series it is natural to ask whether the method can be generalized by simply going to higher order in the Taylor expansion. If we take into account just the next term,

$$y(x + h) = y(x) + hy'(x) + \tfrac{1}{2}h^2y''(x) + O(h^3),$$

we obtain a difference scheme,

$$y_{i+1} = y_i + hf(x_i, y_i) + \tfrac{1}{2}h^2g(x_i, y_i),$$

where
$$g(x, y) = \partial f(x, y)/\partial x + f(x, y)\partial f(x, y)/\partial y.$$

Correspondingly, the "Taylor" increment is given by the function:

```
double taylor(func f, func g, double x, double y, double h)
{       double K = f(x,y);
        double L = g(x,y);
        K += h*L/2;
        return h * K;
}
```

With the usual problem that we studied before $g(x, y) = y - x$, and with $h = 0.1$ and $0 \leq x \leq 1$, we obtain the following results:

x	Exact	Euler	Taylor	Error
0.0	1.0	1.0	1.0	0.0
0.1	1.004837	1.0	1.005	0.00016
0.2	1.018731	1.01	1.019025	0.00029
0.3	1.040818	1.029	1.041218	0.00040
0.4	1.070320	1.0561	1.070802	0.00048
0.5	1.106531	1.09049	1.107076	0.00055
0.6	1.148812	1.131441	1.149404	0.00059
0.7	1.196585	1.178297	1.197210	0.00063
0.8	1.249329	1.230467	1.249975	0.00065
0.9	1.306570	1.287420	1.307228	0.00066
1.0	1.367879	1.348678	1.368541	0.00066

We show the first-order Euler results for comparison. The errors are much reduced, and proportional to h^2. However, in any extension of the Euler method which employs additional terms of the Taylor expansion there is the need to evaluate derivatives higher than the first, which may not always be easy analytically, and if performed numerically involves more function calls and additional errors. For this reason the Taylor expansion is not used directly in this manner, but indirectly, so as to maintain the same truncation error.

First a definition, however. An integration method is said to be *of order p* if p is the largest integer for which

$$y(x + h) = y(x) + h\phi(x, y, h; f) + O(h^{p+1}).$$

For the Euler method, $\phi(x, y, h; f) = f(x, y)$, and the order is one. For the specific "Taylor" method we have just discussed, $\phi(x, y, h; f) = f(x, y) + hf'(x, y)$, and the order is two. The question is whether it is possible to maintain the same order as a Taylor method without introducing higher derivatives. It is. Instead of evaluating f at the values x, $y(x)$ of its parameters, evaluate it at some point $x + \alpha h$, $y + \beta h$, and do so by means of a Taylor expansion about x, y:

$$\phi(x, y, h; f) = f(x + \alpha h, y + \beta h) \approx f(x, y) + \alpha h f_x + \beta h f f_y,$$

where we have used the short-hand $f_x = \partial f(x, y)/\partial x$, $f_y = \partial f(x, y)/\partial y$, and where we have stopped after the lowest terms. This has the same truncation error as the Taylor method of order two if and only if $\alpha = \frac{1}{2}$, $\beta = \frac{1}{2} f(x, y)$. The resulting integration method of order two is therefore

$$y_{i+1} = y_i + hf(x + \tfrac{1}{2}h, \quad y + \tfrac{1}{2}hf(x_i, y_i)),$$

which for obvious reasons is called the *midpoint method*. Notice that it requires *two* function calls at each step.

The midpoint increment may be computed using the function:

```
double midpoint(func f, double x, double y, double h)
{      double K1 = f(x,y);
       double K2 = f(x+h/2, y+h*K1/2);
       return h * K2;
}
```

For our usual problem, the midpoint method gives *exactly* the same result as the second order Taylor method, because if $f(x, y) = x - y + 1$, then $f(x + \frac{1}{2}h, y + \frac{1}{2}hf(x, y))$ reduces to $f(x, y) + \frac{1}{2}hf'(x, y)$. In general, this would not be the case.

18.5 Runge–Kutta methods

The midpoint method is a special case of a whole family of second-order methods for the initial value problem, and there are corresponding families of methods of higher order. We seek a function $\phi(x, y, h; f)$, not involving derivatives of f, the Taylor expansion of which will match the Taylor expansion of the solution up to a certain order. Let us generalize the form of ϕ we used to obtain the midpoint method:

$$\phi(x, y, h; f) = \gamma_1 f(x, y) + \gamma_2 f(x + \alpha h, y + \beta h f(x, y)).$$

This contains four parameters, so it is conceivable that a method of order higher than two might result. Let us therefore expand this expression up to terms of order h^2:

$$\phi(x, y, h; f) = (\gamma_1 + \gamma_2)f(x, y) + \gamma_2 h(\alpha f_x + \beta f f_y) + h^2(\tfrac{1}{2}\alpha^2 f f_{xx} + \alpha\beta f_{xy} + \tfrac{1}{2}\beta^2 f^2 f_y) + O(h^3),$$

where $f_{xx} = \partial^2 f(x, y)/\partial x^2$, etc. But we must also expand $y(x + h)$:

$$y(x + h) = y(x) + hy'(x) + (1/2)h^2 y''(x) + (1/6)h^3 y^{(3)}(x) + O(h^4)$$

$$= y(x) + h[f(x, y) + (1/2)h(f_x + f f_y)$$

$$+ (1/6)h^2(f_{xx} + 2f_{xy}f + f_{yy}f^2 + f_x f_y + f_y^2 f) + O(h^3)].$$

Remembering that $y(x + h) = y(x) + h\phi(x, y, h; f)$, we can attempt to equate coefficients of successive powers of h. For the linear term we have $\gamma_1 + \gamma_2 = 1$; for the quadratic term $\gamma_2\alpha = \gamma_2\beta = \tfrac{1}{2}$; so that $\gamma_2 = 1/2\alpha$, $\gamma_1 = 1 - (1/2\alpha)$, $\beta = \alpha$. There is no way of equating coefficients of the cubic term, so we have failed to increase the order of the method, and we have also failed to determine the four parameters uniquely. What we do have is the *Runge–Kutta family* of methods of order two distinguished by a single parameter α.

If we take $\alpha = \tfrac{1}{2}$, it follows that $\gamma_1 = 0$, and we regain the midpoint method with

$$\phi(x, y, h; f) = f(x + \tfrac{1}{2}h, y + \tfrac{1}{2}hf(x, y)).$$

On the other hand, if we take $\alpha = 1$, so that $\beta = 1$, $\gamma_1 = \gamma_2 = \tfrac{1}{2}$, we have the *improved Euler method* with

$$\phi(x, y, h; f) = \tfrac{1}{2}[f(x, y) + f(x + h, y + hf(x, y))].$$

Other choices of the parameter α are possible; one which (in a certain sense) minimizes the h^3 term is *Heun's method* with $\alpha = 2/3$:

$$\phi(x, y, h; f) \approx \tfrac{1}{4}[f(x, y) + 3f(x + (2/3)h, y + (2/3)hf(x, y))].$$

In future we shall refer to Euler's method of order one as RK1, and the improved Euler's method of order two as RK2, ignoring the other methods of order two. The RK2 increment is computed by the function:

```
double RK2(func f, double x, double y, double h)
{     double K1 = f(x, y);
      double K2 = f(x + h, y + h*K1);
      return h*(K1 + K2)/2;
}
```

The method remained second order because it was impossible to match the the h^3 terms. A match is possible, however, with a different function containing four parameters:

$$\phi(x, y, h; f) = f(x + \alpha_1 h, y + \beta_1 h f(x + \alpha_2, y + \beta_2 h f(x, y))),$$

but it is not much used in practice. The fourth-order family may be derived in exactly the same way; it is a two-parameter family of methods each requiring four function calls. The standard fourth-order method, RK4, which we shall not derive because the algebra is extremely tedious, is a particular case, and the increment may be computed with the function:

```
double RK4(func f, double x, double y, double h)
{       double K1 = f(x, y);
        double K2 = f(x+h/2, y+h*K1/2);
        double K3 = f(x+h/2, y+h*K2/2);
        double K4 = f(x+h, y+h*K3);
        return h * (K1 + 2.0*(K2 + K3) + K4)/6;
}
```

With our usual problem the following results are obtained to eight decimals:

x	Exact	RK4	Error
0.0	1.0	1.0	0.0
0.1	1.00483742	1.0483750	0.00000008
0.2	1.01873075	1.01873090	0.00000015
0.3	1.04081822	1.04081842	0.00000020
0.4	1.07032005	1.07032029	0.00000024
0.5	1.10653066	1.10653093	0.00000027
0.6	1.14881164	1.14881193	0.00000030
0.7	1.19658530	1.19658562	0.00000031
0.8	1.24932896	1.24932929	0.00000033
0.9	1.30656966	1.30656999	0.00000033
1.0	1.36787944	1.36787977	0.00000033

The accumulated error is in the seventh place of decimals. Still higher order methods may be derived, but are not much used because it is better to reduce the step-size h instead. Notice that if the function f does not depend explicitly on y, these methods all reduce to those we have already developed for ordinary integration, RK4 then being equivalent to Simpson's method, for example.

18.

We m
metho

and we
conditio

for all x,
    As a n
establish
ity the mi

$$|\phi(x, y_1,$$

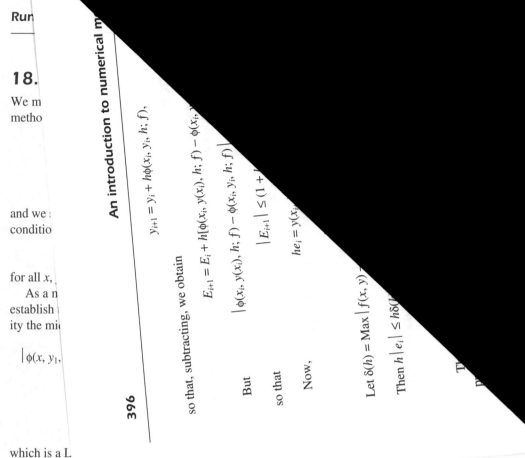

396

$$y_{i+1} = y_i + h\phi(x_i, y_i, h; f),$$

so that, subtracting, we obtain

$$E_{i+1} = E_i + h[\phi(x_i, y(x_i), h; f) - \phi(x_i,$$

But

$$|\phi(x_i, y(x_i), h; f) - \phi(x_i, y_i, h; f)|$$

so that

$$|E_{i+1}| \le (1 +$$

Now,

$$he_i = y(x_i$$

Let $\delta(h) = \mathrm{Max} \, |f(x, y)$

Then $h|e_i| \le h\delta($

which is a L
    As before _____ $ = y(x_i) - y_i$. Then it may be
shown, much _____ ... the first-order Euler method, that the sequence y_i converges to
the exact solution $y(x_i)$ if and only if, in addition, $\phi(x, y, h; f) \to f(x, y)$ as $h \to 0$. This seems
intuitively obvious, and it is a condition we have taken care of by construction in all the
methods we have studied. We must put it on a firmer footing.
    Moreover, it may be shown that $|E_i| = O(h^p)$ for a method of order p. We have already
observed that the accumulated error of the first-order Euler method is $O(h)$. For our usual
initial value problem in fourth order we find

h	0.2	0.1	0.05
Error	0.00000580	0.00000033	0.00000002

successive terms being approximately in the inverse ratio $2^4 = 16$ as expected.
    Define $e(x) = [y(x + h) - y(x)]/h - \phi(x, y, h; f)$. Then $he_i \equiv he(x_i)$ is the local truncation error
of the method. We have

$$y(x_{i+1}) = y(x_i) + h\phi(x_i, y(x_i), h; f) + he_i,$$

$$_i, h; f)] + he_i.$$

$$\leq L_\phi |y(x_i) - y_i| = L_\phi E_i,$$

$$nL_\phi)|E_i| + h|e_i|.$$

$$) - y(x_i) - h\phi(x_i, y(x_i), h; f)$$

$$= hy'(x_i) + \tfrac{1}{2}h^2 y''(\xi_i) - h\phi(x_i, y(x_i), h; f).$$

$$\phi(x, y, h; f)|, \text{ where } a \leq x \leq b, -\infty < y < +\infty.$$

$$) + \tfrac{1}{2}h^2 \|y''\|_\infty; \text{ and if we define } e(h) = \text{Max}|e_i|, M = \|y''\|_\infty, \text{ we find}$$

$$e(h) \leq \delta(h) + \tfrac{1}{2}hM.$$

thus, the local truncation error $e(h) \to 0$, as $h \to 0$, provided $\delta(h) \to 0$ also; that is to say, rovided $\phi(x, y, h; f) \to f(x, y)$. In that case we say that the method ϕ is *consistent*.

Returning now to the accumulated error, we have

$$|E_{i+1}| \leq (1 + hL_\phi)|E_i| + he(h),$$

so that, as before,

$$|E_n| \leq [\exp(L_\phi(b-a) - 1]e(h),$$

The method converges if it is consistent. Moreover, if the method is of order p, then $E_i = O(h^p)$.

18.5.2 Adaptation

When we studied ordinary integration in chapter 16 we found it helpful to adapt the step-size according to the rate of change of the integrand. We may do the same here. The idea is to split the interval $[x_i, x_{i+1}]$ in two, to apply the chosen integration method to each half of the interval separately, and then to compare the resulting increment with that computed for the interval as a whole. Let us denote by RK any of the Runge–Kutta methods. The increment over the whole interval is given by $y_{\text{incl}} = y_{i+1} - y_i = RK(f, x_i, y_i, h)$. The increment over the lower sub-interval is given by $y_{\text{leftinc}} = RK(x_i, y_i, \tfrac{1}{2}h)$ while that over the second is $y_{\text{rightinc}} = RK(x_i + \tfrac{1}{2}h, y_i + y_{\text{leftinc}}, \tfrac{1}{2}h)$. Let the sum $y_{\text{inc}} = y_{\text{leftinc}} + y_{\text{rightinc}}$, and consider the difference $|y_{\text{inc}} - y_{\text{incl}}|$: if it exceeds some predetermined quantity, reduce the step-size; if it does not, by a predetermined margin, increase the step-size for the *next* step. The following routine carries out this idea for a method of order *n*:

```
void   adapt(func f, double &x, double &y, double &xinc, int n)
{      bool adjusted = FALSE;
       while (!adjusted)
       {      double yinc1 = RK(f, x, y, xinc);
              double yinc  = RK(f, x, y, xinc/2);
              y += RK(f, x+h/2, y + yinc, xinc/2);
              double delta = yinc - yinc1;
              double relerr = fabs(delta)/maxerr;
              double nextinc = xinc;
              if (relerr < 1.0)
              {      adjusted = TRUE;
                     if (relerr < 0.95)
                            increase(nextinc, relerr, n);
              }
              else reduce(xinc, relerr, n);
       }
       x += xinc;
       y += yinc;
       xinc = nextinc;
}
```

The quantity `maxerr` is defined globally in the user program. The procedures to reduce or increase the step-size are somewhat arbitrary, but may reasonably be taken to be

```
void   reduce(double &t, double relerr, int n)
{      t /= pow(relerr, 1.0/(n+1)); t *= 0.9; }
void   increase(double &t, double relerr, int n)
{      t /= pow(relerr, 1.0/n); t /= 0.9; }
```

where `double pow(double x, double y)` is an easily written function giving the value of x^y. It is also worth noting that the *y* increments can be slightly improved by higher order adjustments resembling Richardson extrapolation: if $n = 1$, increase the increment `yinc` by `delta`; if $n = 2$, increase it by `delta/3`; and if $n = 4$, increase it by `delta/15`.

To test out this adaptation of the step-size we need a function showing greater variation than our usual one. Consider instead the initial value problem:

$$y' = -y - 2\pi e^{-x}\sin(x), \quad y(0) = 1,$$

for which the exact solution is $y(x) = e^{-x}\cos(2\pi x)$. Setting `maxerr = 0.000001`, and an initial step-size $h = 0.1$, and working with the fourth-order method, RK4, the following results are obtained, which agree with the exact solution to the six figures of decimals shown:

x	y_{calc}
0.0	+1.0
0.055523	+0.889005
0.106442	+0.705348
0.155117	+0.480800
0.202814	+0.238524
0.250476	−0.002330
0.298974	−0.224609
0.349380	−0.412238
0.403537	−0.548976
0.471396	−0.614078
0.550091	−0.548560
0.606308	−0.428159
0.660004	−0.276930
0.712670	−0.113957
0.765529	+0.045307
0.819792	+0.187044
0.877178	+0.298122
0.941551	+0.364016

One drawback with any adaptive method is that additional function calls may be required. Another drawback is that the results are not given at regular intervals. If values are required at regular intervals they must be found from interpolation or adaptation must be abandoned. Results entirely equivalent to those in the table would be obtained with a constant step-size of $h = 0.05$. In that case we would use instead of **adapt** the procedure **uniform** defined by

```
void    uniform(func f, incr rk, double &x, double &y, double &xinc)
{       double yinc = rk(f, x, y, xinc);
        x += xinc;
        y += yinc;
}
```

where **incr** represents any increment function with signature defined in the function type

```
typedef double (*incr)(func, double, double, double);
```

It is perhaps interesting to see how to build all these different methods (and others also if desired) into a single package, which also provides for formatted output in the manner of 6.3.4. Defining yet another function type,

```
typedef void (*step)(func, incr, double&, double&, double&);
```

the required package may be written

```
void  rungekutta(func f, incr rk, step st,
                  double x0, double x1, double y0, double h)
{      long L = ios::left + ios::showpos + ios::fixed;
       format fpw(L, 6, 12);
       cout << fpw << "x"
            << fpw << "y(approx)"
            << fpw << "y(true)\n\n";
       double x = x0, y = y0, xinc = h;
       const double eps = 1.0E-10;
       do
       {      cout << fpw << x
                   << fpw << y
                   << fpw << F0(x) << endl;
              st(f, rk, x, y, xinc);
       }      while (fabs(x - x1) > eps);
}
```

where `eps` is any small positive constant. One simply has to choose for the variable `rk` one of the function names `RK1`, `RK2`, `RK4`, and for the variable `st` either `adapt` or `uniform`, and the routine `rungekutta` does the rest.

18.6 Multiple-step methods

All the Runge–Kutta methods, like the Euler method itself, are single-step methods: the data required for the next estimate is contained within the previous interval. They are a little complicated to formulate in higher order, but the analysis of their behaviour is straightforward. We come now to multiple-step methods, which are easier to formulate but a little more complicated to analyse. There are in fact two types of multiple-step methods: *explicit methods*, which share with single-step methods the virtue that all the data required for a new step are known from earlier steps; and *implicit methods*, where this is not so.

18.6.1 Two-step midpoint method

The simplest two-step method is suggested by the familiar formula

$$y(x + h) - y(x - h) = 2hy'(x) + \tfrac{1}{2}h^3 y^{(3)}(\xi),$$

from which the *two-step midpoint method* is derived:

$$y_{i+1} = y_{i-1} + 2hf(x_i, y_i) \quad \text{(two-step)}$$

where y_0 and y_1 are given. Notice that this is *not* the same as the one-step midpoint method introduced in 18.4:

$$y_{i+1} = y_i + hf(x_i + \tfrac{1}{2}h, y_i + \tfrac{1}{2}hf(x_i, y_i)) \quad \text{(one-step)}$$

where only y_0 is given. Both are of order two, but they differ in higher-order error terms. They also differ in their stability properties: whereas the one-step method is stable for large intervals of integration, where $x_n \equiv b \to \infty$, the two-step method may not be; moreover, the two-step error may be oscillatory. We say that it is *weakly stable*. The following table shows the behaviour of the accumulated error $E_i = y(x_i) - y_i$ for our usual problem, using the one-step and two-step midpoint methods for $0 \le x \le 1$, and $h = 0.1$:

x	One-step	Two-step
0.0	0.0	0.0
0.1	0.000163	0.0
0.2	0.000249	0.000302
0.3	0.000399	0.000213
0.4	0.000482	0.000506
0.5	0.000545	0.000335
0.6	0.000592	0.000642
0.7	0.000625	0.000390
0.8	0.000646	0.000729
0.9	0.000658	0.000394
1.0	0.000662	0.000786

It is clear that the two-step method has an oscillatory element. Moreover, the error becomes large for still higher values of x. To see how this comes about, consider the even simpler problem $y' = \lambda y$, $y(0) = 1$, which has the exact and unique solution $y(x) = \exp(\lambda x)$. We must calculate what solutions there are to the corresponding midpoint equations.

In order to establish a systematic procedure, consider first the simple Euler method of order one, for which the difference equations in this case are

$$y_{i+1} = y_i + hf(x_i, y_i) = y_i + h\lambda y_i = (1 + h\lambda)y_i.$$

We seek a solution of an exponential character; therefore try setting $y_i = \eta^i$, $\eta \neq 0$.

Substitution results in $\qquad\qquad \eta^{i+1} = (1 + h\lambda)\eta^i,$

from which we may cancel a factor η^i, to produce the *characteristic equation*:

$$\eta - (1 + h\lambda) = 0.$$

It is called "characteristic" because it describes qualitatively the response of the method to growth in the solution, and in this case it is linear. Thus, the solution of the difference equations that also satisfies the initial condition $y_0 = 1$ is:

$$y_i = \eta^i = (1 + h\lambda)^i.$$

Now we can suppose that $z \equiv h\lambda \ll 1$, otherwise it would be impossible to obtain even a moderately accurate solution. Moreover, we are seeking a solution approximating the exponential $e^{\lambda x}$. Expanding $(1 + z)^i$, we therefore obtain:

$$y_i = 1 + iz + (1/2)i(i - 1)z^2 + (1/6)i(i - 1)(i - 2)z^3 + \ldots$$

$$= 1 + iz + (1/2)(iz)^2 + (1/6)(iz)^3 + \ldots$$

$$- (1/2)iz^2 - (1/6)(3i - 2)iz^3 - \ldots$$

$$= e^{iz} - \tfrac{1}{2}iz^2 + O(z^3)$$

$$= \exp(\lambda x_i) - \tfrac{1}{2}\lambda^2 x_i h + O(h^2),$$

where in the last line we have set $iz = \lambda x_i$. This is the solution we expected; in particular, the accumulated error is first order:

$$E_i \equiv y(x_i) - y_i = \tfrac{1}{2}\lambda^2 x_i h + O(h^2).$$

Having established how to investigate the nature of the computed solution in the Euler case, consider the same problem using the one-step midpoint method of order two. The difference equations in this case are:

$$y_{i+1} = y_i + hf(x_i + \tfrac{1}{2}h, y_i + \tfrac{1}{2}hf(x_i, y_i))$$

$$= y_i + hf(x_i + \tfrac{1}{2}h, y_i + \tfrac{1}{2}h\lambda y_i)$$

$$= y_i + h\lambda(y_i + \tfrac{1}{2}h\lambda y_i)$$

$$= (1 + h\lambda + \tfrac{1}{2}(h\lambda)^2)y_i.$$

Thus, the characteristic equation for the one-step midpoint method is

$$\eta - (1 + z(1 + \tfrac{1}{2}z)) = 0,$$

which is again linear. The exponential solution for small z may be obtained just as before, and the error is second order: $E_i = (1/6)\lambda^3 x_i h^2 + O(h^3)$.

Now consider the two-step midpoint method:

$$y_{i+1} = y_{i-1} + 2h\lambda y_i, \quad i > 0.$$

This time the characteristic equation is

$$\eta^2 - 2z\eta - 1 = 0,$$

where we have again set $z = h\lambda$; and this is *quadratic*, so there are two solutions,

$$\eta_\pm = z \pm \sqrt{(1 + z^2)},$$

in terms of which the general solution of the two-step midpoint method is

$$y_i = c_+\eta_+{}^i + c_-\eta_-{}^i,$$

where the constants $c_\pm$ must be chosen to accord with the starting values:

$$c_+ + c_- = y_0, \quad c_+\eta_+ + c_-\eta_- = y_1.$$

The starting values may be taken to be the exact values $y_0 = 1$, $y_1 = e^z$. For ease in manipulation, let $G(z) = z + \sqrt{(1 + z^2)}$; then $\eta_+ = G(z)$, $\eta_- = -G(-z)$. Solving for the coefficients c, we find:

$$c_+ = \frac{\exp(z) + G(-z)}{2\sqrt{(1 + z^2)}}, \qquad c_- = \frac{G(z) - \exp(z)}{2\sqrt{(1 + z^2)}}.$$

We seek the exponential approximation to $G(z)$ for small z. Using the binomial expansion,

$$G(z) = z + (1 + z^2)^{1/2}$$

$$= z + 1 + \tfrac{1}{2}z^2 - (1/8)z^4 + \ldots$$

$$= \exp(z) - (1/6)z^3 g(z),$$

where $g(z) = 1 + z + \ldots$ is bounded about $z = 0$. We may now solve for the coefficients c:

$$c_+ = 1 + (1/12)z^3 + O(z^4), \quad c_- = -(1/12)z^3 + O(z^4).$$

To obtain the general solution, we also need to calculate $[G(z)]^i$ and $[-G(-z)]^i$:

$$[G(z)]^i = [e^z - (1/6)z^3 g(z)]^i$$

$$= e^{iz}[1 - (1/6)z^3 e^{-z} g(z)]^i,$$

$$[-G(-z)]^i = (-1)^i e^{-iz}[1 + (1/6)z^3 e^z g(-z)]^i.$$

The general solution for small z is therefore

$$y_i \approx e^{iz}[1 + O(z^3)] - (-1)^i e^{-iz} z^3 / 12.$$

The first term corresponds to the true solution of the differential equation; the second is a spurious product of the particular method of solution. Remembering that $z = h\lambda$, if $\lambda > 0$ the spurious term soon dies out for sufficiently large i, but if $\lambda < 0$, it grows exponentially, eventually swamping the true solution. Note too that the second term alternates in sign: this is characteristic of the weak stability exhibited by the midpoint method. In fact, the error term in this case is highly oscillatory:

$$E_i = y(x_i) - y_i \approx (-1)^i (h\lambda)^3 \exp(-\lambda x_i)/12.$$

The solution of our usual problem $y' = x - y + 1$, $y(0) = 1$, is $y(x) = x + e^{-x}$. It is the decreasing exponential term (for $x > 0$) that gives rise to the oscillatory error observed in the last table when the two-step midpoint method is used, and to the growing error as x increases.

In what follows, we shall be concentrating on the solutions of the characteristic equation. For the two-step midpoint method, as we have seen, it is

$$\eta^2 - 2z\eta - 1 = 0,$$

and the two roots are given as functions of z in the following table:

z	Normal	Spurious
−0.5	0.618	−1.618
−0.4	0.677	−1.477
−0.3	0.744	−1.344
−0.2	0.820	−1.220
−0.1	0.905	−1.105
0.0	1.000	−1.000
0.1	1.105	−0.905
0.2	1.220	−0.820
0.3	1.344	−0.744
0.4	1.477	−0.677
0.5	1.618	−0.618

The spurious term dominates if $z < 0$, and is $O(1)$ even if $z > 0$. Notice that at $z = 0$ the solutions are $\eta_+ = 1$, $\eta_- = -1$. In general, we shall require $|\eta_-| < 1$ if a two-step method is to be reasonably stable.

18.6.2 Other two-step methods

The explicit *Adams–Bashforth* two-step method of order two is obtained by considering the formula:

$$y(x + h) = y(x) + \alpha h f(x, y(x)) + \beta h f(x - h, y(x - h)) + O(h^3).$$

Expanding the terms in $x \pm h$ in a Taylor series, we obtain

$$y + hy' + \tfrac{1}{2}h^2 y'' + O(h^3) = y + (\alpha + \beta)hf - \beta h^2 f' + O(h^3)$$

$$= y + (\alpha + \beta)hy' - \beta h^2 y'' + O(h^3).$$

For consistency we therefore require $\alpha + \beta = 1$, $\beta = -\tfrac{1}{2}$, so that $\alpha = \tfrac{3}{2}$. Thus the method is given by

$$y_{i+1} = y_i + \tfrac{1}{2}h[3f(x_i, y_i) - f(x_{i-1}, y_{i-1})].$$

The characteristic equation is

$$\eta^2 - (1 + 3z/2)\eta + z/2 = 0,$$

which is an immediate improvement over the midpoint method, because at $z = 0$ the solutions are $\eta_+ = 1$, $\eta_- = 0$, instead of $\eta_+ = 1$, $\eta_- = -1$. The solutions as functions of z are given below. It will be seen that $|\eta_-| < 1$ over the range of z shown:

z	Normal	Spurious
−0.5	0.640	−0.390
−0.4	0.690	−0.290
−0.3	0.750	−0.200
−0.2	0.822	−0.122
−0.1	0.905	−0.055
0.0	1.000	0.000
0.1	1.105	0.045
0.2	1.218	0.082
0.3	1.338	0.112
0.4	1.463	0.137
0.5	1.593	0.157

To program the Adams–Bashforth two-step method it will be necessary to carry and update the value of $f(x_{i-1}, y_{i-1})$:

```
double AB2step(func f, double x, double y,
                         double &flast, double h)
{       double K = f(x, y);
        double L = flast;
        flast = K;
        return 0.5*h*(3*K-L);
}
```

We then need functions to format the output and to manage the housekeeping. Formatting may conveniently be handled as in 6.4.5 by defining the structure

```
struct table { double x, y, f, e; };
```

together with a suitable output operator `ostream& operator << (ostream&, table&)`. (We shall not fill in the details again.) Then a housekeeping function which integrates $y' = f(x, y)$ from x_0 to x_1, given that the initial values of y are $y_0 = y(x_0)$ and $y_1 = y(x_0 + h)$, and that the known solution is $y = f(x)$, may be written:

```
void   AB2(func f, double x0, double x1,
                   double y0, double y1, double h)
{       double x = x0, y = y0, flast = f(x0, y0);
        int n = round((x1-x0)/h);
        for (int i = 0; i <= n; ++i)
        {       if (i == 0) y = y0;
                else if (i == 1)
                {       x += h;
                        y = y1;
                }
                else
                {       y += AB2step(f, x, y, flast, h);
                        x += h;
                }
                double fx = F(x);
                table t = {x, y, fx, fx-y };
                cout << t << endl;
        }
}
```

The solution of our usual problem using this method is given in the following table:

x	AB2	$F(x)$	Error
0.0	1.0	1.0	0.0
0.1	1.004837	1.004837	0.0
0.2	1.019112	1.018731	−0.000381
0.3	1.041487	1.040818	−0.000669
0.4	1.071219	1.070320	−0.000899
0.5	1.107611	1.106531	−0.001080
0.6	1.150030	1.148812	−0.001219
0.7	1.197906	1.196585	−0.001321
0.8	1.250722	1.249329	−0.001393
0.9	1.308009	1.306570	−0.001439
1.0	1.369344	1.367879	−0.001464

The error is clearly $O(h^2)$, but is about a factor two larger than that of the various single-step second-order methods. For this particular problem we found that all the second-order methods were identical to the second-order Taylor method, for which the local truncation error is $(1/6)h^3 y^{(3)}(\xi)$. A little more effort shows that the truncation error of the two-step Adams–Bashforth method is $(5/12)h^3 y^{(3)}(\xi)$. The value of ξ need not be the same in both cases, but the ratio of the numerical coefficients is $5/2$, which may be regarded as a sufficient explanation of the larger accumulated error of the Adams–Bashforth method in this case.

Higher order Adams–Bashforth methods may be constructed in exactly the same way by expressing $y(x + h)$ in terms of additional steps. The lowest ones are shown below, where we have included the one-step case to emphasize the development from Euler's method:

Steps	Adams–Bashforth method
1	$y_{i+1} = hy_i + hy'_i + \frac{1}{2}h^2 y''(\xi_i)$
2	$y_{i+1} = hy_i + \frac{1}{2}h[3y'_i - y'_{i-1}] + (5/12)h^3 y^{(3)}(\xi_i)$
3	$y_{i+1} = hy_i + (h/12)[23y'_i - 16y'_{i-1} + 5y'_{i-2}] + (3/8)h^4 y^{(4)}(\xi_i)$
4	$y_{i+1} = hy_i + (h/24)[55y'_i - 59y'_{i-1} + 37y'_{i-2} - 9y'_{i-3}] + (251/720)h^5 y^{(5)}(\xi_i)$

In each case, y'_i is to be replaced by $f(x_i, y_i)$. Note that we have also shown the error term derived from the Taylor expansion of $y(x + h)$; it is, of course, to be dropped when using the method. If s is the number of steps, the single variable `ylast` will have to be replaced by an array of $s - 1$ previously computed values of f. Correspondingly, there will have to be $s - 1$ initial values of f at the points $x, x + h, \ldots, x + (s - 1)h$.

The characteristic equation of an Adams–Bashforth method with s steps is of degree s, and so there are $s - 1$ spurious solutions, η_k, for all of which we shall require $|\eta_k| < 1$ if the method is to be stable.

The implicit *Adams–Moulton* two-step method is of order *three* and may be found in exactly the same way; including the error term we find:

$$y_{i+1} = y_i + (h/12)[5f(x_{i+1}, y_{i+1}) + 8f(x_i, y_i) - f(x_{i-1}, y_{i-1})] - (1/24)h^4 y^{(4)}(\xi_i).$$

Its characteristic equation is:

$$(1 - 5z/12)\eta^2 - (1 + 2z/3)\eta + z/12 = 0,$$

for which again $\eta_-(0) = 0$. The tabulated solutions are:

z	Normal	Spurious
−0.5	0.724	−0.058
−0.4	0.776	−0.043
−0.3	0.830	−0.030
−0.2	0.885	−0.019
−0.1	0.942	−0.009
0.0	1.000	0.000
0.1	1.059	0.008
0.2	1.118	0.015
0.3	1.179	0.021
0.4	1.240	0.027
0.5	1.301	0.032

and it will be observed that $|\eta_-| \ll 1$ over the range shown.

The Adams–Moulton method, being implicit, can only be solved by iteration, using the scheme:

$$y_{i+1}^{(j+1)} = y_i + (h/12)[5f(x_{i+1}, y_{i+1}^{(j)}) + 8f(x_i, y_i) - f(y_{i-1}, y_{i-1})],$$

where at each step a starting approximation $y_{i+1}^{(0)}$ must first be found on which to base the iteration. From the work we have done on iterative methods, we know that the iteration will converge provided $h < 12/5L$, where L is a Lipschitz constant for the function f.

For purposes of illustration, let us take the starting approximation at each step to be that predicted by Euler's method. Then the Adams–Moulton two-step increment function may be written:

```
double AM2Estep(func f, double x, double y, double fa[], double h)
{       double L = fa[0];                       // f(x-h, ylast);
        double K = fa[1];                       // f(x, y);
        double yincr = h*K;                     // Euler predictor
```

```
double t, M;
do                                       // iterate
{       t = yincr;
        M = f(x + h, y + yincr);        // f(x+h, ynext)
        yincr = (h/12)*(5.0*M + 8.0*K − L);// AM 2-step
}       while (fabs(yincr − t) > tol);
fa[0] = K;   fa[1] = M;                 // update fa
return yincr;
}
```

The "past" and "present" values of $f(x, y)$ are stored in the array `fa[2]`, and are updated before the value of the increment is returned. The `do` loop is repeated so long as the change in the increment exceeds a preset value `tol`. The housekeeping procedure is exactly the same as before except that the call to `AB2step` must be replaced by a call to `AM2Estep`. The initial values $f(x_0, y_0)$, $f(x_0 + h, y_1)$ are also required.

If we apply this procedure to our usual problem, setting `tol` $= 10^{-6}$, we find that four iterations are sufficient at each step. The results are shown below:

x	AM2E	F(x)	Error
0.0	1.0	1.0	0.0
0.1	1.004837	1.004837	0.0
0.2	1.018734	1.018731	−0.000004
0.3	1.040825	1.040818	−0.000006
0.4	1.070329	1.070320	−0.000009
0.5	1.106541	1.106531	−0.000010
0.6	1.148823	1.148812	−0.000012
0.7	1.196598	1.196585	−0.000013
0.8	1.249343	1.249329	−0.000014
0.9	1.306584	1.306570	−0.000014
1.0	1.367894	1.367879	−0.000014

The error is about 100 times less than that of the `AB2` method. The Adams–Moulton two-step method is of third order compared with second order for the two-step Adams–Bashforth method. With $h = 0.1$ that accounts for a factor 10, while the numerical coefficients in the error terms account for another factor 10.

We list below the lowest order Adams–Moulton methods with their error terms:

Steps	Adams–Moulton method
1	$y_{i+1} = hy_i + \frac{1}{2}h(y'_{i+1} + y'_i) - (1/12)h^3y^{(3)}(\xi_i)$
2	$y_{i+1} = hy_i + (h/12)[5y'_{i+1} + 8y_i - y'_{i-1}] - (1/24)h^4y^{(4)}(\xi_i)$
3	$y_{i+1} = hy_i + (h/24)[9y'_{i+1} + 19y_i - 5y'_{i-1} + y'_{i-2}] - (19/720)h^5y^{(5)}(\xi_i)$

The one-step Adams–Moulton method is also known as the *trapezoidal* method, from its resemblance to the corresponding method for quadrature. Note that for a given number of steps the Adams–Moulton method is always one order higher than the Adams–Bashforth method. Notice too, that for a given *order* the Adams–Moulton method has a much smaller truncation error, by a factor of about 10 in the numerical coefficient.

18.7 Predictor–corrector methods

This last observation leads to a composite method, where an Adams–Bashforth method of given order is used to *predict* the increment, which is then *corrected* using the corresponding Adams–Moulton method of the same order, and therefore one step less. There should then be no need to iterate the Adams–Moulton method more than once in order to obtain a result almost as good as the fully iterated solution.

Consider a simple case where the two-step Adams–Bashforth method is used as the predictor, with the one-step Adams–Moulton method as corrector. The increment function is

```
double ABM1step(func f, double x, double y, double fa[], double h)
{       double L = fa[0];                    // f(x-h, ylast)
        double K = fa[1];                    // f(x, y)
        double yincr = (h/2)*(3.0*K-L);      // AB2 predictor
        double M = f(x+h, y+yincr);          // f(x+h, ynext)
        yincr = (h/2)*(M + K);               // AM1 corrector
        M = f(x+h, y+yincr);                 // corrected M
        fa[0] = K; fa[1] = M;                // update fa
        return yincr;
}
```

and the following results are obtained for our usual problem:

x	$ABM1$	$F(x)$	Error
0.0	1.0	1.0	0.0
0.1	1.004837	1.004837	0.0
0.2	1.018640	1.018731	+0.000091
0.3	1.040654	1.040818	+0.000165
0.4	1.070097	1.070320	+0.000223
0.5	1.106261	1.106531	+0.000270
0.6	1.148507	1.148812	+0.000305
0.7	1.196254	1.196585	+0.000331
0.8	1.248979	1.249329	+0.000350
0.9	1.306208	1.306570	+0.000362
1.0	1.367511	1.367879	+0.000368

This is about a factor four better than the second order AB2 method by itself. Additional iterations reduce the remaining error by about 30% at the cost of many more function calls.

We have skated over the need to find $s - 1$ initial values of f before any s-step method may be used. We complete this chapter by writing down an increment function of fourth order, in which the initial values are obtained using the standard fourth-order Runge–Kutta method, with the Adams–Bashforth four-step predictor and the Adams–Moulton three-step corrector.

The increment function for the resulting *Adams–Bashforth–Moulton method* may be written:

```
double ABMstep(func f, double x, double &y, double fa[], double h)
{       double N = fa[0], M = fa[1], L = fa[2], K = fa[3];
        double yincr = (h/24)*(55*K - 59*L + 37*M - 9*N);    // AB pred
        double P = f(x+h, y+yincr);
        yincr = (h/24)*(9*P + 19*K - 5*L + M);               // AM corr
        P = f(x+h, y+yincr);                                 // final P
        fa[0] = M; fa[1] = L; fa[2] = K; fa[3] = P;          // renew fa
        return yincr;
}
```

We shall use the standard fourth-order Runge–Kutta increment function RK4. Then the housekeeping function for the Adams–Bashforth–Moulton method becomes:

```
void  ABM(func f, double x0, double x1, double y0, double h)
{       int n = round((x1 - x0)/h);
        double fa[4];
        double x = x0, y = y0;
        for (int i = 0; i <= n; ++i)
        {     if (i == 0) fa[0] = f(x, y);
              else if (i < 4)
              {     y += RK4(f, x, y, h);
                    x += h;  fa[i] = f(x, y);
              }     else
              {     y += ABMstep(f, xz, y, fa, h);
                    x += h;
              }
              double fx = F(x);
              table p = { x, y, fx, fx - y };
              cout << p << endl;
        }
}
```

Applied to our usual problem we get:

x	ABM	$F(x)$	Error
0.0	1.0	1.0	0.0
0.1	1.00483750	1.00483742	−0.00000008
0.2	1.01873090	1.01873075	−0.00000015
0.3	1.04081842	1.04081822	−0.00000020
0.4	1.07031992	1.10653066	+0.00000013
0.5	1.10653027	1.10653066	+0.00000039
0.6	1.14881103	1.14881164	+0.00000060
0.7	1.19658453	1.19658530	+0.00000077
0.8	1.24932806	1.24932896	+0.00000090
0.9	1.30656866	1.30656966	+0.00000100
1.0	1.36787837	1.36787944	+0.00000108

It will be seen that the straightforward fourth-order Runge–Kutta method of 18.5 is somewhat better in this case. The best approach in any particular case may be found by trial and error.

18.8 Domain of existence

The theorem with which we began this chapter gave a sufficient but not necessary condition for the existence of a unique solution. Unique solutions may in fact exist over a more restricted domain. It may be quite difficult to find the actual domain, especially if the problem is non-linear. Consider the initial value problem $y' = y^2$, $y(0) = 1$, for which there is a unique solution $y(x) = 1/(1 - x)$. The range of values of y, however, is not $(-\infty, +\infty)$, as supposed in the theorem, but $(-\infty, 1)$, because y becomes unbounded at $x = 1$.

Consider, next, computing the solution of the problem $y' = x^2 + y^2$, $y(0) = 1$, $0 \le x \le 1$. If we run RK4 over it, with $h = 0.1$, we find that the solution apparently becomes unbounded somewhere between $x = 0.9$ and $x = 1.0$.

Now, $y^2 \le x^2 + y^2 \le 1 + y^2$, if $0 \le x \le 1$. The solution of the problem $y' = y^2$, $y(0) = 1$, we have seen, is $\phi_1(x) = 1/(1 - x)$. Likewise, the solution of $y' = 1 + y^2$, $y(0) = 1$, is $\phi_2(x) = \tan(x + \pi/4)$. But $\phi_1(x) \to \infty$ as $x \to 1$, and $\phi_2(x) \to \infty$ as $x \to \pi/4 \approx 0.785$. We would expect $\phi_1(x) \le y(x) \le \phi_2(x)$ for values of x for which all three functions exist. Therefore $y(x)$ should become unbounded somewhere in the interval $[0.785, 1]$. The table below shows that $y(x)$ becomes unbounded near $x = x_0 \approx 0.97$:

h	$x = 0.96$	0.966	0.968	0.97
0.01	101.160	—	—	899.8
0.005	101.853	—	—	1983.7
0.002	101.925	262.21	546.98	6842.9
0.001	101.927	262.41	551.75	26323.8
0.0005	101.927	262.42	552.25	356794.8

Therefore, there is a unique solution over the range $[0, x_0]$. Note that nothing in the statement of the problem indicates that there will be a singularity at $x = x_0$; indeed, the approximate value of x_0 is itself a result of the computation. It is unwise to attempt to compute the solutions of non-linear differential equations blindfold!

19 More about differential equations

Numerical methods for the solution of differential equations are many and various. So far we have treated only the case of a single differential equation of first order, with one dependent and one independent variable (given an initial condition), but the treatment was fairly rigorous. In this chapter we shall in contrast merely sketch a number of illustrative methods for dealing with more complex problems: initial value problems involving more than one ordinary differential equation, and the related problem of higher order differential equations; boundary value problems; variational methods; and simple eigenvalue problems. Finally, we add a few remarks about partial differential equations. One of the most powerful computational techniques we do not treat at all, namely finite element analysis, because it would take us too far from our main objectives.

19.1 Systems of ordinary differential equations

The most straightforward extension of the work of the last chapter is to the case where there are several dependent variables, each specified at some initial value of the independent variable. For example, we might wish to consider the two simultaneous equations:

$$y' = f(x,y,z), \ z' = g(x,y,z), \ y(0) = y_0, \ z(0) = z_0,$$

where f, g are given functions of x, y, z. In general, it is convenient to state the problem in vector notation. We have a system of n first-order differential equations for the n dependent variables $y_0, \ldots, y_{n-1}$, considered as a single vector $\mathbf{y}$, in terms of a single independent variable x, where $a \leq x \leq b$, with initial conditions $\mathbf{y}(a) = \mathbf{y}_a$, where $\mathbf{y}_a$ is a constant vector:

$$\mathbf{y}' = \mathbf{f}(x, \mathbf{y}), \ \ \mathbf{y}(a) = \mathbf{y}_a.$$

The vector function $\mathbf{f}(x,\mathbf{y})$ is said to satisfy a Lipschitz condition on the domain

$$D = \{ \ (x,\mathbf{y}) \mid a \leq x \leq b, -\infty < y_i < +\infty, i = 0, \ldots, n-1 \ \}$$

if there exists a constant L such that

$$\| \mathbf{f}(x,\mathbf{y}) - \mathbf{f}(x,\mathbf{z}) \|_\infty \le L \| \mathbf{y} - \mathbf{z} \|_\infty, \text{ all } (x,\mathbf{y}),(x,\mathbf{z}) \in D.$$

Using the multi-dimensional Taylor's theorem we can set

$$L = \| \text{Max}_{(x,\mathbf{y}) \in D} J(x,\mathbf{y}) \|_\infty,$$

where $J_{ij}(x,\mathbf{y}) = \partial f_i/\partial y_j$ is the Jacobian matrix. If $\mathbf{f}$ satisfies this Lipschitz condition, then the differential equation $\mathbf{y}' = \mathbf{f}(x,\mathbf{y})$ possesses a unique solution $\mathbf{y}(x)$ in the range $a \le x \le b$.

19.1.1 Runge–Kutta method

The methods introduced in the last chapter for computing approximate solutions to single first-order differential equations may be generalized to systems of differential equations with initial conditions in a fairly obvious manner. In a Runge–Kutta method of order p, for example, the starting point is a function ϕ, linear in f, such that

$$\mathbf{y}(x + h) = \mathbf{y}(x) + h\phi(x,\mathbf{y},h;\mathbf{f}) + O(h^{p+1}).$$

We state without derivation the classic fourth-order Runge–Kutta method for a system of differential equations. It is a straightforward generalization of that given in 18.5. First, we define a function type for the vector function $\mathbf{f}(x,\mathbf{y})$:

```
typedef vector (*funcvec)(double, vector&);
```

The increment function is then:

```
vector RK4sys(funcvec f, double x, vector &y, double h)
{       vector K1 = f(x, y);
        vector K2 = f(x+h/2, y+h*K1/2);
        vector K3 = f(x+h/2, y+h*K2/2);
        vector K4 = f(x + h, y + h*K3);
        return h*(K1 + 2.0(K2 + K3) + K4)/6;
}
```

Notice that the dimension of the vector $\mathbf{f}$, and therefore of `RK4sys`, has not been defined; this is a matter for the user program. Consider the very simple example given by the pair of equations

$$y' = z, \;\; y(0) = 2, \;\; z' = y, \;\; z(0) = 0,$$

for which $L = 1$. The exact solution may readily be verified to be

$$Y(x) = e^x + e^{-x}, \;\; Z(x) = e^x - e^{-x}.$$

The particular vector function of type `funcvec` is therefore given by:

```
vector F(double x, vector &y)
{       int n = y.getsize();            // general prescription
        vector t(n);
        t[0] = y[1];                    // particular case
        t[1] = y[0];
        return t;
}
```

Using `RK4sys` with the function F, and with $h = 0.1$, we find:

x	y	$Y - y$	z	$Z - z$
0.0	2.0	0.0	0.0	0.0
0.1	2.010008	2.8×10^{-9}	0.200333	1.7×10^{-7}
0.2	2.040133	3.9×10^{-8}	0.402672	3.4×10^{-7}
0.3	2.090677	1.1×10^{-7}	0.609040	5.1×10^{-7}
0.4	2.162145	2.1×10^{-7}	0.821504	7.0×10^{-7}
0.5	2.255252	3.6×10^{-7}	1.042190	9.1×10^{-7}
0.6	2.370930	5.4×10^{-7}	1.273306	1.1×10^{-6}
0.7	2.510337	7.7×10^{-7}	1.517166	1.4×10^{-6}
0.8	2.674869	1.0×10^{-6}	1.776210	1.7×10^{-6}
0.9	2.866171	1.4×10^{-6}	2.053031	2.0×10^{-6}
1.0	3.086160	1.8×10^{-6}	2.350400	2.4×10^{-6}

If the results are not accurate enough a smaller value of h may be used. An adaptive version of the increment function may also be devised.

19.1.2 Lotka–Volterra equations

A more interesting example is afforded by early attempts to describe the population dynamics of two competing species. Let x and y be the populations at time t of the prey and predator species, respectively. We assume that the prey has a constant food supply, while the predator feeds only off the prey. Then a plausible model of the population dynamics is given by the pair of equations:

$$dx/dt = Ax - Bxy, \quad dy/dt = Cxy - Dy,$$

where A, B, C, D are constants. These are the Lotka–Volterra equations.

First, we show that there is a fixed point solution at which the two populations do not change with time, and that near that solution the populations perform small oscillations. If we set $x = x_0 + \xi$, $y = y_0 + \eta$, where $\xi \ll x_0$, $\eta \ll y_0$, neglect terms in $\xi\eta$, and denote differentiation by t as usual by a dot, we find

$$\dot{\xi} = A\xi - B(y_0\xi + x_0\eta) + Ax_0 - Bx_0y_0,$$

$$\dot{\eta} = C(y_0\xi + x_0\xi) - D\eta + Cx_0y_0 - Dy_0,$$

which possesses the fixed solution $\dot{\xi} = \dot{\eta} = \xi = \eta = 0$ provided $x_0 = D/C$, $y_0 = A/B$. Giving x_0 and y_0 these values, we obtain more generally

$$\dot{\xi} = -(BD/C)\eta, \quad \dot{\eta} = (CA/B)\xi,$$

so that $\ddot{\xi} + \omega^2\xi = 0$, where $\omega = \sqrt{(AD)}$. This describes simple harmonic motion with period $\tau = 2\pi/\omega$. The trajectories are small circles of radius $\sqrt{(\xi^2 + \eta^2)}$ about the fixed point (x_0, y_0). In general, the solutions will still be oscillatory, but not simple harmonic owing to the non-linear terms in xy.

It is wasteful to make use of the full vector notation when only two simultaneous equations are involved. When the functions f also do not depend upon the independent variable, as in this case, it is preferable to redefine the increment function. Since initial value problems are usually dynamic problems, where the independent variable is time t, and the dependent variables may be represented by x and y, it is natural to define the corresponding two-dimensional Runge–Kutta increment function as a procedure relating the increment in y to the increment in x, thus directly plotting out the "trajectory" in the (x,y)-plane.

If $\qquad\qquad\qquad\qquad \dot{x} = f(x,y), \quad \dot{y} = g(x,y),$

then the increment function is given by:

```
void  RK4traj(func f, func g, double &x, double &y, double h)
{       double s = x, t = y;
        double K1 = f(s, t), L1 = g(s, t);
        s = x + (h/2)*K1; t = y + (h/2)*L1;
        double K2 = f(s, t), L2 = g(s, t);
        s = x + (h/2)*K2; t = y + (h/2)*L2;
        double K3 = f(s, t), L3 = g(s, t);
        s = x + h*K3; t = x + h*L3;
        double K4 = f(s, t), L4 = g(s, t);
        x += h*(K1 + 2.0*(K2 + K3) + K4)/6;
        y += h*(L1 + 2.0*(L2 + L3) + L4)/6;
}
```

Let us compute the solution in a form suitable for displaying in graphics form. The notation and general method were given in chapter 15. We have to shadow the trajectory of a point p in the (x,y)-plane by a pixel s on a screen defined by the points tl and br. If elapsed time is in steps of h, we may follow the trajectory for n steps, displaying it in colour:

```
void   drawtraj(func f, func g, point &p, point &tl, point &br,
                           double h, int n, int color)
{      for (int i = 0; i < n; ++i)
       {      pixel s(p, tl, br);
              putpixel(s, color);
              RK4twofunc(f, g, p.x, p.y, h);
       }
}
```

The trajectory is determined by the starting point p, the time interval h, and the number of steps n.

If we take $$\dot{x} = x(4 - 2y), \ \dot{y} = (x - 3)y,$$

for purposes of illustration, then the fixed point is at (3, 2). It is therefore convenient to trace out the trajectories as a family determined by the initial value of x, keeping the initial value $y = 2$ fixed. The figure reproduced in monochrome shows a few members of this family.

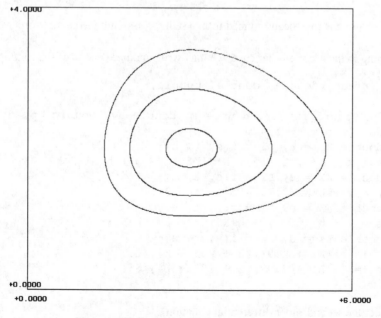

Figure 19.1 Lotka–Volterra trajectories.

19.2 Second-order initial value problems

The second-order initial value problem may be defined by the equations

$$y'' = f(x,y,y'), \quad y(0) = y_0, \quad y'(0) = y'_0.$$

If we define a new variable $p = y'$, where $p(0) = p_0 = y'_0$, the second-order initial value problem can be reformulated in terms of two simultaneous first-order differential equations,

$$p' = f(x,y,p), \qquad p(0) = p_0,$$

$$y' = p, \qquad\qquad y(0) = y_0,$$

which may be solved by the methods of the previous section.

In passing, note that for the $(m+1)$-th order problem, we may define a vector $\mathbf{y} = (y,y', \ldots, y^{(m)})$, and write:

$$y^{(m)'} = f(x,\mathbf{y}),$$

$$y^{(k)'} = y^{(k+1)}, \quad 0 \le i < m,$$

$$\mathbf{y}(0) = \mathbf{y}_0 = (y_0, y'(0), \ldots, y^{(m)}(0)).$$

The problem has thus been reduced to a system of m simultaneous first-order differential equations.

Returning to the second-order case, we may write another version of the routine RK4:

```
double (*func)(double, double, double);

void   RK4secord(func f, double x, double &y, double &p, double h)
{      double s = x, t = y, u = p;
       double K1 = u, L1 = f(s, t, u);
       s = x + h/2; t = y + h*K1/2; u = p + h*L1/2;
       double K2 = u, L2 = f(s, t, u);
       t = y + h*K2/2, u = p + h*L2/2;
       double K3 = u, L3 = f(s, t, u);
       s = x + h; t = y + h*K3; u = p + h*L3;
       double K4 = u, L4 = f(s, t, u);
       y += h*(K1 + 2.0*(K2 + K3) + K4)/6;
       p += h*(L1 + 2.0*(L2 + L3) + L4)/6;
}
```

Consider the second-order differential equation

$$y'' + 2y' + y = e^x, \quad y(0) = 1, \quad y'(0) = -1,$$

for which $f(x,y,p) = e^x - y - 2p$, and $L = 3$.

Ignoring, for a moment, the initial conditions, there is obviously a particular solution

$$y_P(x) = \frac{1}{4} e^x.$$

On the other hand, the general solution of the homogeneous equation

$$y'' + 2y' + y = 0$$

is $y_H(x) = (a + bx)e^{-x},$

where a and b are arbitrary constants. Since the differential equation is linear in y, it follows that the solution we seek is:

$$Y(x) = y_P(x) + y_H(x) = \frac{1}{4} e^x + (a + bx)e^{-x},$$

where the constants a and b are to be determined by the initial conditions. Thus, the exact solution is given by:

$$Y(x) = \frac{1}{4} [e^x + (3 - 2x)e^{-x}].$$

If we compute an approximate solution, for $0 \le x \le 1$, using the routine RK4secord, we may compare it with the exact solution. We find:

x	y	$y - Y$
0	1	0
0.1	0.909679	1.0×10^{-7}
0.2	0.837526	1.5×10^{-7}
0.3	0.781956	1.6×10^{-7}
0.4	0.741632	1.3×10^{-7}
0.5	0.715446	5.7×10^{-8}
0.6	0.702495	-4.2×10^{-8}
0.7	0.702072	-1.7×10^{-7}
0.8	0.713650	-3.2×10^{-7}
0.9	0.736871	-5.0×10^{-7}
1.0	0.771540	-7.0×10^{-7}

Dynamical problems may be expressed in terms of simultaneous differential equations of second order. For example, if the force acting on a particle of mass m at the position $\mathbf{x}$ is $\mathbf{F}(t,\mathbf{x},\dot{\mathbf{x}})$, then by Newton's laws:

$$m\ddot{\mathbf{x}} = \mathbf{F}(t,\mathbf{x},\dot{\mathbf{x}}), \quad \mathbf{x}(0) = \mathbf{x}_0, \quad \dot{\mathbf{x}}(0) = \mathbf{v}_0.$$

It is a straightforward matter to write a variant of the RK4 routine to deal with this problem once the force field has been determined. The real interest then lies in the stability of the solutions, but we shall not carry the matter any further.

19.3 Second-order boundary value problems

Whereas time-dependent problems may naturally be formulated in terms of initial values, space-dependent problems are more usually expressed in terms of *boundary values*. Thus, the static deflection of a horizontal beam supported at the fixed points $x = a$, $x = b$ might be expressed by an equation of the form:

$$y'' = f(x,y), \quad y(a) = y(b) = 0.$$

More generally, one might have a second-order differential equation

$$y'' = f(x,y,y'), \quad y(a) = ya, \quad y(b) = yb.$$

Let the function f, and its partial derivatives $\partial f/\partial y$, $\partial f/\partial y'$, be continuous in the domain

$$D = \{ (x,y,y') \mid a \le x \le b, -\infty < y,y' < \infty \}.$$

Then, if $\partial f/\partial y > 0$ everywhere in D, and if there exists a constant number $M > 0$ such that $|\partial f/\partial y'| \le M$ everywhere in D, it may be shown that the boundary value problem has a unique solution. These conditions are sufficient, but not necessary; many differential equations which do not not satisfy them nevertheless have unique solutions.

There are essentially two different ways to compute a solution of a boundary value problem. In the first approach, called the *shooting method*, one iterates the approximate solution of the initial value problem

$$y'' = f(x,y,y'), \quad y(a) = ya, \quad y'(a) = pa,$$

obtained by any standard method, choosing the value of pa on each iteration so that the computed value of $y(b)$ converges to the desired value yb. This is equivalent to adjusting the elevation of a gun in order to strike the given target, hence the name of the method. The second approach is called the *finite difference method*; it approximates the differential equation by a set of difference equations which satisfy the boundary conditions from the start, but must be solved by matrix methods. We demonstrate these two approaches in turn.

19.3.1 The shooting method

We shall use the RK4 increment function to solve the initial value problem on each iteration, although this is an arbitrary choice, and, equally arbitrarily, the bisect method on the computed value of *y(b)* to choose a better initial value of $pa = y'(a)$. First, then, we require a function which returns the value of *yb*, given the initial values *ya* and *pa*, and the interval *h*:

```
double shoot(func f, double a, double ya, double pa,
                                double b, double h)
{       double x = a, y = ya, p = pa;
        int n = round((b - a)/h);
        for (int i = 0; i < n; ++i)
        {       RK4secord(f, x, y, p, h);
                x += h;
        }
        return y;
}
```

We presented a non-recursive version of the `bisect` function in 4.1.3. It was devised to bracket the root of an equation within preset narrow limits. We adapt it now to bracket that value of *pa* which leads to $y(b) = yb$. The variables *p1*, *p2* are initial guesses of the required initial condition $p = pa$.

```
state  bisectbound(func f, double a, double ya, double b, double yb,
                double p1, double p2, double &pa, double h)
{       state s = ITERATING;
        int iter = 0;
        double yb1 = shoot(f, a, ya, p1, b, h);
        double yb2 = shoot(f, a, ya, p2, b, h);
        if ((sign(yb1 - yb) + sign(yb2 - yb)) != 0) s = SAMESIGN;
        else
        {       double dp;
                pa = (yb1 - yb < 0) ? (dp = p2-p1, p1) : (dp = p1-p2, p2);
                while (s == ITERATING)
                {       double pmid = pa + (dp *= 0.5);
                        double ymid = shoot(f, a, ya, pmid, b, h);
                        if (ymid < yb) pa = pmid;
                        if (fabs(ymid - yb) < tolerance) s = SUCCESS;
                        if (++iter == maxiter) s = WONTSTOP;
                }
        }
        return s;
}
```

The correct initial value of *pa* once determined, we are left with an initial value problem to solve. (Of course, we have already solved it, had we cared to save the computed values of *y* in a vector. It is a balance between extra computing time and extra memory use.)

For example, the boundary value problem

$$y'' + 3y' - 2y = 2x + 3,$$

$$0 \le x \le 1, \ y(0) = 2, y(1) = 1,$$

has $f(x,y,p) = 2(x+y) + 3(1-p)$, so that $\partial f/\partial y = 2 > 0$, $|\partial f/\partial p| = 3$, from which it follows that there is always a unique solution. Using the routine `bisectbound` we obtain the results in the table if $h = 0.1$:

$$p = -7.196164$$

x	y	x	y
0	2	0.6	0.602350
0.1	1.408432	0.7	0.652952
0.2	1.022264	0.8	0.739857
0.3	0.783318	0.9	0.856890
0.4	0.651039	1	1
0.5	0.597228		

The value of p correct to 6 decimals is -7.196069 and is obtained with $h = 0.025$. The values of y shown in the table, obtained with $h = 0.1$, are in error by up to 1 in the fourth decimal place.

Note that with `RK4secord` we compute $p = y'$ as well as y. Therefore we can readily speed up the iterative process by using Newton's method instead of the bisection method, provided that we can first guess sufficiently close values $p1$ and $p2$. We leave that as an exercise for the reader. Note too that the method is applicable to non-linear problems as much as to linear ones. However, in the case of a linear problem, say,

$$y'' = P(x)y' + Q(x)y + R(x), \ y(a) = ya, y(b) = yb,$$

where $Q(x) > 0$, $a \le x \le b$, the boundary value problem may be reduced to the non-iterative solution of two initial value problems: namely, the inhomogeneous problem

$$y_I'' = P(x)y_I' + Q(x)y_I + R(x), \ y_I(a) = ya, y_I'(a) = 0,$$

and the homogeneous problem

$$y_H'' = P(x)y_H' + Q(x)y_H, \ y_H(a) = 0, y_H'(a) = 1.$$

Having computed solutions $y_I(x)$ and $y_H(x)$, the required solution follows from the linearity of the equations:

$$y(x) = y_I(x) + [(yb - y_I(b))/y_H(b)]y_H(x).$$

Much computational effort is thereby saved. This too we leave as an exercise for the reader.

19.3.2 The linear finite difference method

We treat the linear case first because it is simpler and illustrates the method. The boundary value problem may be stated as:

$$y'' = P(x)y' + Q(x)y + R(x), \ \ a \le x \le b, \ \ y(a) = ya, \ y(b) = yb,$$

Take a small interval h, and let $n + 1 = (b-a)/h$. Then the interval $[a,b]$ is divided into $n + 1$ equal subintervals terminated by the points

$$x_0 = a, \ x_i = a + ih, \ 1 \le i \le n, \ x_{n+1} = b.$$

We wish now to approximate the differential equation, which at each of the n interior points is given by

$$y''(x_i) = P(x_i)y'(x_i) + Q(x_i)y(x_i)y(x_i) + R(x_i), \ \ 1 \le i \le n,$$

in terms of finite differences. In 16.1 we found central difference formulae for the first and second derivatives of a function $f(x)$:

$$f'(x) = (1/2h)[f(x + h) - f(x - h)] - (h^2/6)f^{(3)}(\xi),$$

$$f''(x) = (1/h^2)[f(x + h) - 2f(x) + f(x - h)] - (h^2/12)f^{(4)}(\xi),$$

where $x-h \le \xi \le x+h$. Substituting these expressions for the corresponding differentials, replacing $y(x_i)$ by y_i, and dropping the higher-order corrections, we obtain the difference formulae:

$$y_0 = ya, \quad y_{n+1} = yb,$$

$$-[1 + (h/2)P(x_i)]y_{i-1} + [2 + h^2Q(x_i)]y_i - [1 - (h/2)P(x_i)]y_{i+1} = -h^2R(x_i).$$

This is a set of n tridiagonal equations for the approximations y_i to the true values $y(x_i)$, which may be solved in the usual way described in 10.6.

We therefore proceed with the computation by converting the problem into a set of tridiagonal equations. First, it is useful to define a vector the m elements of which vary linearly between given extreme values:

```
vector linvec(double a, double b, int m)
{       vector v(m);                    // m equally spaced points
        double h = (b - a)/(m-1);       // m-2 internal points!
```

```
        v[0] = a;
        for (int i = 0; i < m; ++i)
              v[i] = v[i-1] + h;              // v[m-1] = b
        return v;
}
```

We can now define

```
vector x = linvec(a,  b,  n+2);   vector y = linvec(ya, yb, n+2);
```

of which the second will only be necessary when we treat the non-linear case, but does no
harm here.

A routine may now be written which converts the boundary value problem into a set of
tridiagonal equations:

```
void  maketridiag(tridiag &equ, func p, func q, func r,
                  vector &x, vector &y)
//     here n is the number of internal points
{      int n = equ.getnum();
       double h  = (x[n+1]-x[0])/(n+1);
       double hh = h * h;
       for (int i = 0; i < n; ++i)
       {      double xi = x[i+1];
              double tl, td, tu, tb;
              if (i == 0)
              {      tl = 0.0;
                     td = 2 + hh*q(xi);
                     tu = -1 + h*p(xi)/2;
                     tb = -hh*r(xi) + (1+h*p(xi)/2)*y[0];
              }
              else if (i == n-1)
              {      tl = -1 - h*p(xi)/2;
                     td = 2 + hh*q(xi);
                     tu = 0.0;
                     tb = -hh*r(xi) + (1-h*p(xi)/2)*y[n+1];
              }
              else
              {      tl = -1 - h*p(xi)/2;
                     td = 2 + hh*q(xi);
                     tu = -1 + h*p(xi)/2;
                     tb = -hh*r(xi);
              }
```

```
                    equ[i][L] = tl;
                    equ[i][D] = td;
                    equ[i][U] = tu;
                    equ[i][B] = tb;
        }
}
```

It is also convenient to have a routine which extracts the solution of the set of tridiagonal equations into a vector:

```
void  solution(tridiag &t, vector &v)
{       int n = t.getnum();
        for (int i = 0; i < n; ++i)
                v[i] = t[i-1][B];
}
```

Then the routine which solves the problem is:

```
void  linfd(func p, func q, func r, vector &x, vector &y)
{       int n = x.getsize() - 2;
        tridiag equ(n);
        maketridiag(equ, p, q, r, x, y);
        tridiagonal(equ);
        solution(equ, y);
}
```

Applied to the boundary value problem studied in the last section, namely:

$$y'' = 2(x + y) + 3(1 - p), \ 0 \le x \le 1, \ y(0) = 2, \ y(1) = 1,$$

with 9 internal points (*ie*, $h = 0.1$), we obtain the results in the table:

x	y	x	y
0	2	0.6	0.600150
0.1	1.405352	0.7	0.651452
0.2	1.018097	0.8	0.738961
0.3	0.779036	0.9	0.856494
0.4	0.647367	1	1
0.5	0.594274		

The results are much less accurate than those given by the shooting method; indeed, we have to take $h = 0.025$ even to obtain 4-figure accuracy. The reason is simply that the fourth-order Runge–Kutta method used in the shooting method is $O(h^4)$, whereas the finite difference method is only $O(h^2)$. But the finite difference method, not surprisingly, is usually found to be more stable than the shooting method, and for this reason it is often preferred.

19.3.3 Extrapolated linear finite difference method

We get the best of both worlds if we apply Richardson extrapolation (as explained in 16.2) to the results of the linear finite difference method. It can be shown that the error may be expressed as a series in h^2, so that this is possible. For the above problem we then find:

x	$y(h)$	$y(h/2)$	$y(h/4)$	Extrap 1	Extrap 2	Extrap 3
0	2	2	2	2	2	2
0.1	1.405352	1.407602	1.408157	1.408352	1.408342	1.408341
0.2	1.018097	1.021139	1.021891	1.022153	1.022141	1.022140
0.3	0.779135	0.782187	0.782942	0.783204	0.783193	0.783192
0.4	0.647367	0.650044	0.650707	0.650937	0.650928	0.650927
0.5	0.594274	0.596426	0.596959	0.597144	0.597137	0.597137
0.6	0.600150	0.601751	0.602148	0.602285	0.602280	0.602280
0.7	0.651452	0.652543	0.652814	0.652907	0.652904	0.652904
0.8	0.738961	0.739612	0.739773	0.739829	0.739827	0.739827
0.9	0.856494	0.856781	0.856852	0.856877	0.856876	0.856876
1	1	1	1	1	1	1

19.3.4 The non-linear finite difference method

If the boundary value problem is non-linear, the finite difference method leads to a tridiagonal system of equations which is non-linear and therefore has to be solved by iteration. The method resembles that used to solve non-linear systems of ordinary equations by Newton's method described in 12.6. We consider the general second-order boundary value problem:

$$y'' = f(x,y,y'), \ a \leq x \leq b, \ y(a) = ya, y(b) = yb.$$

As in the last section, we replace the derivatives by finite differences, and obtain the tridiagonal system:

$$y_0 = ya, \quad y_{n+1} = yb,$$

$$y_{i+1} - 2y_i + y_{i-1} = h^2 f(x_i, y_i, (y_{i+1} - y_{i-1})/2h),$$

which we have to solve by iteration.

Let $\qquad P(x,y,p) = \partial f(x,y,p)/\partial p, \quad Q(x,y,p) = \partial f(x,y,p)/\partial y.$

Then the tridiagonal Jacobean matrix for the $n \times n$ system is given by:

$$J_{i,\,i-1}(\mathbf{y}) = -1 - (h/2)P(x_i,y_i,(y_{i+1} - y_{i-1})/2h),$$

$$J_{i,\,i}(\mathbf{y}) = 2 + h^2 Q(x_i,y_i,(y_{i+1} - y_{i-1})/2h),$$

$$J_{i,\,i+1}(\mathbf{y}) = -1 + (h/2)P(x_i,y_i,(y_{i+1} - y_{i-1})/2h).$$

Starting with a suitable vector $\mathbf{y}^{(0)}$, we must then construct a sequence of vectors

$$\mathbf{y}^{(k+1)} = \mathbf{y}^{(k)} + \mathbf{u}^{(k)},$$

where $J\mathbf{u} = \mathbf{b}$ is the tridiagonal system to be solved at each iteration, and

$$b_i = -y_{i-1} + 2y_i - y_{i+1} + h^2 f(x_i,y_i,(y_{i+1} - y_{i-1})/2h).$$

(Of course, these are the "internal" terms; obvious modifications are required when $i = 0$ or $i = n$, which bring in the boundary values.)

The programming is much the same as before, with only obvious changes to the routine `maketridiag`. The main modification required is to `linfd` to include the iteration; it becomes:

```
state nonlinfd(func f, func p, func q, vector &x, vector &y)
{       int n = x.getsize() - 2;
        state s = ITERATING;
        while (s == ITERATING)
        {       static int iter = 0;
                vector u(n+2);
                tridiag equ(n);
                maketridiag(equ, f, p, q, x, y);
                tridiagonal(equ);
                solution(equ, u);
                y += u;
                if (norminf(u) < tolerance) s = SUCCESS;
                if (++iter == maxiter) s = WONTSTOP;
        }
        return s;
}
```

The boundary value problem

$$y'' = y^3 - yy', \; 0 \le x \le 1, \; y(0) = 1, \; y(1) = \tfrac{1}{2},$$

has the unique solution $y(x) = 1/(x+1)$. Using $h = 0.1$, and just one stage of Richardson extrapolation, we obtain the results shown in the table:

x	$y(h)$	$y(h/2)$	Extrap
0	1	1	1
0.1	0.909214	0.909122	0.909091
0.2	0.833513	0.833379	0.833334
0.3	0.769429	0.769281	0.769231
0.4	0.714479	0.714334	0.714286
0.5	0.666841	0.666711	0.666667
0.6	0.625147	0.625037	0.625000
0.7	0.588350	0.588264	0.588235
0.8	0.555634	0.555575	0.555556
0.9	0.526356	0.526326	0.526316
1.0	0.5	0.5	0.5

The results are exact to the accuracy shown, with the exception of $x = 0.4$ which is too large by 1 in the last place of decimals.

19.4 Variational methods

Let $P(x)$, $Q(x)$, $R(x)$, $\phi(x)$ be continuous on the interval $a \le x \le b$; furthermore, let $P(x)$ be differentiable on that interval, and $\phi(x)$ twice differentiable; and let

$$\phi(a) = \phi(b) = 0.$$

Now consider the integral:

$$I[\phi] = \int_a^b \{P(x)[\phi'(x)]^2 + Q(x)[\phi(x)]^2 - 2R(x)\phi(x)\}\,dx.$$

The quantity $I[\phi]$ is referred to as a *functional* of ϕ: its argument is the complete function $\phi(x)$ which occurs within the integral, rather than an ordinary variable.

Let us consider the effect of changing the function ϕ by a small amount:

$$\phi(x) \rightarrow \phi(x) + \delta\phi(x),$$

where $\delta\phi(x)$ satisfies the same conditions as $\phi(x)$, and in particular $\delta\phi(a) = \delta\phi(b) = 0$, but is otherwise arbitrary. We shall assume that the square of $\delta\phi(x)$ may be neglected. Then the change induced in the functional is:

$$\delta I = \delta\int_a^b \{P(x)[\phi'(x)]^2 + Q(x)[\phi(x)]^2 - 2R(x)\phi(x)\}\,dx$$

$$= \delta I_0[\phi] + \int_a^b \delta\phi(x)\{2Q(x)\phi(x) - 2R(x)\}\,dx,$$

where
$$I_0[\phi] = \int_a^b P(x)[\phi'(x)]^2 dx.$$

Now
$$\delta I_0 = \delta \int_a^b P(x)[\phi'(x)]^2 dx$$

$$= 2\int_a^b P(x)\phi'(x)(d/dx)\delta\phi(x)dx$$

$$= 2\delta\phi(x)P(x)\phi'(x)\Big|_a^b - 2\int_a^b \delta\phi(x)(d/dx)[P(x)\phi'(x)]dx,$$

where the first term vanishes because $\delta\phi(a) = \delta\phi(b) = 0$.

Collecting terms, we finally obtain:

$$\delta I[\phi] = 2\int_a^b \delta\phi(x)\{-(d/dx)[P(x)d\phi/dx] + Q(x)\phi(x) - R(x)\}dx.$$

We obtain a variational principle by noting that $\delta\phi(x)$ is arbitrary inside the interval of integration. The variation $\delta I[\phi]$ therefore vanishes if and only if the bracketed expression multiplying $\delta\phi$ also vanishes, a condition that is conventionally written $\delta I/\delta\phi = 0$. Thus, the solution of the rather general linear boundary value problem

$$-(d/dx)[P(x)y'] + Q(x)y = R(x), \quad a \le x \le b, \quad y(a) = y(b) = 0,$$

is a function $y(x)$ which makes the functional $I[y]$ stationary. It may further be shown that if $P(x) > 0$ and $Q(x) > 0$ on the interval of integration, this solution is unique.

It will be a rare case in which the variational principle may be used to find the exact solution directly; complete freedom to vary the function $\phi(x)$ within the interval of integration is not usually achievable. Instead, it is customary to restrict ϕ to a set of functions which may more readily be handled. In what follows, we shall assume that ϕ may be expressed as a linear combination of n orthonormal functions $\phi_k(x)$ satisfying the same boundary conditions as ϕ:

$$\phi(x) = \Sigma_k c_k \phi_k(x),$$

where the n coefficients a_k are to be determined by the set of n variational conditions

$$\partial I/\partial c_k = 0.$$

This version of the variational method is called the *Rayleigh–Ritz method*, and it leads directly to a system of linear equations for the unknown coefficients c_k. If we assume that the functions ϕ_k satisfy the orthonormality relations

$$\int_a^b \phi_j(x)\phi_k(x)dx = \delta_{j\,k},$$

then the variational equations become

$$Ac = b,$$

where the matrix A has matrix elements

$$a_{jk} = \int_a^b [P(x)\phi_j'(x)\phi_k'(x) + Q(x)\phi_j(x)\phi_k(x)]dx,$$

and the vector **b** has elements

$$b_k = \int_a^b R(x)\phi_k(x)dx.$$

Once these elements have been calculated, the system of linear equations may be solved by any convenient method. In general, the calculation of the matrix and vector elements involves quadrature, and that is where much of the effort of computation lies.

In simple cases, however, it is possible to obtain closed formulae for these elements. Consider, for example, the second-order problem:

$$-(d/dx)[(x^2 + 1)dy/dx] + 2y = 4(x^2 + 1), \quad -1 \le x \le +1, \quad y(-1) = y(+1) = 0.$$

where $P(x) = x^2 + 1$, and $Q(x) = 2$. The conditions for a unique solution hold, and it is $y(x) = 1 - x^2$. A suitable set of orthonormal functions is $\phi_k(x) = \cos(k+\frac{1}{2})\pi x$, for which it is an easy matter to calculate all the elements involved:

$$a_{jk} = (5/2) + (4/3)(k + \tfrac{1}{2})^2\pi^2, \qquad\qquad\qquad (j = k),$$

$$= 2(-1)^{j-k}(j+\tfrac{1}{2})(k+\tfrac{1}{2})[1/(j - k)^2 + 1/(j - k + 1)^2], \qquad (j \ne k),$$

$$b_k = 16(-1)^k[1 - 1/(k+\tfrac{1}{2})^2\pi^2]/(k+\tfrac{1}{2})\pi.$$

The unknown coefficients in the approximation $y^{(n)}(x) = \sum_{k=0}^{n-1} c^{(n)}_k \cos(k + \frac{1}{2})\pi x$ may be found using Crout's method, and the results are given below for $1 \le n \le 6$:

n	c_0	c_1	c_2	c_3	c_4	c_5
1	1.0463	–	–	–	–	–
2	1.0331	−0.0407	–	–	–	–
3	1.0324	−0.0385	0.0090	–	–	–
4	1.0322	−0.0383	0.0083	−0.0034	–	–
5	1.0321	−0.0383	0.0083	−0.0030	0.0016	–
6	1.0321	−0.0382	0.0083	−0.0030	0.0014	−0.0009

Using these coefficients the corresponding approximations to the solution may readily be computed, and are given in the table below:

x	$n = 1$	2	3	4	5	6	Exact
0.0	1.0463	0.9924	1.0030	0.9989	1.0007	0.9996	1.00
0.1	1.0334	0.9842	0.9918	0.9897	0.9900	0.9901	0.99
0.2	0.9951	0.9586	0.9593	0.9611	0.9594	0.9604	0.96
0.3	0.9322	0.9141	0.9075	0.9111	0.9101	0.9097	0.91
0.4	0.8464	0.8484	0.8381	0.8396	0.8408	0.8399	0.84
0.5	0.7398	0.7593	0.7508	0.7487	0.7500	0.7505	0.75
0.6	0.6150	0.6459	0.6434	0.6400	0.6392	0.6398	0.64
0.7	0.4750	0.5092	0.5131	0.5118	0.5103	0.5096	0.51
0.8	0.3233	0.3521	0.3592	0.3610	0.3612	0.3608	0.36
0.9	0.1637	0.1801	0.1853	0.1878	0.1890	0.1897	0.19
1.0	0	0	0	0	0	0	0

Successive approximations are not very rapidly convergent, and they depend upon the first approximation, in this case the function cos $\frac{1}{2}\pi x$, being itself a fair approximation to the exact solution. Of course, had we chosen Legendre polynomials instead of cosines the exact result would have been obtained immediately in this case because $1 - x^2 = (2/3)(P_0(x) - P_2(x))$. The interest of the method is that it obtains an approximate solution by working with the whole function rather than by proceeding stepwise. It is also possible to take a stepwise approach in the Rayleigh–Ritz method, however. The functions ϕ_k do not have to be orthonormal; they merely have to be linearly independent. A set of functions each of which differs from zero only in its own sub-interval of the interval of integration has this property, and Rayleigh–Ritz equations may be written down accordingly, and more rapid convergence thereby obtained (see the book by Burden and Faires).

19.5 Eigenvalue problems

Normally, a differential equation of order n, if it has a solution at all, will require just n conditions to determine a unique solution. We have seen, for example, that a first-order differential equation requires just one initial condition, while a second-order equation requires two initial conditions or two boundary values, depending how the problem is posed. In general it is not possible to demand that a first-order differential equation satisfies two conditions, or that a second-order equation satisfies three conditions. However, if the differential equation contains an adjustable parameter, then it may be possible to find solutions which satisfy an extra boundary condition for certain values of the parameter. These values, if there are any, are called *eigenvalues*, and the corresponding solutions are called *eigenfunctions*.

To take a very simple example, consider the second-order initial value problem:

$$y'' + k^2 y = 0, \quad 0 \le x \le 1, \quad y(0) = 0, \quad y'(0) = 1,$$

which has the unique solution $y(x) = (1/k)\sin kx$. Now let us demand that the solution satisfies the additional boundary condition $y(1) = 0$. This requires that $\sin k = 0$, or $k = n\pi$, where n is an integer. Only for these values of k may solutions be found which satisfy the extra boundary condition. These are the eigenvalues of the problem.

In passing, we may note that if we formally set $A \equiv d^2/dx^2$, $\lambda \equiv -k^2$, then the above equation may be written

$$Ay = \lambda y,$$

a form that is familiar from our study of matrix eigenvalue problems. Many practical eigenvalue problems are of this form, except that the operator A may be of greater complexity, for example:

$$A \equiv P(x)d^2/dx^2 + Q(x)d/dx + R(x).$$

The problem is then determined by the boundary conditions, say $y(a) = ya$, $y(b) = yb$, together with a third condition which could take many forms. A common one is a normalization condition such as

$$\int_a^b [y(x)]^2 dx = 1.$$

As a very simple illustration we develop a routine derived from the shooting method of 19.3.1 which computes the eigenvalues of the variable k in the problem

$$y'' = f(k, x, y, p),$$

where as usual $p = dy/dx$ and $y(a) = ya$, $y(b) = yb$, $p(a) = pa$. Instead of varying pa until $y(b) = yb$, we fix pa and vary k until the same condition is met, starting with guessed values $k1$, $k2$ which are supposed to bracket the true eigenvalue.

We shall need a new function type in order to include the variable k:

```
typedef double (*func)(double k, double x, double y, double p);
```

The functions `RK4secord` and `shoot` become quite trivially

```
void RK4secord(func f, double k,
               double x, double &y, double &p, double h);
```

and

```
double shoot(func f, double k,
       double a, double ya, double pa, double b, double h);
```

We may now define a new `bisect` function to achieve our goal:

```
state   bisecteigen(func f, double &k, double a, double ya, double pa,
                        double b, double yb, double k1, double k2, double h)
{       state s = ITERATING;
        int iter = 0;
        double yb1 = shoot(f, k1, a, ya, pa, b, h);
        double yb2 = shoot(f, k2, a, ya, pa, b, h);
        if ((sign(yb1 - yb) + sign(yb2 - yb)) != 0) s = SAMESIGN;
        else
        {       double dk;
                k = (yb1 - yb < 0) ? (dk = k2-k1, k1) : (dk = k1-k2, k2);
                while (s == ITERATING)
                {       double kmid = k + (dk *= 0.5);
                        double ymid = shoot(f, kmid, a, ya, pa, b, h);
                        if (ymid < yb) k = kmid;
                        if (fabs(ymid - yb) < tolerance
                                && fabs(dk) < tolerance) s = SUCCESS;
                        if (++iter == maxiter) s = WONTSTOP;
                }
        }
        return s;
}
```

If we apply this routine to the eigenvalue problem

$$y'' + k^2 y = 0, \quad 0 \le x \le 1, \quad y(0) = y(1) = 0, \quad y'(0) = 1,$$

we obtain the following values of k/π:

$h =$	0.1	0.05	0.025
k/π	1.000078	1.000005	1.000000
	2.002253	2.000157	2.000010
	3.014158	3.001139	3.000076

which approximate to the first three positive eigenvalues 1, 2, 3. Approximate eigenfunctions may also be printed out if desired.

19.6 Partial differential equations

Many of the laws of the physical sciences may be expressed in terms of partial derivatives, and a great deal of applied mathematics and mathematical physics therefore consists of devising methods for solving partial differential equations. The difficulties may be formidable. A simple illustration of a typical difficulty is afforded by the equation

$$x(\partial u/\partial x) + y(\partial u/\partial y) = 0,$$

the solution of which may readily be seen to be $u(x, y) = f(y/x)$, where $f(z)$ is an arbitrary function of z. Similarly, the *wave equation*

$$\partial^2 u/\partial x^2 = (1/c^2)\partial^2 u/\partial t^2$$

possesses solutions of the form

$$u(x, t) = f_+(x - ct) + f_-(x + ct),$$

where the first term represents an arbitrary waveform travelling with velocity c in the direction of increasing x, and the second another arbitrary waveform travelling with velocity c in the direction of decreasing x. The wave equation is an example of a *hyperbolic equation*, and it may be generalized to three dimensions, when it becomes

$$\nabla^2 u \equiv (\partial^2 u/\partial x^2) + (\partial^2 u/\partial y^2) + (\partial^2 u/\partial z^2) = (1/c^2)(\partial^2 u/\partial t^2).$$

There are also *parabolic equations*, of which the prototype is the *diffusion equation*

$$\partial u/\partial t = D\nabla^2 u,$$

and *elliptic equations* such as *Poisson's equation*

$$\nabla^2 u = f(x,y,z),$$

where f is a given source distribution. Each of these three families of equations requires somewhat different computational treatment because they have different stability properties.

As an indication of what may be done we offer a means of computing approximate solutions of the two-dimensional Poisson equation defined on the square

$$S = \{ (x,y) \mid a \le x,y \le b \},$$

in which we replace the partial second derivatives by central differences according to Taylor's Theorem:

$$\frac{\partial^2 u}{\partial x^2}(x_i, y_j) = \frac{u(x_{i+1}, y_j) - 2u(x_i, y_j) + u(x_{i-1}, y_j)}{h^2} - \frac{h^2}{12} \cdot \frac{\partial^4 u}{\partial x^4}(\xi_i, y_j)$$

$$\frac{\partial^2 u}{\partial y^2}(x_i, y_j) = \frac{u(x_i, y_{j+1}) - 2u(x_i, y_j) + u(x_i, y_{j-1})}{h^2} - \frac{h^2}{12} \cdot \frac{\partial^4 u}{\partial y^4}(x_i, \eta_j)$$

Here, $u(x, y) = g(x, y, a, b)$, a given function, whenever the point (x, y) is on the boundary of S defined by (a, b). If we define the integer $n = (b - a)/h$, the grid-points (x_i, y_j) are given by:

$$x_i, y_i = a + ih, \quad i = 0, 1, \ldots, n.$$

Dropping the truncation terms, we then arrive at the central difference equations

$$4u_{i,j} - (u_{i-1,j} + u_{i+1,j}) - (u_{i,j+1} + u_{i,j-1}) = -h^2 f(x_i, y_j),$$

for the interior points for which $i, j = 1, \ldots, n - 1$, while on the boundary,

$$u_{0,j} = g(a, y_j, a, b), \quad u_{n,j} = g(b, y_j, a, b), \quad j = 0, 1, \ldots, n,$$
$$u_{i,0} = g(x_i, b, a, b), \quad u_{i,n} = g(x_i, b, a, b), \quad i = 0, 1, \ldots, n.$$

It is apparent that in the central difference approximation the value of $u_{i,j}$ at each interior point is determined by its four nearest neighbours $u_{i\pm1,j}$, $u_{i,j\pm1}$, together with the values $f_{i,j} = f(x_i, y_j)$, and the boundary values, which suggests a relabelling of elements to achieve an appropriate set of linear equations. There are $(n - 1)^2$ interior points; denote them by $z_k \equiv (x_i, y_j)$, where $k = (i - 1) + (n - 1)(j - 1)$ for each $i, j = 1, 2, \ldots, n - 1$. Then $0 \leq k < (n - 1)^2$. The original labelling may always be recovered since $i = 1 + k\%(n - 1)$, $j = 1 + k/(n - 1)$. The two labellings are illustrated below for the case $n = 4$:

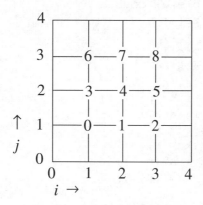

Our task is now to set up the system of $(n - 1)^2$ linear equations at the points z_k,

$$A\mathbf{u} = \mathbf{b},$$

where we have to determine the non-zero elements of the matrix A in terms of the coefficients of the u_k, and the elements of the vector $\mathbf{b}$ in terms of the boundary values and the source terms f_k. We have to distinguish three types of points: first, *internal* points which involve no boundary points, of which $k = 4$ is the only example in the diagram; the four *corner* points which involve two boundary points each, 0, 2, 6, 8 in the diagram; and *border* points which involve one boundary point each, which in the diagram are 1, 3, 5, 7. Of course, with a fine grid (when

n is large) the interior points will be the most numerous. For the case $n = 4$, the equation corresponding to the internal point $k = 4$ is

$$4u_4 - u_3 - u_5 - u_1 - u_7 = b_4 = -h^2 f_4 \, ;$$

the corner point $k = 0$ yields

$$4u_0 - u_1 - u_3 = b_0 = -h^2 f_0 + g_{0,1} + g_{1,0} \, ;$$

and the border point $k = 1$ gives

$$4u_1 - u_0 - u_2 - u_4 = b_1 = -h^2 f_1 + g_{0,2} \, .$$

Thus the matrix A is constant, symmetric and tridiagonal, with two additional symmetric diagonals the placing of which is determined by n. The diagonal elements all equal 4, the non-zero off-diagonal elements all equal -1. The following routine constructs the matrix A and the vector **b** for the general case:

```cpp
typedef double (*func)(double, double);
typedef double (*bndfunc)(double, double, double, double);
void  poisson(func f, bndfunc g, matrix &A, vector &u,
              double a, double b)
{       int N = u.getsize();
        int n = sqrt(N) + 1;
        vector x(n+1), y(n+1);
        x[0] = a; y[0] = a;
        double h = (b - a)/n;
        for (int i = 1; i <= n; ++i)
                x[i] = y[i] = x[i-1] + h;
        double hh = h*h;
        for (int k = 0; k < N; ++k)
        {       int i = 1 + k%(n-1);
                int j = 1 + k/(n-1);
                A[k][k] = 4;
                if (i < n-1) A[k][k+1]   = -1;
                if (i > 1)   A[k][k-1]   = -1;
                if (j < n-1) A[k][k+n-1] = -1;
                if (j > 1)   A[k][k-n+1] = -1;
                u[k] = -hh*f(x[i],y[j]);
                if (i == 1)   u[k] += g(x[0], y[j], a, b);
                if (i == n-1) u[k] += g(x[n], y[j], a, b);
                if (j == 1)   u[k] += g(x[i], y[0], a, b);
                if (j == n-1) u[k] += g(x[i], y[n], a, b);

        }
}
```

Note that `poisson` has been designed to solve Poisson's equation *on a square*, with a square grid; it is not of general applicability, although it may readily be generalized.

By way of example, consider the very simple equation

$$(\partial^2 u/\partial x^2) + (\partial^2 u/\partial y^2) = 0, \ 0 \leq x,y \leq 0.5,$$

where $u(0, y) = u(x, 0) = 0$, $u(0.5, y) = 200x$, $u(x, 0.5) = 200y$. If we take $n = 5$ (so that the interval is $h = 0.1$), the solution obtained with the routine `poisson` is:

$k =$	0	1	2	3	4	5	6	7	8	9	10	11	12	13	14	15
$x =$	.1	.2	.3	.4	.1	.2	.3	.4	.1	.2	.3	.4	.1	.2	.3	.4
$y =$	.1	.1	.1	.1	.2	.2	.2	.2	.3	.3	.3	.3	.4	.4	.4	.4
$u =$	4	8	12	16	8	16	24	32	12	24	36	48	16	32	48	64

The exact solution, as may easily be verified, is $u(x, y) = 400xy$. The computed solution is also exact, because the truncation terms involving derivatives of higher than second order in this case vanish identically.

For a further discussion of partial differential equations, see the book by Burden and Faires.

20 Recursive data types – lists

In this chapter we shall be concerned with *arrays of objects* of given type, and with how to handle these arrays in a variety of applications in which the ordering of the elements of the array (*ie*, of the individual objects) matters. We may wish, for example, to sort them according to some measure which will determine their order; or we may wish to shuffle them, or reorder them according to some other measure; or add to or subtract from their number while preserving the given order. The result can be a highly developed theory of data handling, much of which is of little relevance to numerical analysis. We dealt briefly with the sorting of an array in 14.4. Here, we shall consider more elaborate methods, but shall limit ourselves to the construction, manipulation and numerical applications of *lists*, both *linear lists* and *circular lists*. We shall not develop the beautiful theory of *trees*, useful though they are for certain applications.

First we note that an array is itself a list in the very limited sense that it has an order. If A is an array of n elements, each of type `type`, so that we can write

```
type A[n];
```

then the elements A[i], as we have seen, are ordered according to i = 0,1, . . . , n − 1. This ordering, however, merely reflects the order in which the various elements were inserted into the array. If we wish to change that order, for instance to re-order the elements according to their value (as determined by some measure), this may involve a great deal of memory manipulation if the individual elements are themselves substantial structures.

Suppose that we have five objects of given type, possibly complex structures or class objects, and suppose each has a measure associated with it, which for simplicity we shall take to be an integer, each object having a different measure. Let the five measures be 1,5,3,2,7 in that order. In the simplest case, the objects could simply be the integer measures themselves. In any case, we denote this situation by the diagram:

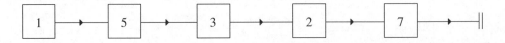

where the arrowed lines imply "followed by" and the double line indicates that the object of measure 7 is followed by nothing at all. This diagram represents a linear list which begins with the object of measure 1 and is terminated by the *null* object. The recursive nature of a list defined by such a diagram is clear: we may say that a list is either *empty*, that is to say it consists of the null object, or it consists of a single object, which we call the *head*, in this case object 1, followed by another list, which we call the *tail*.

Now suppose we prefer to have the objects in rising order. We could do this by interchanging the positions of objects 2 and 5. That would mean copying object 5 into a temporary object declared to be of the same type, copying object 2 into the position at present occupied by 5, and then copying the temporary object into the position at present occupied by 2. Finally, we would have to destroy the temporary object. We are describing in words the `swap` operation we have already met in a number of contexts. When these (possibly memory intensive) operations have been completed we would have the sorted list:

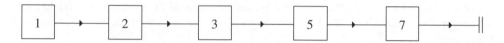

We might save ourselves a great deal of copying activity if, alternatively, we could leave the objects where they are but change the "followed by" relationships. In that case we obtain the list

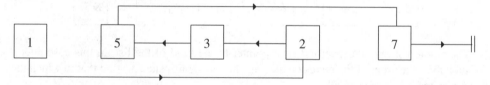

where we merely have to follow the arrows to find the order of the objects in the list.

Another typical list operation is the insertion of a further object into a given list while preserving the order. In the above example, suppose that we want to add a new object of measure 4 in its correct arithmetical position. If we move objects around this might entail a good deal of copying. But if we work on the "followed by" lines instead we obtain the augmented list:

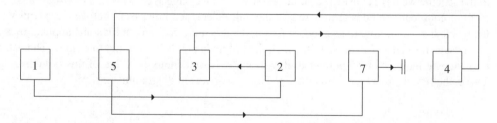

In spite of the position of object 4 on the page, it correctly follows object 3 and precedes object 5 according to the "followed by" lines.

To represent the "followed by" lines in C++, all we have to do is to augment each of our objects of type `type` with a pointer, and the pointer must point to the following augmented object. We must imagine the square boxes, or *nodes*, in the diagram as augmented structures containing this additional pointer. Schematically, we define a node as a structure with two members:

```
struct node {
        type object;
        node *next;
};
```

The structure so defined contains an object of type `type`, as required, and a pointer to the node which is to follow. If no node is to follow we take `next = 0`, the value of the null pointer, by convention the unique address of no object at all. Otherwise, it is the address of the next node. If `n1` and `n2` are consecutive nodes, then the objects we wish to order are `n1.object` and `n2.object`, respectively, and the value of the pointer from `n1` to `n2` is `n1.next = &n2`. Thus, the last box diagram may be redrawn in terms of nodes, and becomes:

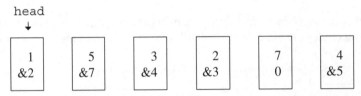

The whole list may be specified by a pointer to the head of the list, in this case node 1. Written thus, the order of the boxes on the page has no significance beyond that in which they happened to be first presented.

Notice that the node definition we have given is recursive: it contains a pointer to the structure which it defines. It could not contain an *object* of that structure (*ie*, a node) because the object would not have been fully defined and so would be of indeterminate size. It would be an error to try to do so. But it may contain a pointer to a node, because that is of fixed size.

Before offering a definition of the class `list`, it is convenient to make two modifications to the scheme we have considered so far. Firstly, we have envisaged a list as a *linear* arrangement of consecutive nodes terminated by the null object, and have seen that we can identify the whole list by pointing to its head, or first node. It is more convenient for some applications to envisage instead a *circular list*, in which the last node points back to the first node. The list may then be indicated by a pointer to its *last* node. If we denote the value of this pointer by `last`, the first node is then pointed to by `last->next`. In diagrammatic form:

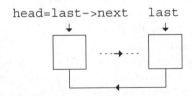

Of course, the last node of the list may not be the last shown on the line; the example we were treating before becomes in terms of nodes:

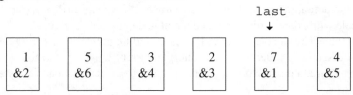

Second, it is convenient to define `list` as a *base class*, from which may be derived lists of objects of different types: lists of integers, lists of complex numbers, lists of named geometrical points such as towns on a map, and so on. In that case, we must replace the data member of the structure `node` by a *pointer* to the data type, as we did for arrays in 14.4. But since for the base class we do not wish to declare the data type, we must declare the pointer to be a *void pointer* (*cf* 1.5.1). On derivation, pointers to real objects, such as elements of an array which provides the required data storage, will be converted to void pointers in order to access the base class (see 6.1). The list itself provides only the storage required for void pointers.

20.1 The base class `list`

It is convenient to define three pointer types, which we do by the following statements:

```
class node;        // incomplete
class list;        // declarations
typedef void *vptr;
typedef node *nptr;
typedef list *lptr;
```

It is in terms of these pointer types that we shall define our classes. First, we redefine `node` as an entirely private class. However, its private members should be available to the class `list` when we define it, and also to a class which provides us with the facility to scan through a list. A single node, not part of a list, should point to the null node. Thus:

```
class node {
friend class list;
friend class listscan;
private:
      vptr data;        // "data" pointer
      nptr next;        // node pointer
      node(vptr, nptr=0);
};

inline node::node(vptr vp, nptr np) { data = vp; next = np; }
```

We have referred to the void pointer member as a "data" pointer because on derivation its value will be obtained by conversion from a real data pointer.

Thanks to the default declaration of the constructor, if we call node `n(vp);` we obtain an isolated node with "data" pointer `vp` and a null node pointer (*ie*, `n.next = 0`).

We may now declare the class `list`. It is essentially a pointer to the last node, protected so that it is accessible to a derived class, and the current length of the list, also protected, accompanied by three constructors, a destructor, and a set of functions for assembling and manipulating lists:

```
class    list {
friend   class listscan;
protected:
        nptr last;         // last->next is head of cyclic list
        int  len;          // length of list, defined dynamically
private:
        void clearlist();       // remove all nodes
public:
        list();                 // empty list constructor
        list(list&);            // copy list constructor
        ~list();                // destructor
        list& operator=(list&); // assignment
        list& operator+(list&); // concatenation
        void prepend(vptr);     // insert at head of list
        void append(vptr);      // insert at foot of list
        vptr get();             // return and remove head
        int  length();          // returns current length
        vptr operator[](int);   // data ptr of numbered node
        void swap(int, int);    // swap given items in list
        vptr foot();            // data ptr at foot
        vptr head();            // data ptr at head
};
```

The private function `clearlist` is an auxiliary function required for defining the list destructor, and for re-initializing a list before assignment to it; we shall define it presently.

The class `listscan` must be declared a friend of `list` as well as of `node`, because its task is to scan through all the nodes of a given list. We may therefore declare it to be a pair of pointers, one to the list and one to a node of the list, together with a constructor and an operator which on iteration scans through the list:

```
class    listscan {
private:
        nptr np;
        lptr lp;
```

```
public:
      listscan(list&);        // constructor
      vptr operator()();      // scanner
};
```

We now have to define all the functions which appear in these declarations. The first list constructor defines the null list and is very simple:

```
inline list::list() { last = 0; len = 0; }
```

There is a certain redundancy here: an empty list is sufficiently defined by `last = 0`; we do not need to know that `len = 0` also. However, it is useful to carry the current length of a list as a data member, just as we did in 5.2 for the class `string`.

Next, we consider the insertion of a new node into a circular list, and thereby the construction of lists beginning with the empty list. To insert a node with data pointer `vp` into an empty list, thus creating a list with one node, we merely have to create that node and then cause it to point to itself:

```
last = new node(vp);    last->next = last;
```

If, however, the list already contains one or several nodes, we have to decide in particular whether to insert the new node at the beginning of the list or at the end. In either case, since we are dealing with a circular list, we have to insert the new node between the last node and the head node. The operation we are performing in diagrammatic form is:

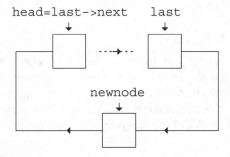

We carry out this operation by creating a new node which is pointed to by the last node, and which in turn points to the head node:

```
last->next = new node(vp, head=last->next);
```

In a circular list, the only difference between inserting the new node at the front of the list and at the end is that in the latter case we have to redefine `last` to point to the new node:

```
last = last->next;
```

In all these cases, we also have to increase the length of the list by one. Allowing for error conditions, we thus obtain the two alternative procedures already declared: the first we shall call `prepend` (from the Latin *praependere*), and the second `append`. Their definitions are:

```
void  list::prepend(vptr vp)
{      nptr p = new node(vp);
       if (!p) error("allocation error in list::insert(vptr)");
       if (!last)
       {      last = p; last->next = last;
       }      else
       {      p->next = last->next; last->next = p;
       }
       ++len;
}

void  list::append(vptr vp)
{      nptr p = new node(vp);
       if (!p) error("allocation error in list::append(vptr)");
       if (!last)
       {      last = p; last->next = last;
       }      else
       {      p->next = last->next; last->next = p;
              last = p;
       }
       ++len;
}
```

Note that all nodes are created with the aid of the `new` operator. This means that when a node is removed from a list a pointer to it must be deleted, otherwise the memory it occupies remains allocated to it. But the node must first be decoupled from the rest of the list or great confusion and error may result. For example, a function which removes a node from the front of a list and then returns the value of the data pointer it contained is:

```
vptr  list::get()
{      if (!last) error("can't get from empty list!");
       nptr head = last->next;
       vptr hv = head->vpt;
       if (head==last) last = 0;
       else last->next = head->next;            // free head
       delete head;
       --len;
       return hv;
}
```

Likewise, the private function `clearlist`, needed for the list destructor, requires the successive deletion of all nodes:

```
void  list::clearlist()
{       if (!last) return;
        nptr np = last;
        do
        {       nptr nnp = np;
                np = np->next;
                delete nnp;
        }       while (np != last);
        last = 0; len = 0;
}
```

In terms of this function the list destructor is simply

```
inline list::~list() { clearlist(); }
```

The copy list constructor requires both `append` and `get`:

```
list::list(list &vl)
{       last = 0; len = 0;
        while (vl.last) append(vl.get());
}
```

Note that as each call to `append` adds to the length of the list, so each call to `get` reduces the length of the list being copied; lengths are taken care of automatically. Note, too, that had we used `prepend` instead of `append`, the list would have been copied in reverse order. If required, we could use this fact to construct a procedure for reversing a list.

Assignment also calls upon the procedure `clearlist`, because before we can assign one list to another list, we first have to clear the list assigned to:

```
list&   list::operator=(list &vl)
{       clearlist();
        while (vl.last) append(vl.get());
        return *this;
}
```

Clearance is not required for the concatenation operator; indeed, to include it would reduce the operation to mere assignment:

```
list&   list::operator+(list &vl)
{       while (vl.last) append(vl.get());
        return *this;
}
```

The remaining list functions are conveniently provided with the aid of the list scanner, so we turn to that next. Remember, a list is specified by a pointer to its last node, while the scan should begin at the first node. The `listscan` constructor (which is merely an initializer) is therefore given quite simply by:

```
inline listscan::listscan(list &vl) { lp = &vl; np = lp->last; }
```

The scan operator, a member function of the class `listscan`, is conveniently provided by overloading the function call operator `( )`. It returns the value of the data pointer of the node pointed to, moves on to the next node, and conventionally returns a zero at the end of the scan to terminate any iterative procedures in which it is used:

```
vptr   listscan::operator()()
//     return 0 at end of scan
{      vptr vp = np ? (np = np->next)->data : 0;
       if (np == lp->last) np = 0;
       return vp;
}
```

This is the operator we use in applications for printing out the contents of a list, and for many other purposes. A very simple but revealing illustration of its use is in constructing a function that returns the number of nodes in a list:

```
int    list::length()                // unnecessary!
{      listscan thisnode(*this);
       int count = 0;
       while(thisnode()) ++count;
       return count;
}
```

(Remember that in the class `list`, the self-referencing pointer `this` is a pointer to the list itself. The list pointed to is given by `*this`.) However, although the function so defined certainly works, it is quite unnecessary because we have included the current list length as a data member in its own right. The simpler definition of the member function `length` is therefore:

```
inline int list::length() { return len; }
```

Although, whenever possible, we shall avoid referring to nodes in the numbered order in which they occur in a list, it is sometimes necessary, or more efficient, to do so. We therefore include an indexing function which returns the "data" pointer of a given numbered node. This can be obtained using the list scanner, but we prefer to compute it directly:

```
vptr   list::operator[](int i)
{      if (i < 0 || i > len-1)
       error("int out of range in list::numtoptr(int)");
```

```
        nptr np = last;
        for (int k = 0; k <=i; ++k) np = np->next;
        return np->data;
}
```

A similar problem occurs if we wish to interchange two given nodes. It is in fact quite a complex operation, as a few diagrams will quickly demonstrate. Fortunately, it is not usually necessary to interchange the nodes, only the data pointers contained in the nodes. The following procedure effects the interchange of two data pointers according to the numbers of the corresponding nodes in the list:

```
void  list::swap(int i, int j)
{     if (i < 0 || i >= len-1 || j < 0 || j >= len-1)
      error("indices out of range in list::swap(int,int)");
      if (i == j) return;
      if (j < i)                        // ensure j > i
      { int t = i; i = j; j = t; }
      nptr ip, jp = last;
      for (int k = 0; k <= j; ++k)
      {     jp = jp->next;
            if (k == i) ip = jp;
      }
      vptr tp = ip->data;               // swap pointers
      ip->data = jp->data;
      jp->data = tp;
}
```

Strictly speaking, these functions are enough for most applications. However, it is sometimes efficient to be able to refer to the last data item pointed to by a list, and also the first. Both may be obtained using the list scanner, but it is wasteful to scan right through a list unnecessarily. We therefore define the additional member functions:

```
vptr list::foot()
{     if (!last) return 0;
      else return last->data;
}
```

```
vptr list::head()
{     if (!last) return 0;
      else return last->next->data;
}
```

Note that if the list is empty we take the returned data pointer to be null.

20.2 List class template

The list we have just constructed is a base class only; its "data" are pointers to void. In practice, we shall be concerned with data which are of any given type: integers, floating point numbers, complex numbers, structures, and so on. The operations we shall perform on these lists, whatever the type of their elements, will mostly be those already fully considered for the base class. To take advantage of all that work we should clearly derive our lists from list as the base class. That would mean that we had to convert pointers to int, say, if we were considering a list intlist of integers, to pointers to void, which are what the base class understands. We should have to do that afresh whenever we were considering a list of elements of a different type. The conversions would be carried out in the manner described in 6.1.1, and stated more formally in 6.2, but clearly there would be much repetitive work.

We introduced the concept of a class template in 5.4.3 in the context of arrays. Using a list class template we may perform these wearisome pointer conversions once and for all, so that thereafter we may introduce lists of elements of any type as particular instances of the template. The class template Tlist, and its scanner Tlistscan, will be derived from list as its base class as follows:

```
template <class T> class Tlist : public list {
public:
      Tlist() {}
      Tlist(Tlist<T> &tl) : list(tl) {}
      ~Tlist() {}
      void prepend(T* tp) { list::prepend((void*)tp); }
      void append(T* tp) { list::append((void*)tp); }
      Tlist& operator=(Tlist<T> &tl) \
          { return *(Tlist<T>*)&list::operator=(tl); }
      Tlist& operator+(Tlist<T> &tl) \
          { return *(Tlist<T>*)&list::operator+(tl); }
      T* get()     { return (T*)list::get(); }
      T* operator[](int i) \
                { return (T*)list::operator[](i); }
      T* foot() { return (T*)list::foot(); }
      T* head() { return (T*)list::head(); }
};


template <class T> class Tlistscan : public listscan {
public:
      Tlistscan(Tlist<T> &tl) : listscan(*(list*)&tl) {}
      T* operator()() \
                { return (T*)listscan::operator()(); }
};
```

The template device is implemented as a macro. If we declare

```
#define intlist Tlist<int>
```

not only may all occurrences of `Tlist<T>` be replaced by `intlist`, but `T` may be replaced by `int`, and `T*` by `int*`, throughout the class template, if we wish to see what is happening in the real world of integers. Of course, we must also declare

```
#define intlistscan Tlistscan<int>
```

Continuing with a list of integers as an example, we shall need to make provision for an array to hold up to 20 integers (of course, this number could be changed):

```
typedef int array[20];
```

Suppose, therefore, that we wish to construct a list of *n* integers, where *n* < 20, by filling the first *n* elements of the array with integers of our choice. The following procedure achieves this, having first invited us to declare the number of integers in the list:

```
void  makelist(array a, intlist &il, char *listname)
{      int n;
       cout << "number of integers in " << listname << " = ";
       cin >> n;                       // define length of list
       for (int i = 0; i < n; ++i)
       {    cout << "enter a[" << i << "] = ";
            cin >> a[i];               // construct array
            il.append(&a[i]);          // link to list
       }
}
```

Note the automatic conversion of the "real" data pointer `&a[i]` to the void pointer expected by `append` in the base class. (Had we wished the list to appear in the reverse order of this array, we would have used the member function `prepend` instead of `append`.)

We can use the scanner to print out the list with suitable messages:

```
void  printlist(intlist &il, char *listname)
{      intlistscan thisptr(il);
       int* ip;
       int length = il.length();
       cout << listname << ":\n";
       while ((ip=thisptr()) != 0)
           cout << *ip << "\t";
       cout << "\n" << listname << " has "
           << length << " member(s)\n\n";
}
```

We must now give `intlist` something testing to do. Let us define a function which splits a given list into two lists, the first integer into the "even" list, the second into the "odd" list, and so on, alternately:

```
void   split(intlist &il, intlist &ileven, intlist &ilodd)
{      while (il.length())
           {   ileven.append(il.get());
               if (il.length()) ilodd.append(il.get());
           }
}
```

We may take the testing process a little further by means of recursion. Having split the original list into an "even" list and an "odd" list, we may split each of the new lists in the same way, and so on. Finally, we may reassemble the lists in revised order by concatenation:

```
void   eosort(intlist &il)
{      if (il.length() < 2) return;
       intlist ileven, ilodd;
       split(il, ileven, ilodd);
       eosort(ileven);
       eosort(ilodd);
       il = ileven + ilodd;
}
```

This is an interesting recursive function which merits careful study by the reader. A suitable user program is the following:

```
main()
{      clear;
       char *listname1 = "original list";
       char *listname2 = "reformed list";
       array a;
       intlist il;
       makelist(a, il, listname1);
       newline;
       printlist(il, listname1);
       eosort(il);
       printlist(il, listname2);
}
```

This may seem a rather contrived example, but it has the immediate advantage of testing most of the member functions of `intlist` and of the base class `list`. (It will reappear in the next chapter in the context of fast Fourier transforms.) If we enter the 10 consecutive integers:

1 2 3 4 5 6 7 8 9 10

this program gives us the rearranged list:

1 9 5 3 7 2 10 6 4 8

In a similar way, we can manipulate lists of objects of any type. All we have to do is to define "pointer to type" by means of `typedef type *tptr;`, create an array of `type`, and then convert to void pointers when calling upon base class members. The only substantial differences are naturally in the input and output procedures, where the exact nature of "type" must be taken into account. It is a trivial matter, for example, to construct a class `complist` for lists of complex numbers, and we shall do it in the next chapter.

We end this section by giving the list version of the quicksort routine introduced in 14.4.1. It closely resembles `eosort`, with `partition` taking the place of `split`:

```
void   partition(intlist &L, intlist &less,
                 intlist &pivot, intlist &more)
{
       intlistscan thisptr(L);
       int* ip, hp = thisptr();
       pivot.append((void*)hp);
       while ((ip=thisptr()) != 0)
       {     if (*ip <= *hp) less.append((void*)ip);
             else more.append((void*)ip);
       }
}


void   quicksort(intlist &L)
{
       if (L.length() < 2) return;
       intlist less, pivot, more;
       partition(L, less, pivot, more);
       quicksort(less);
       quicksort(more);
       L = less + pivot + more;
}
```

20.3 The travelling salesman

Finding the minimum of a function can be very simple or very difficult indeed. We consider here the problem of finding the path of minimum length which has to be travelled by a "salesman" who visits each of a number of given "towns" once only before returning to the starting point. Clearly, the path is a circular list of towns. Our first task is therefore to construct such a list; the second is to choose a path which minimizes a suitable measure of the distance traversed, given the distances between all pairs of towns; the third is to try to find a path for which the distance traversed is an absolute minimum (or infimum)—for if there are many towns there may be many paths each of which is a "local" minimum with respect to minor variations of path, and we seek the minimum of these local minima. We shall not usually achieve our goal exactly, but we may find a useful approximation to it.

We introduced the classes `point` and `town` in 6.1 as classes derived from a base class `duple`. Since, however, we are intending to display in graphic form the paths taken, we shall use the definition of `point` given as a structure in 15.2.1 and the header file `stdgraph.h` of Appendix F. As in chapter 6 we shall continue to name the towns by means of an integer, although we could use their real names as strings if we so wished. It is natural to regard the name and the location as public variables; we therefore redefine `town` as a structure, first defining a "town pointer":

```
struct  town;           // incomplete definition
typedef town *tptr;     // for pointer type
```

A town may be regarded as a named point, so we derive `town` from `point`:

```
struct  town : public point {
     int name;
     town() : point()                      { name = 0; }
     town(int n, point &p) : point(p)   { name = n; }
};
```

We have already defined an output operator for the class `point`. We may extend that to towns quite simply:

```
ostream& operator<<(ostream &s, town &t)
{ return s << t.name << tab << *(point*)&t; }
```

We define `path` as a list of towns using our list template:

```
#define path Tlist<town>
```

That takes care of all the usual list functions, but there are others that we shall have to consider also, which will enable us to construct a path through a set of towns, measure the length of the possible paths between them, and so on.

Using the member function `append` we can now construct a set of n towns, storing their data in an array of type `town` which we shall call `map`. It would be perfectly possible to take real towns from a real map; but for simplicity and ease of changing data we shall take the euclidean coordinates of the towns to be random numbers between 0 and 1, using the function `uniform` declared in `random.h`; and we shall construct a path through the towns, assembling them in the order in which they are defined. This will therefore be a random path through randomly placed towns:

```
void    makepath(path &p, int n, tptr map, long &s)
{       for (int i = 0; i < n; ++i)
        {   double x = (double)uniform(s);
            double y = (double)uniform(s);
            point pt(x, y);
            map[i] = town(i, pt);
            p.append(&map[i]);
        }
}
```

Note that the index i in map[i] really is an integer, whereas the variable i in town(i, pt) stands for a name, which might in principle be a string rather than an integer, in which case we should also need a conversion table. We ignore that complication. The variable s is the usual long integer we use to generate random numbers and it must be initialized at the beginning of the user program.

Using the path scanner we may arrange to print out the towns in the order in which they occur on the path:

```
void    printpath(path &p)
{       pathscan thistown(p);
        tptr tp;
        while ((tp = thistown()) != 0)
            cout << *tp << endl;
}
```

We must now discuss the distances between towns, and for simplicity we shall take them to be the euclidean distances between points. Had we taken real towns from a real map, we could have read real distances from the table which accompanies a good road map. Here we shall take them to be the euclidean distances between our random and unreal towns.

If there are n towns there are $K = \frac{1}{2}n(n-1)$ distances, and these may be stored in an array declared as double distance[K]. Notice that K could be quite large; we therefore define it to be a long integer. It would be simpler to define a matrix (D_{ij}) giving the distance between towns i, j, but this would be wasteful of memory. We therefore have to relate the town "names" to the index k of the distance array. If we take $j > i$, and arrange the distances in increasing order of j given i, and then increasing i, we find:

$$k(i, j) = ni + j - \frac{1}{2}(i + 1)(i + 2).$$

In the case of $n = 4$, for example, we obtain the numbering:

	$j = 0$	1	2	3
$i = 0$	—	0	1	2
1	—	—	3	4
2	—	—	—	5

That gives us k in terms of i, j. If instead we wish to know the value of k for the pair of towns pointed to by townpointers tp1, tp2, it is given by the function:

```
long  index(path &p, tptr tp1, tptr tp2
{       int i = tp1->name, j = tp2->name;
        if (j<i)              // ensure j > i
        {       int t = i;
                i = j;
                j = t;
        }
        long k = p.length()*i - (i+1)*(i+2)/2 + j;
        return k;
}
```

We shall also need to be able to determine i and j in terms of k. Given that i and j are integers satisfying $0 \le i, j < n$, and also $j > i$, we can determine them uniquely by means of a simple algorithm. Note first that the leading term in each row, for which $j = i + 1$, satisfies $k = ni - \frac{1}{2}i(i + 1)$, so that in this case

$$i^2 - (2n - 1)i + 2k = 0 .$$

The value of i satisfying this relation is therefore given by

$$i = \frac{1}{2}\{(2n - 1) \pm \sqrt{[(2n - 1)^2 - 8k]}\} ,$$

where we have to take the negative sign in order that $k = 0$ corresponds to $i = 0$. In general, however, the equation has a solution in integers if and only if the quantity $(2n - 1)^2 - 8k$ is a perfect square. This is sufficient to determine the starting point of each row. The following procedure computes a table of euclidean distances between the map points of towns:

```
void  makedistance(tptr map, double *distance, int n)
{       const double eps = 1.0E-6;
        long nn = long(n);
        long K = nn*(nn-1)/2;
        long N = (2*nn-1)*(2*nn-1);
        int i = 0, j = 0;
        for (long k = 0; k < K; ++k)
        {       double d = sqrt(N - 8*k);
                if (k && fabs(d - round(d)) < eps) j = ++i;
                ++j;
                double x = map[i].x - map[j].x;
                double y = map[i].y - map[j].y;
                distance[k] = sqrt(x*x + y*y);
        }
}
```

The total length of a path, given the table of distances, may now be computed using the path scanner:

```
double pathlength(path &p, double *distance)
{       double d = 0.0;
        pathscan nexttown(p);
        tptr head = nexttown();
        tptr tp1 = head, tp2;
        while ((tp2 = nexttown()) != 0)
        {       long k = index(p, tp1, tp2);
                d += distance[k];
                tp1 = tp2;
        }
        tptr foot = tp1;
        long k = index(p, foot, head);
        d += distance[k];
        return d;
}
```

We are now ready to approach the core of the travelling salesman problem. We wish to find a permutation of the path for which the total path length is a minimum. Many strategies are possible. The simplest is to suppose that two towns chosen at random are interchanged: if the path length is thereby decreased that interchange is accepted, otherwise it is rejected. The process continues until a given number of random interchanges has been effected. If that number is large enough it may be assumed that a "local" minimum has been found. This minimization process may be repeated a number of times and the lowest path length chosen as an approximation to the "global" minimum.

A more sophisticated strategy is the "annealing" method of Metropolis. (It is described in the book by Press *et al.*) Here we accept any interchange which lowers the path length, but we also accept certain interchanges which increase the path length following a concept suggested by statistical mechanics. A system near equilibrium may undergo fluctuations of energy of magnitude ΔE with a probability proportional to $\exp(-\Delta E/T)$, where T is a measure of the temperature of the system in energy units. If the temperature is high, large positive fluctuations of energy, comparable with T, are probable; as the temperature falls, fluctuations of the same magnitude become less probable. While the temperature is high, the various local minima are explored; as the temperature falls, the tendency is to fall into a deep local minimum, although not necessarily the deepest.

Let us define a boolean function which depends upon whether we choose the first strategy or the second according to the value of a boolean variable mode. In the second strategy, we determine the probability of a positive increment incr of path length according to whether or not uniform(s) < exp(-incr/temp), where temp plays the role of temperature:

```
bool  improve(double incr, double temp, long &s, bool mode)
{     double t = incr/temp;    // avoid overflow of exp at temp = 0
      double z = (t > overflow) ? 0.0 : exp(-t);
      if (mode) return (incr < 0.0 || uniform(s) < z) ? TRUE : FALSE;
      else return (incr < 0.0) ? TRUE : FALSE;
}
```

The constant `overflow`, which we have arbitrarily taken to be 500, is included to prevent large positive increments when the "temperature" is very small.

For simplicity, we present a minimization procedure designed for the annealing strategy, merely switching off the possibility of positive increments if we prefer the other mode:

```
void  minimize(path &p, double *distance, long &s, bool mode)
{     int n = p.length();
      int maxtries = 50 * n;      // max tries allowed
      int maxswaps = 7 * n;       // max successful swaps
      int maxtempsteps = 75;      // max temp steps
      double temp = 5.0;          // starting `temperature'
      for (int k = 0; k < maxtempsteps; ++k)
      {     int numswaps = 0;
            int numtries = 0;
            while (++numtries < maxtries && numswaps < maxswaps)
            {     int i = int(uniform(s) * n);
                  int j = int(uniform(s) * n);
                  if (i && i<j)
                  {     double incr = swapincr(p, distance, i, j, n);
                        if (improve(incr, temp, s, mode))
                        {     p.swap(i, j);
                              ++numswaps;
                        }
                  }
            }
            if (numswaps == 0) break;
            temp *= coolfactor;
      }
}
```

The procedure examines a large number of interchanges at each of a number of temperature steps. If the interchange would be an "improvement" the towns are swapped, otherwise the iteration proceeds. Note that the interchange does *not* take place unless there would be an improvement. The function which computes what the increment would be is `swapincr`; we shall define it presently. Each time an interchange is tried the integer variable `numtries` is increased by one; each time an interchange is successful the integer variable `numswaps` is increased by one. At each temperature step, the number of iterations is determined by the requirement that both `numtries` < `maxtries` and `numswaps` < `maxswaps`, where values of these limits are suggested above, but may be changed by the user – not least, to adapt the program to the speed limitations of the machine used.

The outer loop of this procedure is the annealing process. The starting temperature should be chosen somewhat larger than the largest increments obtained, so that there is a reasonable probability of their taking place at high temperatures. In our case, most positive increments are less than about 2, so we set the initial temperature `temp` = `5.0`. This also may be changed by the user. We have allowed for a maximum of 75 temperature steps, at each of which the temperature is reduced by 10%. Again, these parameters may be changed; there is much scope for experimentation. However, this procedure would take a great deal of time to complete had we not added the statement `if (numswaps == 0) break;`. If no successful interchanges have been found at some temperature step the procedure is terminated, because it is most unlikely that they would be found at a lower temperature.

We state the function `swapincr`, which returns the value of the path increment that would result from an interchange, without detailed explanation. It is simple in principle, but complicated in practice, because one has to take into account all the distances involved in every possible interchange:

```
double swapincr(path &p, double *distance, int i, int j, int n)
{      double d = 0.0;                // swap increments path by d
       if (i == j) return d;
       tptr pi = p[i];
       tptr pj = p[j];
       tptr ai, bi, aj, bj;          // before and after pointers
       tptr head = p.head(), foot = p.head();
       pathscan nexttown(p);
       tptr tpb = foot, tp = head;   // initialise
       tptr tpa = nexttown();        // start tpa running
       while ((tpa = nexttown()) != 0)
       {      if (tp == pi) { ai = tpa; bi = tpb; }
              if (tp == pj) { aj = tpa; bj = tpb; }
              tpb = tp; tp = tpa;    // move on
       }
       if (pi == foot) { ai = head; bi = tpb; }
       if (pj == foot) { aj = head; bj = tpb; }
       if (abs(i-j) == 1)
       {      d += distance[index(bi, pj)];
              d += distance[index(pi, aj)];
              d -= distance[index(bi, pi)];
              d -= distance[index(pj, aj)];
       }
       else if (n-abs(i-j) == 1)
       {      d += distance[index(bj, pi)];
              d += distance[index(pj, ai)];
```

```
                    d -= distance[index(bj, pj)];
                    d -= distance[index(pi, ai)];
          }
          else
          {       d += distance[index(bi, pj)];
                  d += distance[index(pj, ai)];
                  d += distance[index(bj, pi)];
                  d += distance[index(pi, aj)];
                  d -= distance[index(bi, pi)];
                  d -= distance[index(pi, ai)];
                  d -= distance[index(bj, pj)];
                  d -= distance[index(pj, aj)];
          }
          return d;
}
```

We have here used the member function `numtoptr` to convert from the name of a town to its pointer, because the `list` function `swap` takes the integer positions of nodes on the list as parameters.

The user program may now be constructed, except that we wish to show the path graphically. For this we need graphics functions to represent each path by straight lines drawn between the spots which represent towns on the map. Using the functions presented in `stdgraph.h`, explained in chapter 15, and other obvious TurboC++ facilities, the required functions are:

```
void    spot(pixel &s)
{       const int SPOTSIZE = 4;
        int color = s.getcol();
        setfillstyle(SOLID_FILL, color);        // TurboC++
        setcolor(color);                        // TurboC++
        int sx = s.getx(), sy = s.gety();
        circle(sx, sy, SPOTSIZE);               // TuboC++
        floodfill(sx, sy, color);               // TurboC++
}

void    showtowns(path &p, point &tl, point &br, int color)
{       pathscan nexttown(p);
        tptr tp;
        while((tp = nexttown()) != 0)
        {       pixel s(*(point*)tp, tl, br, color);
                spot(s);
        }
}
```

```
void   drawpath(path &p, point &tl, point &br, int color)
{      setcolor(color);
       tptr head = p.head(), foot = p.foot(), tp1, tp2;
       pathscan nexttown(p);
       tp1 = head;
       tp2 = nexttown();        // = head, but starts tp2 running!
       while ((tp2 = nexttown()) != 0)
       {      pixel s1(*(point*)tp1, tl, br, color);
              pixel s2(*(point*)tp2, tl, br, color);
              line(s1, s2);
              tp1 = tp2;
       }
       pixel s1(*(point*)head, tl, br, color);
       pixel s2(*(point*)foot, tl, br, color);
       line(s1, s2);
}
```

Finally, we offer a user program, with which the reader may experiment at will:

```
main()
{      clear;
       int n;
       input("number of towns = ", n);
       long s;
       input("large negative seed = ", s);         // defines map
       bool mode = getyes("annealing mode? - (Y/N): ");
       newline;
       tptr town = new towntype[n];
       path p;
       makepath(p, n, town, s);
       long nn = long(n);
       long K = nn*(nn-1)/2;
       double *distance = new double[K];
       makedistance(town, distance, n);
       statement("\n\ninitial path:\n");
       printpath(p);
       newline;
       output("distance = ", p.pathlength(distance));
       newline; newline;
       input("new negative seed = ", s);         // defines path
       if (getyes("\nenter graphical mode? - (Y/N): "))
       {      opengraph();
              cleardevice();
              point tl(0.0, 1.0), br(1.0, 0.0);
              showtowns(p, tl, br, LIGHTRED);
```

```
            drawpath(p, tl, br, CYAN);
            char *msg = "Wait ... ";
            setcolor(WHITE);
            outtextxy(0, 0, msg);
            minimize(p, distance, s, mode);
            setcolor(BLACK);
            outtextxy(0, 0, msg);
            drawpath(p, tl, br, YELLOW);
            msg = "Press any key...";
            setcolor(WHITE);
            outtextxy(0, 0, msg);
            getch();
            closegraph();
        }
    else minimize(p, distance, s, mode);
    clear;
    statement("\nminimized path:\n");
    printpath(p);
    output("\ndistance = ", p.pathlength(distance));
    newline;
}
```

The annealing process is not much use for small numbers of towns; beyond about 15 towns, however, it produces the lowest minima. For 100 towns, this program gave the

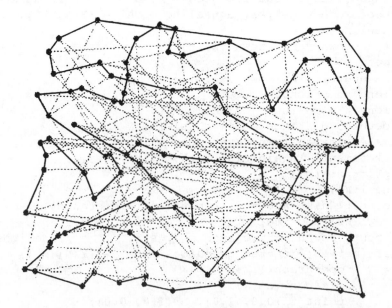

Figure 20.1 Travelling salesman's minimum path for 100 towns.

following results. For a particular map (*ie,* for a given choice of the first seed), the total length of the initially random path was 55.59. Ten minimum paths were then computed with different values of the second seed. Without annealing, the path lengths varied from 12.42 to 16.51, from which we deduce that the infimum $\leq$ 12.42. With annealing, they varied from 10.48 to 12.31. The infimum is therefore $\leq$ 10.48. The path which gives this value is shown in monochrome in Figure 20.1.

Of course, more runs might in general produce a still lower path length; so might larger values of some of the fixed parameters. Moreover, the only permutations we have investigated have been simple interchanges of pairs of towns. It is also possible that an admixture of more complex permatations might lower the lowest value so far obtained.

20.4 Multiple byte integers

The largest integers we have been able to work with have been confined to an `int` or a `long`. In TurboC++, for example, these are two 8-bit bytes and four 8-bit bytes long, respectively. For both types, advanced arithmetical operations are standard provisions of the language, which makes their use overwhelmingly desirable wherever possible. In principle, however, there is no reason to confine an integer to a specific number of bytes: the concept of a list allows us to work with any number of bytes, dynamically determined according to the transient needs of the program.

The lists we have discussed so far have been singly linked: each node contains a single pointer to the succeeding node in the list. Multiple byte integers, however, are not efficiently dealt with by means of single linkages: we shall find it necessary to scan a list both forwards and backwards, and this is most easily done if each node contains a pointer to the previous node as well as a pointer to the the succeeding node. Morever, it is no longer desirable to work with pointers to the elements of a fixed data array: since we do not know in advance the number of bytes required by an integer, it would be most inconvenient to work with fixed arrays; we need to work with the bytes themselves. For this application, therefore, we abandon our list template and start afresh.

An 8-bit byte, unsigned, can accommodate the integers 0, 1, . . . , 255. If we set $B = 256$, we may represent any non-negative integer N as a polynomial in B:

$$N = a_m B^m + \ldots + a_1 B + a_0,$$

or

$$N = (a_m, \ldots, a_1, a_0)_B,$$

where, in the second formula, we have recognised B as the *radix,* and the integer coefficients a_i, which must obey the restriction $0 \leq a_i < B$, as the *digits* of the integer to the base B. There may in principle be any number of digits. If B were equal to 10, we should have the usual decimal representation, but we shall take B to be 256.

If we define a digit to be a node containing a byte b and two pointers, a multiple byte integer (which we shall call a `mult`) may be represented by a list of digits. In diagrammatic form:

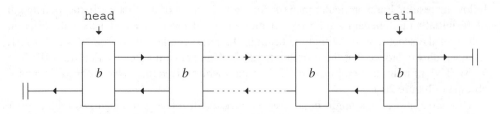

We shall use linear lists instead of circular lists, because in this case it is more natural to mark both the head and the tail of the list. These ideas we must now write in code. First, we define a digit as a class with no public members:

```
class digit;
typedef digit *dgtp;
class digit {
friend class mult;
friend class multscan;
private:
      static int radix;
      byte dgt;
      dgtp next;
      dgtp last;
      digit(int);                   // constructor
friend ostream& operator<<(ostream&, digit&);
friend istream& operator>>(istream&, digit&);
};
const int digit::radix = 256;
```

The constructor for an isolated digit checks that the data member may be confined within a byte:

```
digit::digit(int i)
{      if (i < 0 || i >= radix)
       error("bytes only in digit::digit(int)");
       dgt = (byte)i; next = last = 0;
};
```

We have provided input and output operators for digits:

```
ostream& operator<<(ostream &s, digit &d)
{      s << int(d.dgt); return s; }
```

```
istream& operator>>(istream &s, digit &d)
{      int i; s >> i;
       if (i < 0 || i >= digit::radix)
              error("enter a byte in op>>(os&, dig&)");
       d.dgt = (byte)i; return s;
}
```

It is convenient to introduce the constants,

```
const int POSV = +1;
const int NEGV = -1;
```

Then we may define the class of multiple byte integers as a doubly-linked list of digits:

```
class mult {
friend class multscan;
private:
       int  sgn;                // sign of integer
       dgtp head;               // most significiant digit
       dgtp tail;               // least significant digit
       int  numdgt;             // number of digits
       clearmult();
public:
       mult()        { sgn = POSV; head = tail = 0; numdgt = 0; }
       mult(int);
       mult(mult&);
       ~mult()                  { clearmult(); }
       void setsgn(char);
       char getsign();
       void chsgn();
       void prepend(int);
       void append(int);
       int gethead();
       int gettail();
       int lastdec();           // find last decimal
       mult& subdec(int);       // subtract last decimal
       mult& divten();          // integer division by 10
       char *decimalize();      // output as decimal string
};
```

Note that by definition a digit is non-negative; the sign belongs to the number as a whole, and is therefore included as a data member of the class mult. It is unnecessary, but convenient, also to include the current number of bytes in a mult integer.

As noted above, we shall need to be able to scan a list in either direction. Therefore we define the class multscan to contain an additional boolean data member which determines whether the scan is to be forward (in the "next" direction) or backward (in the "last" direction). For this purpose, and to avoid confusion, we define alternative names for the boolean values:

```
#define NEXT TRUE
#define LAST FALSE
```

The declaration of `multscan` is then

```
class multscan {
private:
      mult *mp;
      dgtp dp;
      bool dir;
public:
      multscan(mult&, bool);       // forward or backward scan
      dgtp operator()();           // returns pointer to digit
};
```

where the constructor is simply an initializer:

```
multscan::multscan(mult &m, bool direction)
{     mp = &m;
      dir = direction;   // NEXT for forward, LAST for backward
      dp = dir ? m.head : m.tail;
}
```

The scanning operator is a little simpler than that for a circular list:

```
dgtp  multscan::operator()()
{     dgtp dp = dp;
      if (dir) dp = p ? dp->next : 0;
      else     dp = p ? dp->last : 0;
      return p;
}
```

We must now develop the member functions of the class `mult`. The private auxiliary function required for the destructor is:

```
void  mult::clearmult()
{     if (!head) return;         // empty list!
      dgtp dp = head;
      while (dp)
      {     dgtp ddp = dp;
            if ((dp = dp->next) != 0) dp->last = 0;
            delete ddp;
```

```
        }
        head = tail = 0;
        numdgt = 0;
}
```

There are three constructors: the first is the trivial empty list constructor stated inline; notice, however, that the empty list has to have a sign. The second is required in order to invoke a multiple byte integer which in fact contains only one byte. In particular, we may wish to invoke the zero integer. We cannot write that `mult m;`, because that invokes the *empty* list, not the list with only one digit the value of which is zero. The second constructor enables us to write `mult m(0);` for the zero integer:

```
mult::mult(int i)
{       if (i < 0 || i >= digit::radix)
                error("enter a byte!");
        sgn = POSV;
        head = tail = new digit(i);
        numdgt = 1;
}
```

The sign functions are simple. In the first, we wish to indicate by means of an input character whether the number is positive or negative:

```
void   mult::setsign(char ch)
{       if (ch == '+') sgn = POSV;
        else if (ch = '-') sgn = NEGV;
        else error("enter '+' or '-' only!");
}
char mult::getsign() { return (sgn == POSV) ? '+': '-'; }

void mult::chsgn() { sgn = (sgn == POSV) ? NEGV : POSV; }
```

The insertion functions are, for head insertion:

```
void   mult::prepend(int i)
{       dgtp p = new digit(i);
        if (!p) error("allocation error in mult::prepend(int)");
        if (!head) head = tail = p; else
        {       head->last = p;
                p ->next = head;
                head = p;
        }
        ++numdgt;
}
```

and for tail insertion:

```
void   mult::append(int i)
{      dgtp p = new digit(i);
       if (!p) error("allocation error in mult::append(int)");
       if (!tail) head = tail = p; else
       {      tail->next = p;
              p->last = tail;
              tail = p;
       }
       ++numdgt;
}
```

Correspondingly, there are the two "get" functions:

```
int    mult::gethead()
{      if (!head)
          error("can't get from empty list in mult::gethead()");
       dgtp p = head;
       int i = int(p->dgt);
       if ((head = head->next) != 0) head->last = 0;
       delete p;
       --numdgt;
       return i;
}
```

```
int    mult::gettail()
{      if (!tail)
          error("can't get from empty list in mult::gettail()");
       dgtp p = tail;
       int i = int(p->dgt);
       if ((tail = tail->last) != 0) tail->next = 0;
       delete p;
       --numdgt;
       return i;
}
```

The copy constructor may be written in terms of prepend or append; we choose the latter:

```
mult::mult(mult &m)
{      head = tail = 0; sgn = m.sgn; numdgt = 0;
       while (m.head) append(m.gethead());
}
```

Next, we need input and output functions; these need not be member functions. The input function is one which assembles a list of digits:

```
void   makemult(mult &m)
{      char ch; input("enter sign: ", ch);
       m.setsign(ch);
       int n; input("number of bytes? : ", n);
       for (int i = 0; i < n; ++i)
       {      int b; input("enter byte: ", b);
              m.append(b);
       }
}
```

and the output function should be one which prints out a multiple byte integer in some suitable format, for example:

```
void   printmult(mult &m)
{      multscan thisptr(m, NEXT);
       cout << "{" << m.getsign();
       dgtp p;
       while ((p = thisptr()) != 0)
              cout << *p << sep;
       cout << "\b\b}";
}
```

Here, `sep` is defined in `standard.h` to be the string constant `", "`, and `'\b'` represents the backspace character.

The heart of the problem of representing large integers as lists of bytes, however, is the conversion of the multiple byte integer so constructed into a normal integer. We have to convert an integer N to the base $B = 256$ into the same integer to the base 10. Let us recall that to base B we are working with polynomials:

$$N = a_m B^m + \ldots + a_1 B + a_0.$$

If we convert this into ordinary decimals, the last decimal digit will be $N \bmod 10$. It is easy to establish the identity:

$$(uv + w) \bmod m \equiv [(u \bmod m)(v \bmod m) + w \bmod m] \bmod m \,.$$

With the aid of this identity, we find:

$$N \bmod 10 = [(\ldots (a_m B + a_{m-1})B + \ldots + a_1 B + a_0] \bmod 10$$
$$= [(\ldots ((a_m \bmod 10)c + a_{m-1} \bmod 10)c + \ldots + (a_1 \bmod 10)c + (a_0 \bmod 10)] \bmod 10,$$

where $c \equiv B \bmod 10 = 6$. The following member function computes the last decimal digit of a multiple byte integer, using Horner's algorithm (see 4.10.1) for the evaluation of a polynomial:

```
int    mult::lastdec()
{      int B = digit::radix;
       int c = B % 10;                          // c = 6
       multscan thisptr(*this, NEXT);
       dgtp p;
       while ((p = thisptr()) != 0)
              d = (c*d + (p->dgt%10))%10;    // Horner
       return d;
}
```

Let the decimal digit so obtained be denoted by d_0: then $N - d_0$ is divisible by 10. We therefore need a member function which subtracts a single decimal digit from a multiple byte integer. The only difficulty arises from the possible need to "carry" during the usual backward scan required for subtraction, and to delete the head of the list if its resulting byte value is zero:

```
mult& subdec(int i)
{      int B = digit::radix;
       bool carry = FALSE;
       multscan thisptr(*this, LAST);
       dgtp p = thisptr();                      // last digit
       if (!p) error("can't subtract from empty list!");
       int j = int(p->dgt) - i;
       if (j < 0) { j += B; carry = TRUE; }
       p->dgt = (byte)j;
       while ((p = thisptrt()) != 0)        // remaining digits
       {      j = int(p->dgt);
              if (carry) { --j; carry = FALSE; }
              if (j < 0) { j += B; carry = TRUE; }
              p->dgt = (byte)j;
       }
       if (head->dgt) gethead();
       return *this;
}
```

We may now divide by 10, knowing that there will be no overall remainder, carrying the stepwise remainder, and deleting the head digit if its data value is zero:

```
mult& mult::divten()
{      int B = digit::radix;
       int c = 0;
```

```
        multscan thisptr(*this, NEXT);
        dgtp p;
        while ((p = thisptr()) != 0)
        {       int i = int(p->dgt);
                i += (c*B);
                p->dgt = (byte)(i/10);
                c = i % 10;
        }
        if (!head->dgt) gethead();
        return *this;
}
```

If we now repeat the process we obtain the second to last decimal digit; and so on. However, if we print out the decimals as they arrive, they will be in reverse order. It is therefore necessary to output the character value of each decimal digit in turn to the *head* of a character string. For a C string:

```
char    *chstr(char ch, char *str)
{       int n = strlen(str);
        char *newstr = new char[n+2];
        newstr[0] = ch; newstr[1] = '\0';
        str = strcat(newstr, str);
        newstr = 0;
        return str;
}
```

Printing out the string which results yields decimal digits in the normal order with the aid of the decimal output member function (which also precedes the number with its sign):

```
char    *mult::decimalize()
{       char *number = "";        // empty string
        while (head)
        {       int d = lastdec();
                char ch = char(d + '0');
                number = chstr(ch, number);
                subdec();
                if (head) divten();
        }
        char s = (sgn == POSV) ? '+' : '-';
        number = chstr(s, number);
        return number;
}
```

The routine is extremely fast and is limited only by the memory available. We find, for example, for a particular 16-byte number having 37 decimal digits, that:

$$\{+1, 2, 3, 4, 5, 6, 7, 8, 9, 10, 11, 12, 13, 14, 15, 16\}$$

$$= +1339673755198158349044581307228491536.$$

It is now a relatively easy matter to add the usual assignment and arithmetic operators to the class `mult`, and to proceed to do arithmetic on multiple byte integers. We shall not do this, but shall note that it is helpful first to write routines for the addition, subtraction, multiplication and integer division of polynomials, and to take care with the sign of zero. It is also possible to define a class of multiple precision floating point numbers.

21 Elements of Fourier analysis

In this last chapter we return to the subject of chapter 17, in which we omitted to discuss the most common of all orthogonal functions, namely the trigonometric functions. This will lead us to introduce Fourier series, which are used to describe many physical phenomena, especially those which are periodic in time or space and may therefore be described in terms of a spectrum of frequencies or wavenumbers. Finally, we offer a rather compact treatment of the Fast Fourier Transform in terms of lists.

21.1 Fourier series

Following the work of chapter 17, we may say that for each positive integer n the $2n$ functions $\{\phi_0, \phi_1, \ldots, \phi_{2n-1}\}$ form an orthonormal set, where:

$$\phi_0(x) = (2\pi)^{-1/2},$$
$$\phi_k(x) = \pi^{-1/2} \cos kx, \quad k = 1, 2, \ldots, n$$
$$\phi_{n+k}(x) = \pi^{-1/2} \sin kx, \quad k = 1, 2, \ldots, n-1.$$

The interval is $[-\pi, +\pi]$, and the weighting function is $w(x) \equiv 1$. The orthogonality properties follow from considering integrals of the kind

$$\int_{-\pi}^{\pi} \sin hx \cos kx \, dx = \tfrac{1}{2}\int_{-\pi}^{\pi} [\sin(h+k)x + \sin(h-k)x] \, dx$$

and

$$\int_{-\pi}^{\pi} \cos hx \cos kx \, dx = \tfrac{1}{2}\int_{-\pi}^{\pi} [\cos(h+k)x + \cos(h-k)x] \, dx$$

from which it is clear that only terms with $h = k$ can be non-zero.

The linear combination

$$T_n(x) = \sum_{k=0}^{2n-1} a_k \phi_k(x), \quad k = 0, 1, \ldots, n,$$

may be called a *trigonometric polynomial* of degree not more than n. We may choose the coefficients a_k in any way we like. Here we shall choose them so that $T_n(x)$ is the least-squares approximation to a function $f(x)$ over the interval $[-\pi, +\pi]$, namely:

$$a_k = \int_{-\pi}^{\pi} f(x)\phi_k(x) \, dx.$$

Of course, we could change to an arbitrary interval $[a, b]$ by means of a linear transformation $z = (b - a)x/2\pi + (a + b)/2$.

Consider, for example, the inverted triangle function defined by:

$$f(x) = |x|, \quad -\pi < x < +\pi.$$

It is an even function of x, and to degree n only coefficients a_k with $k = 0, 1, \ldots, n$ could be non-zero. We have

$$a_0 = (2\pi)^{-1/2}\int_{-\pi}^{\pi} |x|\, dx = \pi^{3/2}/\sqrt{2}, \quad a_k = -(4/k^2\sqrt{\pi}), \quad k = \text{odd}, \quad a_k = 0, \quad k = \text{even},$$

so that the trigonometric polynomial approximations to the triangle function are given by

$$T_n(x) = (\pi/2) - (4/\pi) \sum_{\text{odd } k \le n} (\cos kx)/k^2.$$

The first few polynomials in this case are:

$$T_0(x) = (\pi/2)$$
$$T_1(x) = (\pi/2) - (4/\pi)\cos x$$
$$T_2(x) = T_1(x)$$
$$T_3(x) = (\pi/2) - (4/\pi)\cos x - (4/9\pi)\cos 3x$$
$$T_4(x) = T_3(x)$$
$$T_5(x) = (\pi/2) - (4/\pi)\cos x - (4/9\pi)\cos 3x - (4/25\pi)\cos 5x$$

and they may be seen to approximate the function quite closely.

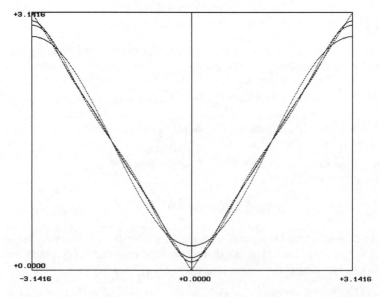

Figure 21.1 Trigonometric polynomial fit to inverted triangle function.

We shall consider the convergence of the sequence $\{T_n(x)\}$ of polynomial approximations to the function $f(x)$ in due course. In the meantime, we note that if it converges for all values of x, it converges to the function f. That is to say, if the limit exists, then

$$T(x) \equiv \lim_{n \to \infty} T_n(x) = f(x),$$

and $T(x)$ is called the *Fourier series* for the function $f(x)$. Thus, the Fourier series for the triangle function is

$$T(x) = (\pi/2) - (4/\pi) \sum_{k=0}^{\infty} \cos(2k+1)x/(2k+1)^2,$$

where we have written $2k+1$ instead of k in order to guarantee that it is odd.

21.2 Periodic functions

Fourier series are particularly useful in analysing periodic phenomena. A function $f(x)$ is said to be *periodic* if there exists a constant ξ such that

$$f(x + \xi) = f(x), \quad \text{for all } x.$$

The constant ξ is called a *period* of the function f. If ξ is a period, then so is any integral multiple of ξ. The smallest such period is called the *fundamental period*, or simply *the period*, of f. If x represents a space coordinate, then the period may instead be called the *wavelength* and is often written λ; if it represents the time coordinate, the period might instead be denoted by τ.

If two functions, f_1 and f_2, are periodic with the same period ξ, then the product $F = f_1 f_2$ and the linear combination $G = \alpha_1 f_1 + \alpha_2 f_2$, where α_1, α_2 are arbitrary constants, are also periodic with period ξ.

For
$$F(x + \xi) = f_1(x + \xi)f_2(x + \xi) = f_1(x)f_2(x) = F(x),$$

and
$$G(x + \xi) = \alpha_1 f_1(x + \xi) + \alpha_2 f_2(x + \xi) = \alpha_1 f(x) + \alpha f_2(x) = G(x).$$

Clearly, the functions $\sin kx$ and $\cos kx$ are periodic with period $\lambda = 2\pi/k$. If x represents a space coordinate, k is called the *wavenumber*; if x represents time, k is the *angular frequency*, usually written $\omega = 2\pi\nu$, where $\nu = 1/\tau$ is the ordinary frequency. The Fourier series for the triangle function, which we considered above, has period 2π, because the function $\cos x$ has that fundamental period, and all the other terms have periods which are integral multiples of it, and therefore share 2π as a common period.

It follows that if the Fourier series of the periodic function $f(x)$, with fundamental period ξ, on the interval $[-\infty, +\infty]$ is $T(x)$, then the Fourier series of $f(x + \xi)$ is $T(x + \xi) = T(x)$. If we regard the triangle function as periodic, it becomes a triangular waveform. The coefficients a_k of its Fourier series are the *amplitudes* of the cosine components of wavenumber k of that waveform. In general, the coefficients a_k of the Fourier series representation of a given waveform provide the wavenumber *spectrum* of the waveform, or its *Fourier transform*. (Note that these coefficients may be obtained from a single period of the waveform.) The Fourier series is a good representation of the triangular waveform because the amplitudes of successive terms of the series vary as $(2k + 1)^{-2}$.

Let us now consider another example, the square wave defined by:

$$f(x) = 0, \quad -\pi < x < 0, \quad f(x) = \pi, \quad 0 < x < \pi, \quad f(x + 2\pi) = f(x).$$

Note that $f(x)$ is not yet defined at the points of discontinuity, $x = 0, \pm\pi, \ldots$ Carrying out the integrals involved, we find

$$a_0 = \pi, \quad a_k = 0, \quad 0 < k \le n, \quad a_{n+k} = 0, \quad k \text{ even}, a_{n+k} = 2/k, \quad k \text{ odd}.$$

The Fourier series representation of the square wave is therefore

$$T(x) = (\pi/2) + 2\sum_{k=0}^{\infty}\sin(2k + 1)x/(2k + 1).$$

The Fourier series for the square waveform does not converge as rapidly as that for the triangular waveform, because successive amplitudes vary as $(2k + 1)^{-1}$ only.

At the points of discontinuity, $x = 0, \pm\pi, \ldots$, the sinusoidal terms vanish, so that T has the value $\pi/2$. The Fourier series $T(x)$ therefore does not converge to $f(x)$ everywhere unless $f(x)$ also is defined to have the value $\pi/2$ at these points.

More generally, the function $f(x)$ is said to be *piecewise continuous* on the interval $a \le x \le b$ if the interval can be partitioned by a finite number of points,

$$a = x_0 < x_1 < \ldots < x_n = b$$

into a number of open sub-intervals $x_{k-1} < x < x_k$ on which f is continuous; and if f approaches a finite limit at both endpoints of each sub-interval. Denote by $f(c+)$ the limit of $f(x)$ as x approaches c from above. If $\varepsilon > 0$, we write

$$f(c+) = \lim_{\varepsilon \to 0} f(c + \varepsilon).$$

Similarly, $$f(c-) = \lim_{\varepsilon \to 0} f(c - \varepsilon).$$

If f is continuous at $x = c$ it follows that $f(c+) = f(c-) = f(c)$. More generally, we can say that the Fourier series defined above for the function f converges to $\frac{1}{2}[f(x+) + f(x-)]$ for all x. (We have excluded from consideration functions f which possess more complicated discontinuities.) This is the *Fourier theorem*. The solutions of many physical problems may be expressed in terms of Fourier series.

21.3 Discrete Fourier transforms

Suppose that we are given $2m$ data points (x_j, y_j), $0 \leq j < 2m$, and for simplicity suppose that the x_j are equally spaced on the interval $[-\pi, +\pi]$, so that $x_j = -\pi + j\pi/m$. We wish to find the trigonometric polynomial of degree n which minimizes the expression

$$E(a_0, \ldots, a_{2n-1}) \equiv \sum_{j=1}^{2m-1} [y_j - T_n(x_j)]^2.$$

It is the polynomial

$$T_n(x) = \Sigma_k a_k \phi_k(x), \qquad 0 \leq k < 2n,$$

where

$$a_0 = (2\pi)^{-1/2} \Sigma_j y_j, \qquad 0 \leq j < 2m,$$

$$a_k = \pi^{-1/2} \Sigma_j y_j \phi_k(x_j), \qquad 0 < k \leq 2n.$$

The highest degree polynomial for which the coefficients are uniquely determined is $n = m$, and this is the case we shall study. It is more convenient in what follows, however, to work in the complex plane, where we shall as usual set $i = \sqrt{(-1)}$. Let

$$f(x) = \Sigma_k c_k e^{ikx}, \qquad 0 \leq k < 2n,$$

where

$$c_k = (2/n) \Sigma_j y_j e^{-\pi i jk/n}, \qquad 0 \leq j < 2n.$$

The real coefficients a_k may easily be recovered from the complex coefficients c_k:

$$a_k + i a_{n+k} = \pi^{-1/2} c_k e^{-i\pi k}, \quad 0 \leq k < n.$$

Let $\omega = \exp(-2\pi i/2n)$ be a constant complex number; it is one of the $(2n)$-th roots of unity, but the parameter n will be taken as understood. (It should not be confused with the angular frequency of the previous section.) Then the coefficients c_k are given by:

$$c_k = \sum_{j=0}^{2n-1} y_j \omega^{jk}.$$

Having computed these coefficients, the function $f(x)$ may be calculated. Clearly, this is a process for which the computation time is $O(n^2)$. The following routine carries out the transformation from the y_j to the c_k. However, the factor $\sqrt{\pi}$ has been dropped, and we have written n instead of $2n$. We should remember, therefore, that n is supposed to be *even*.

```
void  fourier(int n, complex y[], complex c[])
{     double theta = 2.0 * PI / n;
      for (int k = 0; k < n; ++k)
      {     double kth = k*theta;
            c[k] = 0;
```

```
            for (int j = 0; j < n; ++j)
            {      double jkth = j * kth;
                   complex z(0, jkth);
                   c[k] += y[j] * exp(-z);
            }
            c[k] /= n;
      }
}
```

There is a very simple test for this routine: apply to it a sinusoidal trial function,

$$y(x) = \exp(ihx),$$

where h is an integer wavenumber. Setting $x_j = -\pi + 2j/n$, the corresponding values of y are computed by:

```
void  sinusoidal(int n, int h, complex y[])
{      double phi = 2.0 * PI * h / n;
       double hpi = h*PI;
       complex z0(0, hpi);
       for (int j = 0; j < n; ++j)
       {      double jphi = j * phi;
              complex z(0, jphi);
              y[j] = exp(z-z0);
       }
}
```

When a particular value of h is chosen, and the y_j set by sinusoidal, the routine fourier yields, as expected, $c_k = \delta_{hk}e^{i\pi h} = (-1)^h\delta_{hk}$, where δ is Kronecker's delta.

21.4 Fast Fourier transforms

The discrete Fourier transform just described is $O(n^2)$, which is much too slow for many practical applications, where we may wish to fit many thousands of points. The method we are about to explain is $O(n \log n)$, which is much more practicable.

We shall again assume that the number of points to be fitted is n, instead of $2n$, and further that n is a power of 2, $n = 2^p$, say. The relation between the y_j and the c_k is then

$$c_k = \sum_{j=0}^{n-1} y_j \omega_n{}^{jk}, \qquad 0 \le k < n,$$

where $\omega_n = e^{-2\pi i/n}$ is one of the nth roots of unity. But since n is even, we may split this sum into equal numbers of even and odd terms:

$$c_k = \sum_{j=0}^{n/2-1} y_{2j}\,\omega_n^{2jk} + \sum_{j=0}^{n/2-1} y_{2j+1}\,\omega_n^{(2j+1)k}$$

$$= \sum_{j=0}^{n/2-1} y_{2j}\,\omega_{n/2}^{jk} + \omega_n^k \sum_{j=0}^{n/2-1} y_{2j+1}\,\omega_{n/2}^{jk}.$$

Therefore,
$$c_k = c_k^{(\text{even})} + \omega_n^k c_k^{(\text{odd})}, \qquad 0 \le k < n/2,$$

and
$$c_{n/2+k} = c_k^{(\text{even})} - \omega_n^k c_k^{(\text{odd})}, \qquad 0 \le k < n/2.$$

This relates the discrete Fourier transform c_k of all the y_j to the discrete Fourier transforms $c_k^{(\text{even})}$ of the even terms $y_0, y_2, \ldots,$ and $c_k^{(\text{odd})}$ of the odd terms $y_1, y_3, \ldots,$ through the roots of unity. But $n/2$ is also even (unless it is 1). Therefore the same process may be repeated recursively. We shall now demonstrate that its execution time is $O(n \log n)$.

To be definite, let us suppose that we can measure the computational time required by the number of complex multiplications involved. For given n, we shall denote this by $T(n)$. Then, since $c^{(\text{even})}$ and $c^{(\text{odd})}$ are both of length $n/2$, $T(n) = \frac{1}{2}n + 2T(n/2)$, where the first term comes from the multiplication of $c^{(\text{odd})}$ by powers of ω. The solution of this functional equation is easily seen to be

$$T(n) = \alpha n + \frac{1}{2}n \log_2 n = O(n \log n),$$

where α is a constant.

The problem of computing a discrete Fourier transform, as now stated, is similar to the recursive splitting procedure `eosort` we introduced in 20.2 to test the operations of the class `list` in the context of a list of integers. This strongly suggests that we ought to be able to express the solution in terms of lists, although we must expect complications to result from the appearance of multiplications. Of course, these multiplications are the heart of the matter. To give effect to the above considerations, we shall need routines which add and subtract corresponding terms of a list, and a routine which multiplies the terms of a list by the corresponding terms of another list.

First we define a pointer to a complex number,

```
typedef complex *cptr;   // pointer to complex data
```

and then use our list template to give us a list of complex numbers,

```
#define complist Tlist<complex>
```

and similarly for the scanner,

```
#define complistscan Tlistscan<complex>
```

The nth roots of unity may be stored as an array of complex numbers, and arranged into a list. It will be observed that we need only *half* the number of roots for the appropriate multiplications:

```
complist omega(int n, cptr wp)
//      half list of roots
{       complist wl;
        double theta = -2.0*PI/n;
        for (int m = 0; m < n/2; ++m)
        {       double t = theta * m;
                wp[m] = complex(cos(t), sin(t));
                wl.append(&wp[m]);
        }
        return wl;
}
```

Next, we have to be able to split a list L into an even list and an odd list:

```
void    split(complist &L, complist &Leven, complist &Lodd)
{       while (L.length())
        {       Leven.append(L.get());
                if (L.length()) Lodd.append(L.get());
        }
}
```

Note that the length of L reduces to zero as Leven and Lodd are built up. We must also be able to reverse the splitting process:

```
void    restore(complist &L, complist &Leven, complist &Lodd)
{       while (Leven.length())


        {       L.append(Leven.get());
                if (Lodd.length())
                        L.append(Lodd.get());
        }
}
```

It is convenient at this point to postpone some of the detail and write down the procedure fastfour which encodes the strategy of the problem as described above:

```
void   fastfour(complist &F, complist &w)
{      if (F.length() < 2) return;
       complist Feven, Fodd;
       split(F, Feven, Fodd);
       complist weven, wodd;
       split(w, weven, wodd);
       fastfour(Feven, weven);
       fastfour(Fodd,  weven);
       recombine(F, Feven, Fodd, w, weven, wodd);
}
```

The procedure takes as arguments a list F of length n, and a list w of length $n/2$. It defines two sub-lists, Feven and Fodd, each of length $n/2$, and likewise two sub-lists, weven and wodd, each of length $n/4$. The splitting procedure is then called recursively on Feven and Fodd, in each case replacing w with weven. Finally, we have to reconstitute the lists F and w, having carried out the required arithmetic operations.

We carry out all the arithmetic, as well as concatenation, in the procedure recombine. The arithmetic has to be carried out on the array to which F points, so that the original data are lost. It is assumed that they can always be reconstructed by calculation or by reading from disc.

```
void recombine(complist &F, complist &Feven, complist &Fodd,
               complist &w, complist &weven, complist &wodd)
{      restore(w, weven, wodd);
       complistscan Wp(w), Fep(Feven), Fop(Fodd);
       cptr wp, fep, fop;
       while ((fep = Fep()) != 0)
       {      wp  = Wp();
              fop = Fop();
              *fop *= *wp;
              complex t = *fep + *fop;
              *fop = *fep - *fop;
              *fep = t;
       }
       F = Feven + Fodd;
}
```

The while loop carries out all the arithmetic, replacing the appropriate array elements as it goes along. The final step concatenates the two sub-lists, using the concatenation operator defined for the base class list. However, we have neglected to normalize the final list, and this we may do with the function:

```
void    normalize(complist &F)
{       int n = F.length();
        complistscan fptr(F);
        cptr fp;
        while ((fp = fptr()) != 0)
        *fp /= n;
}
```

That is really the end, except that it is more elegant to draw together all the various functions into a single procedure which carries out the fast Fourier transform on a given list F:

```
void    fastfourier(complist &F)
//      on input F is list of data
//      on output it is the Fourier Transform
{       int n = F.length();
        cptr wp = new complex[n/2];
        complist w = omega(n, wp);
        fastfour(F, w);
        normalize(F);
}
```

Applied to the sinusoidal function of the last section, fastfour yields the same result as fourier, but as expected is very much faster. There is a certain overhead due to the amount of memory occupied by list pointers, exacerbated by the fact that fastfour is not tail-recursive. Non-recursive methods exist which do all the work on the data array itself, and therefore avoid this overhead. They are very difficult to understand, however, because the arithmetic has to be performed in exactly the right order, and that order is exceedingly complicated to describe in non-recursive terms. Using lists takes care of the order automatically, and that is a great advantage.

Addendum: Programming in Windows

In this addendum we offer an alternative treatment of graphics to that presented in chapter 15. There we sketched out how to write a graphics package in TurboC++ running in DOS, which would suffice for the limited purposes of this book. We wanted to be able to sketch out functions of a single variable and to map out the screen according to some mathematical prescription. Here we indicate how to achieve the same objectives in TurboC++ for Windows, at the level of Windows 3.1. Later versions of C++ running under Windows 95 and 98 require differences of detail, but the essential techniques described in this Addendum survive intact.

First, however, let us note that all the programs presented in this book, with the exception of those requiring graphics, will run under Windows without modification. Windows runs these programs automatically using the normal library functions. Indeed, the presence of the function `main` ensures that this will happen. The graphics programs, however, will not work without major modification, and it is the purpose of the addendum to illustrate these modifications.

Programming in Windows is notoriously complicated, as a glance at the classic book by Petzold will confirm. There is a two-way interaction between applications and windows which has to be catered for systematically. Borland have made things somewhat easier, however, by providing with TurboC++ (and also with Borland C++) an extensive Object Windows Library, OWL, of functions which perform automatically much of the more routine coding that is necessary for any Windows program. The book by Perry provides an introduction to the process. The programming we perform in this addendum employs OWL throughout.

About the simplest thing we can do with OWL is to create a special window called a "message box". We do this in the following program:

```
#define WIN31
#define STRICT


#include <owl.h>
```

```
#pragma argsused
int    PASCAL WinMain(HINSTANCE hInstance,
                      HINSTANCE hPrevInstance,
                      LPSTR lpCmdLine, int nCmdShow)
{
       MessageBox(NULL, "Hello, Windows!", "A Message Box",
           MB_ICONEXCLAMATION | MB_OK);
       return 0;
}
```

The program begins as usual with some defines and an include: these ensure that the header files required both for Windows 3.1 and OWL are present, and that type checking will be followed strictly. There is then a compiler directive which we have not come across before: #pragma argsused. This prevents the compiler printing unnecessary warnings to the effect that not all variables passed to the function which immediately follows are used in the function body.

We might then have expected to find the function main. Not so; under the OWL convention, main is replaced by WinMain. This tells the compiler that the full OWL treatment required for graphics is needed. Like main, WinMain returns an integer. Note the partial capitalization of the identifier: in Windows, variables and functions tend to have rather long and descriptive names, for which a degree of capitalization adds clarity. We shall follow that convention for Windows variables and functions, following our previous use of lower case identifiers when referring to purely mathematical manipulations. Similarly, type names, such as HINSTANCE, of which there are many in Windows, are often written entirely in capitals so that one may readily recognise them as such. In some cases, lower case letters precede the main part of an identifier, as in lpCmdLine, which indicates that the corresponding variable is a far (long) pointer—its type being denoted by LPSTR, it is a long pointer to a string; indeed, somewhere in the header files there is the statement

```
typedef char far *LPSTR
```

WinMain takes four parameters. The first two are "handles" (unsigned integers, or words) to the present instance of the application, and the previous instance, respectively. If there is no previous instance the handle is zero, or NULL. The pointer variable lpCmdLine points to a command line of parameters, if any, much as described for the function main in 2.10. Finally, nCmdShow is an integer which indicates how the window is to be displayed on the Windows desktop—in full, or as an icon, for example. We shall not have occasion to do more than somewhat laboriously write down these parameters.

Note that WinMain is preceded by the modifier PASCAL. This instructs the compiler in the interests of efficiency to call parameters in the Pascal order rather than the C order.

The body of the function WinMain, apart from the trivial return statement, consists of a single call to the standard function MessageBox, which creates a window of predetermined dimensions and position, with the caption "A Message Box", and the message "Hello, Windows!". (The NULL value of its first parameter indicates that it has no parent window.) It is a very primitive window which can neither be moved, scrolled, nor resized, as most windows can. It is embellished, however, by an exclamation mark and an OK button, brought about by the predefined integer parameter

```
MB_ICONEXCLAMATION | MB_OK
```

and we shall come across other such parameters later. The resulting window is shown in Figure A.1, where it appears superimposed upon the working windows of TurboC++ for Windows. If the mouse is clicked, the window vanishes. It cannot be otherwise affected.

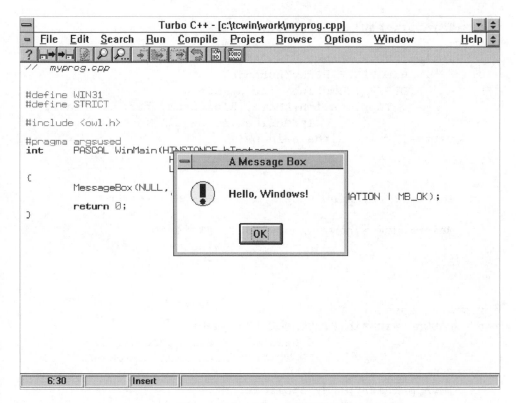

Figure A.1 A message box

A1 The Window classes TApplication and TWindow

In general, we shall be concerned to run an application in a window. TurboC++ for Windows supplies classes with the aid of which windows may be defined by the user in accordance with the needs of a specific application. In Windows it is usual to begin the name of a class with a capital 'T' indicating that it is a type name. Windows provides us with the classes TApplication and TWindow, from which classes may be derived to suit the purpose in hand. In particular, the class `TWindow` is a class of windows which may be moved, scrolled, resized, and replaced by an icon, by simple operations of the mouse.

The following program creates a named window with no content:

```
#define WIN31
#define STRICT


#include <owl.h>


class TMyApp : public TApplication {
public:
      TMyApp(LPSTR AName, HINSTANCE hInstance,
             HINSTANCE hPrevInstance,
             LPSTR lpCmdLine, int nCmdShow)
             : TApplication(AName, hInstance, hPrevInstance,
                            lpCmdLine, nCmdShow) {};
      virtual void InitMainWindow();
};


void  TMyApp::InitMainWindow()
{
      MainWindow = new TWindow(NULL, "This is my \
                                    window!");
}


#pragma argsused
int   PASCAL WinMain(HINSTANCE hInstance,
                     HINSTANCE hPrevInstance,
                     LPSTR lpCmdLine, int nCmdShow)
{
      TMyApp ThisApp("My Application",
                 hInstance, hPrevInstance,
                 lpCmdLine, nCmdShow);
      ThisApp.Run();
      return ThisApp.Status;
}
```

The program first of all creates a class `TMyApp` derived from `TApplication`. It requires only two member functions (in addition to data members and functions derived from `TApplication`). The first is the constructor, which simply creates an object of `TApplication`. The second is a *virtual function* which initializes the window.

We have not met virtual member functions before, and a few words of explanation are called for. (A more complete discussion will be found especially in the book by Ellis and Stroustrup.) By declaring a base class member function `virtual`, a redefinition of a function of the same name and type in a derived class overrides the base class definition even when access is obtained through a pointer to the base class.

For example, suppose we have a class B containing virtual functions, from which a class D is derived:

```
class B {
public:
        ..................
        virtual void vf(int);
        void f();
};


class D : public B {
public:
        ..................
        virtual void vf(int);
        void f();
};
```

Then the function F below behaves as indicated:

```
void   F()
{      D d;
       B *p = &d;
       p->vf(7);            // calls D::vf(7)
       p->f();             // calls B::f()
}
```

The use of virtual functions in a base class allows polymorphism in derived classes by redefining these functions according to the job they have to do. Note that the use of the specifier `virtual` in the derived class D is unnecessary: once a virtual function, always a virtual function. However, the repetition is allowable, and serves to remind the user that it is indeed a virtual function. Note that it is often helpful to declare the destructor, if any, of a base class `virtual`; any derived class is then guaranteed to have a destructor.

Returning to our program, the virtual function `InitMainWindow`, already defined in the base class, must now be redefined in the circumstances of the derived class `TMyApp`. This means that we have to understand a little of the makeup of the two classes `TApplication` and `TWindow`. Each is, in fact, a derived class. It is sufficient if we note the following features of `TApplication`:

```
_CLASSDEF(TApplication)
class TApplication : public TModule {
public:
        . . . . . . . . . . . . . . . . . . . . . . .
        HINSTANCE hPrevInstance;
        int nCmdShow;
        PTWindowsObject MainWindow;
        TApplication(LPSTR AName, HINSTANCE AnInstance,
                     HINSTANCE APrevInstance,
                     LPSTR ACmdLine, int ACmdShow);
        ~TRApplication();
        virtual void Run();
protected:
        virtual void InitMainWindow();
};
```

Note first the macro `_CLASSDEF(TApplication)`. It is little complicated to write down, but it has the effect, given any class, of defining far pointers and references automatically. Thus, the effect of `_CLASSDEF(classname)` is to make the following type definitions (among others):

```
typedef classname far * Pclassname
typedef classname far & Rclassname
```

The class `TApplication` is derived from a class `TModule` which provides much of the functionality required for any application. For the present purpose it is sufficient to represent it as follows:

```
class TModule {
public:
        . . . . . . . . . . . . . . . . . . .
        HINSTANCE hInstance;
        LPSTR lpCmdLine;
        int Status;
        LPSTR Name;
};
```

Note that the `WinMain` parameters `hInstance`, `hPrevInstance`, `lpCmdLine`, and `nCmdShow` appear as data members of one or other of the classes `TApplication` and `TModule`. In addition `TModule` provides the data members `Status` and `Name`. The first, an integer, if non-zero, denotes the error status of the application; the second, a string, a name for the application, if desired.

The class `TApplication`, it will be seen, has a constructor and a destructor; a function `Run` which, when called, causes the application to execute; and the virtual function

InitMainWindow which we have already come across. The class also includes a data member MainWindow which is a far pointer (remembering the _CLASSDEF convention) to the not yet defined class TObjectWindows. We shall not in fact define TObjectWindows, except to say that it provides the basic functionality of a window, and is the base class from which TWindow is derived.

We shall come to the class TWindow presently. For the present, it is sufficient to note that our program initializes the main window by setting the TApplication data member MainWindow to point to a newly created object of the class TWindow, the constructor of which has two parameters: the first, NULL, indicating that the window has no parent; the second, a caption which names the window. It is in this manner that the application creates its own window.

Our program ends with the function WinMain; it contains just three statements. The first declares an object ThisApp of the derived class TMyApp, in which the parameter AName has been replaced by the string "My Application". The second is a call to the function Run; the third, the usual return statement involving the data member Status. If we always refer to the instance of TApplication as TMyApp, and if we always name the application "My Application", the TMyApp class definition and the function WinMain as written above will suffice for all OWL applications. In the programs which follow we shall omit the OWL include and define statements, the application class definition, and the function WinMain, regarding them as standard. In any particular case, however, we may choose to use a different name for the TApplication instance, and a different name for the application, in which case the application class and the function WinMain will have to redefined accordingly. Note also that if we had written the window initialization statement in the form

```
MainWindow = new TWindow(NULL, Name);
```

then the parameter Name would have instead been given the value "My Application", inherited from TModule, and that would then have been the caption of the window.

The empty window resulting from this program is shown in Figure A.2, again superimposed on the working windows of TurboC++. It may readily be verified that this window is moveable and resizeable by the usual mouse operations.

A2 The class TWindow and the function Paint

Having succeeded in creating our own window, we are well on the way to our final objective. An empty window, however, is of rather limited use! We must learn how to put into it text or graphics at will. The class TWindow possesses a virtual function Paint which provides the necessary facilities.

It suffices for the present to observe that the class TWindow may be partially declared to be:

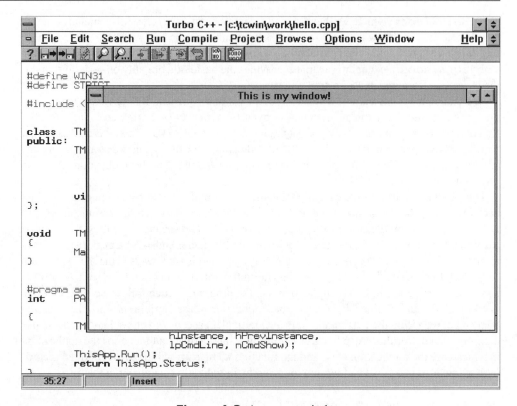

Figure A.2 An empty window

```
class TWindow : public TWindowsObject {
public:
        TWindow(PTWindowsObject AParent, LPSTR ATitle,
                               PTModule AModule=NULL);
        virtual ~TWindow();
        ........................................
protected:
        virtual void Paint(HDC PaintDC,
                               PAINTSTRUCT &PaintInfo);
};
```

It contains a constructor with three parameters: the first a pointer to the parent window, if any; the second a title, or caption, for the window; and the third, a pointer to a TModule, which is given the default value NULL. There follows a destructor, specified to be virtual as already discussed. And then there is the virtual function Paint: we have to redefine this function. For future reference we note also that the class TWindowsObject from which TWindows is derived contains as a data member a handle (unsigned integer) HWindow by which the window may be accessed, and which is initialized when the constructor is called.

As we did with `TApplication`, let us therefore construct a derived class from `TWindow` in terms of which we can redefine `Paint`:

```
class TMyWindow : public TWindow {
public:
        TMyWindow(PTWindowsObject AParent, LPSTR ATitle) :
              TWindow(AParent, ATitle) {};
        virtual void Paint(HDC PaintDC,
                             PAINTSTRUCT &PaintInfo);
};
```

Accordingly, we shall initialize the main window so as to create a window of the class `TMyWindow`:

```
void   TMyApp::InitMainWindow()
{
        MainWindow = new TMyWindow(NULL, Name);
}
```

It remains to define the function `Paint`, called internally by `Run()` in `WinMain`. It is a function of two parameters. The first is a handle to the "device context", or DC for short. The DC possesses attributes, together with functions for changing these attributes. We shall come across a number of these later, but for the moment suffice it to say that it is through the DC that the mode of writing to a window is defined or altered: the background colour, for example, the font and colour of text, and the colouring of individual pixels. The handle `PaintDC` is found automatically in OWL programs through the second parameter, which is a reference to a structure which contains information required for painting to a window, including the handle.

In order to define the function `Paint`, we first have to decide just what it is to do: in future this will be seen as the heart of an OWL program. We could simply write some text to the window using a standard function `TextOut`. Let us be a little more enterprising, however. Let us write a first statement to the window; then a message box asking whether a second statement should be written; and finally, if the answer is affirmative, a second statement:

```
#pragma argsused
void TMyWindow::Paint(HDC PaintDC, PAINTSTRUCT
                        &PaintInfo)
{
        TextOut(PaintDC, 30, 30, "First Statement", 15);
        if (MessageBox(HWindow, "Add Statement?",
              "The Question",
              MB_ICONQUESTION | MB_OKCANCEL) == IDOK)
        {
```

```
        TextOut(PaintDC, 30, 50,
                        "Second Statement", 16);
    }
}
```

The standard function `TextOut` has five parameters. The first is the DC handle; the second and third are respectively the x and y pixel coordinates of where in the window the statement string is to commence; the fourth is the string itself; and the fifth is the length of the string. There follows a second call to `TextOut`, which prints a second statement to the window, but it is conditional. The standard function `MessageBox`, which we have met before, returns a boolean value. This time we have given the message box two buttons—OK and `Cancel`. If the OK button is clicked, `MessageBox` returns the predetermined number `IDOK`, the condition is fulfilled, and the second statement is printed; if on the other hand the `Cancel` button is clicked, the condition is not fulfilled, and no further action takes place. In Figure A.3 we show the screen at the instant of decision.

Thus, we have shown that the problem of programming in Windows falls into two parts. The first is to define a class of windows suitable for the application in mind; the second is to redefine the virtual function `Paint` so that it writes to the chosen window whatever the

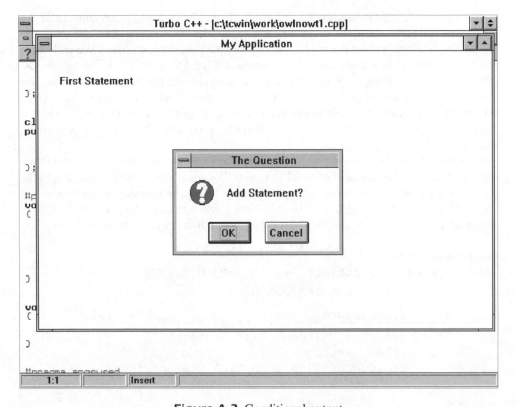

Figure A.3 Conditional output

application requires. The remainder of this addendum is concerned with making these two important choices.

A3 Programming mouse buttons

In Windows much use is normally made of the mouse as an input device. The various mouse operations are defined by virtual functions with default values. However, the mouse buttons can be redefined to carry out whatever task we wish. We illustrate the technique by programming the right mouse button to create the message box of the last section. We begin by redefining the class TMyWindow, as follows:

```
class TMyWindow : public TWindow {
private:
      BOOL ButtonDown;
public:
      TMyWindow(PTWindowsObject AParent, LPSTR ATitle) :
      TWindow(AParent, ATitle) { ButtonDown = FALSE; }
      virtual void WMRButtonDown(RTMessage)
                        = [WM_FIRST+WM_RBUTTONDOWN];
      virtual void Paint(HDC PaintDC,
                              PAINTSTRUCT &PaintInfo);
};
```

Note that we have included as a private data member the boolean variable ButtonDown, which is initialized to FALSE in the constructor. The other modification is the inclusion of the virtual function WMRButtonDown, initialized to the predetermined integer value WM_FIRST + WM_RBUTTONDOWN contained in square brackets. (Of course, these numbers are defined somewhere in the include files.) The prefix WM indicates that it is a function for processing a "Window Message", in this case a message sent by the mouse. Indeed, the function WMRButtonDown takes as a parameter a reference (remember the _CLASSDEF convention) to a structure with the typename TMessage. It contains information which enables the window to interpret the message.

In our particular case we can define the message function to be:

```
void   TMyWindow::WMRButtonDown(RTMessage)
{
      if (MessageBox(HWindow, "Add Statement?",
                        "The Question",
                        MB_OKCANCEL) == IDOK)
      {
```

```
            if (!ButtonDown)
            {
                  ButtonDown = TRUE;
                  InvalidateRect(HWindow, NULL, TRUE);
            }
      }
}
```

This function reintroduces the message box of the last section. If the OK button of the box is clicked, using the left mouse button as usual, then if ButtonDown is still FALSE, the window will be invalidated or, in a simpler word, cleared. Having once been cleared, a further mouse operation will be ineffective because by this point ButtonDown is TRUE. Thus, the effect of calling the WMRButtonDown function by pressing the right mouse button is to clear the window and to make ineffective the right button.

Finally, we have to redefine Paint. We take it to be:

```
#pragma argsused
void  TMyWindow::Paint(HDC PaintDC,
                                PAINTSTRUCT &PaintInfo)
{
      if (!ButtonDown)
      {     TextOut(PaintDC, 30, 30,
                        "First Statement", 15);
      }
      else
      {
            TextOut(PaintDC, 30, 30,
                        "First Statement", 15);
            TextOut(PaintDC, 30, 50,
                        "Second Statement", 16);
      }
      ButtonDown = FALSE;
}
```

This has exactly the same effect as the program in the last section. Note, however, that to make it quite equivalent it is necessary to restore ButtonDown to FALSE before leaving Paint.

A4 An Input Dialogue Box

In Windows, just as in DOS, we have much need for keyboard input, of integers, floating point numbers, and strings. In Windows this is achieved by means of Input Dialogue Boxes. We shall illustrate the general technique by building a dialogue box for strings. We begin by declaring a function SetString in the class TMyWindow:

```
class TMyWindow : public TWindow {
public:
      TMyWindow(PTWindowsObject AParent, LPSTR ATitle) :
           TWindow(AParent, ATitle) {};
      void SetString(LPSTR Str, LPSTR Caption);
      virtual void Paint(HDC PaintDC,
                             PAINTSTRUCT &PaintInfo);
};
```

The function takes two parameters: a pointer to the string that is to be entered; and a caption for the dialogue box. We define it as follows:

```
void  TMyWindow::SetString(LPSTR Str, LPSTR Caption)
{
      char Buffer[20];
      strcpy(Buffer, "");
      if (GetApplication()->ExecDialog(new TInputDialog
                 (this, Caption, "Enter String:",
                 Buffer, sizeof(Buffer))) == IDOK)
            sscanf(Buffer, "%s", Str);
}
```

For the most part this function is exactly as might be expected. A buffer is declared and initialized to the null string; the input string is then read into the buffer, and finally the buffer is copied into the string `Str` as desired. But the reading is conditional, and the condition involves the creation of an appropriate dialogue box.

Now the class `TWindowsObject` (from which `TWindows` is derived) contains a member function `GetApplication` which returns a pointer to `TApplication`, itself a data member of `TWindowsObject`:

```
PTApplication Application;
PTApplication TWindowsObject::GetApplication()
{
      return Application;
}
```

Of course, `Application` points to the application in hand. In turn, `TApplication` inherits from its base class `TModule` a function

```
void TModule::ExecDialog(PTWindowsObject ADialog);
```

which takes as its parameter a pointer to `TWindowsObject`, for the dialogue box we wish to create is itself a window. For the dialogue box we shall choose an object of the predefined class `TInputDialog`, itself derived from `TWindowsObject`. We set

```
ADialog = new TInputDialog(this, Caption,
                    "Enter String", Buffer, sizeof(Buffer));
```

The `TInputDialog` constructor takes five parameters: the first is a pointer to the window in which the dialogue box is to appear, and it suffices to represent it by `this`; the second is the caption of the dialogue box; the third is the chosen prompt; the fourth and the fifth the buffer and its size, respectively. As for `GetApplication`, it returns a boolean variable which is `TRUE` if the memory allocation implied by the operator `new` is successful, otherwise `FALSE`. The construction of the function `SetString`, as we have explained it, shows very clearly the interplay between window and application that we remarked upon earlier.

It is straightforward to construct `Paint`:

```
#pragma argsused
void  TMyWindow::Paint(HDC PaintDC,
                              PAINTSTRUCT &PaintInfo)
{
    char YourName[15];
    strcpy(YourName, "");
    SetString(YourName, "What is your name?");
    char *Str = "Your name is ";
    strcat(Str, YourName);
    TextOut(PaintDC, 30, 50, Str, strlen(Str));
}
```

Figure A.4 shows the resulting window with the dialogue box awaiting input. If we enter "`Peter`", say, and click the OK button, there appears in the main window the statement "`Your name is Peter`". However, the program cannot be run as it stands. It has first to be linked to a "resource file" which provides the detailed instructions for constructing the dialogue box, the parameters of which are to be found in a header file `inputdia.h` which must also be included. TurboC++ for Windows provides special facilities for constructing resource files; they are also discussed in the book by Perry. In a precisely similar way we may construct input dialogue boxes for other types.

A5 A multiple document interface

We shall sometimes wish to display simultaneously more than one aspect of a problem—the phase trajectory and the time dependence of a solution of a differential equation, for example. Thus, we are interested in having more than one window open at once. This is commonplace in word processors, where many document windows may be simultaneously open; indeed, the TurboC++ for Windows working environment normally involves several windows. We have to learn how to bring about a *multiple document interface*, or MDI—for which a better name might be multiple window environment.

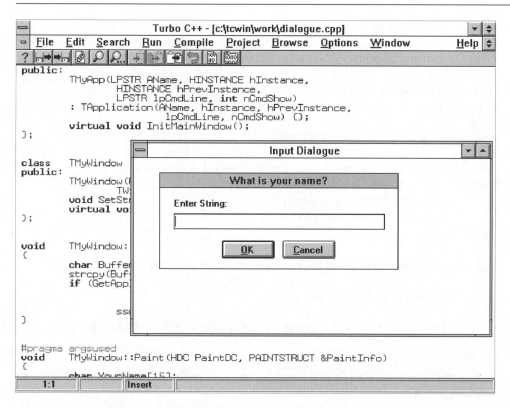

Figure A.4 Input dialogue box

In OWL there has been defined a class of windows, called *MDI frames* and derived from TWindow, which supports any number of *child windows*. It is sufficient for our purpose to note that the class definition includes the following:

```
class TMDIFrame : public TWindow {
public:
        ...................................
        TMDIFrame(LPSTR ATitle, LPSTR MenuName,
                                 PTModule AModule = NULL);
        virtual PTWinowsObject CreateChild();
        virtual BOOL CloseChildren();
        virtual void TileChildren();
        virtual void CascadeChildren();
        void CloseWindow();
};
```

The constructor has three parameters, of which the third is given its default value. The first is a name for the frame window, which we shall give the value "OWL-MDI DEMO"; the second is a name for the menu convention which the frame is to carry, for which we shall supply the value "MAINMENU", indicating that menu bars are to carry the usual items, in particular File and Window, although the precise details depend upon the chosen resource file—a matter discussed by Perry.

We have to redefine some of these virtual functions. We therefore declare our own class of frames:

```
class TFrame : public TMDIFrame {
public:
    TFrame(LPSTR Atitle, LPSTR MenuName);
    virtual PTWindowsObject CreateChild();
    virtual BOOL CloseChildren();
    virtual void SetupWindow();
};
```

Correspondingly, we initialize the main window in terms of `TFrame`:

```
void TMyApp::InitMainWindow()
{
    MainWindow = new TFrame("OWL-MDI DEMO", "MAINMENU");
}
```

The `TFrame` constructor is a little complicated simply because we would like the frame window to be of maximum size:

```
TFrame::TFrame(LPSTR ATitle, LPSTR AMenuName) :
    TMDIFrame(ATitle, AMenuName)
{
    HDC hDC = GetDC(NULL);
    int w = GetDeviceCaps(hDC, HORZRES);
    int h = GetDeviceCaps(hDC, VERTRES);
    Attr.X = 0;
    Attr.Y = 0;
    Attr.W = w;
    Attr.H = h;
    ReleaseDC(NULL, hDC);
}
```

The first statement in this function obtains a handle to the device context for the frame window by calling `GetDC`. Since only a few handles may be active simultaneously, it is a wise precaution to release the device context as soon as it has been used, which we do in the last statement by calling the function `ReleaseDC`. The remaining statements set the size of the window to the maximum possible. The function `GetDeviceCaps` obtains the "capability" of the device context for a range of parameters, in this case height and width.

TWindow includes a public data member `Attr`, of type `TWindowAttr`, a structure which includes the following members:

```
struct TWindowsAttr {
      ................
      int X, Y, W, H;           // Top left, bottom right
      LPSTR Menu;               // Menu name
      int Id;                   // Child identifier
};
```

The top left and bottom right corners of the window are therefore deposited with the parameter `Attr`. Note that `Attr` is also the repository of the menu name, which we met in the last section, and of a child identifier which we shall need presently.

The remaining `TFrame` functions require us first to define a class of child windows:

```
_CLASSDEF(TChild)
class TChild : public TWindow {
private:
      BOOL Active;
public:
      TChild(PTWindowsObject AParent, LPSTR ATitle) :
            TWindow(AParent, ATitle)      { Active = FALSE; }
      void SetActive(BOOL b)              { Active = b; }
      virtual void WMRButtonDown(RTMessage)
      = [WM_FIRST + WM_RBUTTONDOWN];
      virtual void Paint(HDC, PAINTSTRUCT&);
};
```

Note that we have included a private boolean data member `Active`; in the constructor this is set to `FALSE`, so that the child window is inactive, but we also provide a function `SetActive` enabling us to change its value. Again, we include a redefinitions of the virtual functions `WMRButtonDown`, the duty of which is to make a child window active when the right mouse button is clicked within it, and of course `Paint`. We may define `WMRButtonDown` immediately:

```
void   TChild::WMRButtonDown(RTMessage)
{
      Active = TRUE;
      InvalidateRect(HWindow, NULL, TRUE);


}
```

In addition to making the window active it also clears it.

We may now consider the remaining TFrame functions. First we want to be able to create a child window:

```
PTWindowsObject TFrame::CreateChild()
{
      return GetApplication()->MakeWindow(new TChild
                                (this, "Child Window"));
}
```

The TWindowsObject function GetApplication we have already met. With it we call a TModule function MakeWindow, which in turn creates a window of type TChild entitled simply "Child Window". It is a child of the frame window itself, hence the first parameter value this. The predefined class TMDIFrame includes a boolean function which closes all child windows. Accordingly, we may write:

```
BOOL   TFrame::CloseChildren()
{
      return TMDIFrame::CloseChildren();
}
```

These last two functions, together with the menu facilities provided in the initialization of the (main) frame window, enable us to create child windows at will, and to close them all, simply by using the <u>W</u>indow menu.

To complete the functionality of TFrame we have to redefine the virtual function SetupWindow, and this will determine the shape our program takes. We define a representative function first, then describes what it does:

```
void   TFrame::SetupWindow()
{
      TMDIFrame::SetupWindow();
      PTChild Child1, Child2, Child3;
      Child1 = new TChild(this, "Child One");
      Child2 = new TChild(this, "Child Two");
      Child3 = new TChild(this, "Child Three");
      Child1->Attr.Id = 1;
      Child2->Attr.Id = 2;
      Child3->Attr.Id = 3;
      Child1->SetActive(TRUE);
      GetApplication()->MakeWindow(Child3);
      GetApplication()->MakeWindow(Child2);
      TileChildren();
      GetApplication()->MakeWindow(Child1);
}
```

First, we call the TMDIFrame function SetupWindow; then we declare three pointers to TChild and reserve the corresponding memory space for three child windows. We set the identifying member of Attr to 1,2,3 respectively, and declare that Child1 will be the active window initially. Using the familiar functions GetApplication and MakeWindow, we then create child windows 3 and 2 in that order, and tile them. Finally, we create child window 1. As a result we may expect to see within the frame window, child windows 2 and 3 side by side, with child window 1 superimposed. This is the somewhat contrived environment within which we must now construct the function Paint.

Let us decide to write some distinctive text to each child window in turn, closing the first window after having written to it in order to show windows 2 and 3 fully. To do this we must use the attribute identifier. The function seems otherwise self-explanatory:

```
#pragma argsused
void   TChild::Paint(HDC PaintDC, PAINTSTRUCT &PaintInfo)
{
       char String[25];
       strcpy(String, "");
       int StrLen;
       HDC hDC = GetDC(NULL);


       if (Attr.Id == 1 && Active)
       {
             Active = FALSE;
             MessageBox(HWindow, "For First Window?",
                   "Ready", MB_OKCANCEL);
             strcpy(String, "This is the First Window");
             StrLen = strlen(String);
             TextOut(PaintDC, 5, 5, String, StrLen);
             MessageBox(HWindow, "First Window?",
                                "Delete", MB_OKCANCEL);
             CloseWindow();
       }


       if (Attr.Id == 2 && Active)
       {
             Active = FALSE;
             MessageBox(HWindow, "For Second Window?",
                   "Ready", MB_OKCANCEL);
             strcpy(String, "This is the Second Window");
             StrLen = strlen(String);
             TextOut(PaintDC, 5, 5, String, StrLen);
       }
```

```
if (Attr.Id == 3 && Active)
{
        Active = FALSE;
        MessageBox(HWindow, "For Third Window?",
                "Ready", MB_OKCANCEL);
        strcpy(String, "This is the Third Window");
        StrLen = strlen(String);
        TextOut(PaintDC, 5, 5, String, StrLen);
}
ReleaseDC(NULL, hDC);
}
```

The program must now be completed in the usual way, including definitions of the class
TMyApp and the function WinMain. The header files mdi.h and inputdia.h must also
be included with the more usual header files. When the program is run, linked to the owlmdi
resource file, it opens in the manner displayed in Figure A.5. If the OK message box button is
clicked the statement "This is the First Window" appears in child window 1,
accompanied by another message box asking whether this window should be deleted. If the

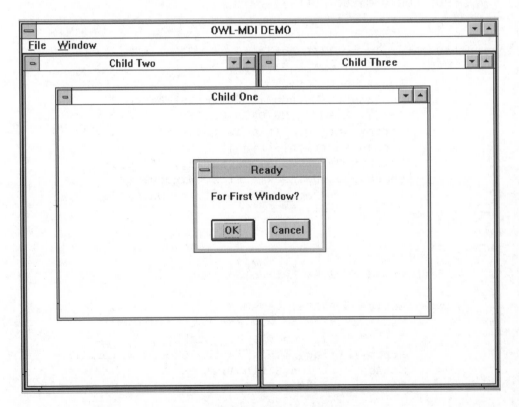

Figure A.5 Multiple documents

OK button is again clicked, window 1 disappears, leaving windows 2 and 3 side by side, both empty.

We may now use the right mouse button to activate one of the remaining windows, whereupon another message box will appear asking whether we are ready to activate that window. If we click the OK button the message "This is the Second Window" appears, if that is the one we have chosen, otherwise "This is the Third Window". We can, of course, activate them in turn.

Having completed those operations, we may now create further windows by calling up New in the File menu. They will all be entitled "Child Window". After the creation of three such windows the screen appears as in Figure A.6. Finally, the whole family of windows may be cleared by calling Exit in the File menu. These last operations are brought about by the functions CreateChild and CloseChildren.

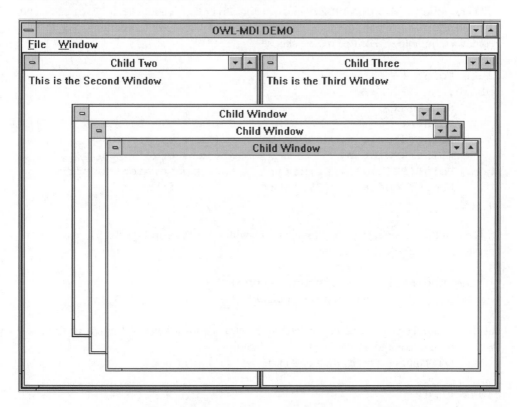

Figure A.6 Writing to multiple documents

A6 The class TScreen

With those preliminaries behind us we may advance more quickly towards our objectives. We have reached a point in Windows programming equivalent to that from which in DOS we began chapter 15, but from now on we can draw upon many of the details of that chapter. In particular, we want to derive from `TWindow` a class `TScreen` which will enable us to draw a function and map the screen with rather little extra work.

The construction of `TScreen` is summarized is Appendix H, which shows the header file `screen.h`. It begins with the usual header files, including those appropriate to OWL. It continues with a set of standard colour values of the form

```
const COLORREF AColor = RGB(r, g, b);
```

where `COLORREF` is another name for a long unsigned integer, and `RGB` is a Windows function which constructs, subject to the capabilities of the output device, colours according to their red, green and blue components, each given as a byte.

There follow the `_CLASSDEF` macros for `Point`, `Pixel`, `TScreen` and `TScreenApplication`, the four classes in terms of which we translate chapter 15 into Windows terminology. For `Point` we choose

```
class Point {
public:
      double x, y;
      Point()                       { x = y = 0; }
      Point(double a, double b)     { x = a; y = b; }
      Point(RPoint p)               { x = p.x; y = p.y; }
      Point(PTScreen, RPixel);      // conversion constructor
      RPoint operator=(RPoint);     // assignment
};
```

The first three constructors are defined inline within the class declaration. The assignment operator may also be defined inline,

```
inline RPoint Point::operator=(RPoint p)
{ x = p.x; y = p.y; return *this; }
```

but the remaining pixel to point conversion constructor must await definition of the class `TScreen`, in relation to which the point is defined.

Next there comes the declaration of `Pixel`:

```
class Pixel {
friend class Point;
friend class TScreen;
private:
      int x, y;
      COLORREF c;
```

```
public:
      Pixel();       // top left client rect, default color
      Pixel(int, int, COLORREF);
      // image of point in screen:
      Pixel(PTScreen, RPoint p, COLORREF color);
      Pixel(RPixel);
      RPixel operator=(RPixel);
      int GetX();
      int GetY();


      COLORREF GetCol();
      void SetCol(COLORREF);
      friend COLORREF SetPixel(HDC, RPixel);
      friend void MoveTo(HDC, RPixel);
      friend void LineTo(HDC, RPixel);
};
```

There are four constructors, three of which are defined inline (see Appendix H). The remaining one is the point to pixel conversion constructor which awaits the definition of TScreen. The assignment operator is also defined inline, as are the functions GetX, GetY, GetCol and SetCol. There remain three friend functions, SetPixel, MoveTo and LineTo, all defined inline. The first is the Windows equivalent of the DOS putpixel, and needs no further comment. The second and third enable a straight line to be drawn between an initial pixel p1 and a final pixel p2 simply by writing

```
MoveTo(hDC, p1);
LineTo(hDC, p2);
```

Taken together, they are the equivalent of the DOS function line.

The class TScreen itself contains all the data and functions necessary for setting offsets, making axes and writing to the window in various ways:

```
class TScreen : public TWindow {
friend      class Point;
friend      class Pixel;
private:
      int tsox, tsoy;
      double Slope(double);    // convert tangent
protected:
      double tsxinc, tsyinc;
      Point TopLeft, BottomRight;
      RECT showrect;
```

```
        void SetPoint(RPoint, LPSTR);
        void SetOffSets();
        void SetShowRect(RECT&);
        void SetInteger(int&, LPSTR);
        void SetLong(long&, LPSTR);
        void SetDouble(double&, LPSTR);
        double GetAspect();
public:
        TScreen(PTWindowsObject, LPSTR);      // max window
        BOOL InShowRect(RPixel);
        void FloatOut(int, int, float);
        void DigitOut(HDC, RPixel, int);
        void Axes(HDC, int, COLORREF);
        void ShowRect(HDC, int, COLORREF);
        void SetScreen(RECT&);
        void Spot(HDC, RPixel, int, COLORREF);
        void Arrow(HDC, double, double, double,
                                   BOOL, COLORREF);
};
```

The private data members are the offsets `tsox`, `tsoy`, measured in pixels. It is also useful to include a private function `Slope` which converts a tangent in point space into a tangent in pixel space:

```
inline      double TScreen::Slope(double t)
{ return t*tsxinc/tsyinc; }
```

Here the parameters `tsxinc` and `tsyinc` are protected data members giving the increments in point space corresponding to an increment of one pixel, horizontally and vertically, respectively, in pixel space. Their values will be determined by the function `SetScreen` when we come to it.

The other protected data members of `TScreen` define the rectangle to be shown, first in point space using the parameters `TopLeft` and `BottomRight`, then in pixel space. `RECT` is the Windows structure defined by:

```
typedef struct {
        int left;                  // pixel coordinates
        int top;
        int right;
        int bottom;
} RECT;
```

In terms of `RECT` we may define the function `SetShowRect`, which determines the region of the window within the offsets, the "show rectangle", to be:

```
void    TScreen::SetShowRect(RECT &rect)
{
        showrect.top = rect.top + tsoy;
        showrect.left = rect.left + tsox;
        showrect.bottom = rect.bottom - tsoy;
        showrect.right = rect.right - tsox;
}
```

There then follows a list of protected functions which make it possible to read into the window values for a `Point`, the offsets, an `int`, a `long`, and a `double`. These functions are constructed in strict analogy to the function `SetString` which we introduced as an input dialogue box in section A4. There is also a function `GetAspect` which returns the aspect ratio of the window:

```
double TScreen::GetAspect()
{
        int nox = showrect.right - showrect.left;
        int noy = showrect.bottom - showrect.top;
        return double(nox)/noy;
}
```

After those come the public functions, beginning with the constructor. We define it so as to obtain a maximum window, as we did for `TFrame` in A4:

```
#pragma argsused
TScreen::TScreen(PTWindowsObject AParent, LPSTR ATitle) :
                          TWindow(AParent, ATitle)
{
        HDC hDC = GetDC(NULL);
        int w = GetDeviceCaps(hDC, HORZRES);
        int h = GetDeviceCaps(hDC, VERTRES);
        Attr.X = 0;
        Attr.Y = 0;
        Attr.W = w; // 640, say
        Attr.H = h; // 480, say
        ReleaseDC(NULL, hDC);
}
```

We may also write a boolean function which determines whether or not a particular pixel lies within the show rectangle:

```
BOOL    TScreen::InShowRect(RPixel s)
{
        int b = TRUE;
```

```
    if (s.x <= showrect.left || s.x >= showrect.right)
        b = FALSE;
    if (s.y <= showrect.top || s.y >= showrect.bottom)
        b = FALSE;
    return b;
}
```

The pixel to point and point to pixel conversion constructors, respectively, may now be defined in the TScreen context by:

```
Point::Point(PTScreen pS, RPixel s)
{
    x = pS->TopLeft.x + pS->tsxinc*(s.x - pS->tsox);
    y = pS->TopLeft.y - pS->tsyinc*(s.y - pS->tsoy);
}


Pixel::Pixel(PTScreen pS, RPoint p, COLORREF color)
{
    x = int(round((p.x - pS->TopLeft.x)/pS->tsxinc))
                                            + pS->tsox;
    y = int(round((pS->TopLeft.y - p.y)/pS->tsyinc))
                                            + pS->tsoy;
    c = color;
}
```

We then come to the public functions which allow us to write floating point numbers and digits to the window, draw axes, and display the rectangle within the offsets. Floating point output first, where we have to select the font in which the number is to be written:

```
void  TScreen::FloatOut(int x, int y, float f)
{
    char str[15];
    strcpy(str, "");
    sprintf(str, "%-+8.4f", f);

    HDC hDC = GetDC(HWindow);

    static LOGFONT lf;
    lf.lfHeight = 16;
    lf.lfWidth = 6;
    lf.lfPitchAndFamily = DEFAULT_PITCH | FF_SWISS;

    HFONT hFont = CreateFontIndirect(&lf);
    HFONT hOldFont = SelectFont(hDC, hFont);
```

```
      TextOut(hDC, x, y, str, strlen(str));

      SelectFont(hDC, hOldFont);
      DeleteFont(hFont);

      ReleaseDC(HWindow, hDC);
}
```

This calls for some explanation, and we have left some blank lines for added clarity. The number f is first written to the string str using the C output function sprintf, and the device context is sought. LOGFONT is a structure holding in "logical units", *ie* pixels, the full description of the font to be chosen for writing. There is, of course, a default font which we have been using so far, but now we would like to switch to a swiss font of the same pitch, having chosen character height and width. We then create the desired font using the Windows function CreateFontIndirect, which returns a handle hFont to that font, and select it using the function SelectFont, which conveniently returns a handle hOldFont to the previous (in our case, default) font. Finally, we print out the string str representing the number using our old friend TextOut, reselect the previous font, delete the swiss font, and release the device context. The function TScreen::DigitOut is constructed in exactly the same fashion.

We next draw axes with extreme values written adjacent to them:

```
void   TScreen::Axes(HDC hDC, int Width, COLORREF Color)
{
      Point O(0.0, 0.0);

      if (O.y > TopLeft.y)      O.y = TopLeft.y;
      if (O.y < BottomRight.y)  O.y = BottomRight.y;
      if (O.x < TopLeft.x)      O.x = TopLeft.x;
      if (O.x > BottomRight.x)  O.x = BottomRight.x;

      Point A(TopLeft.x, O.y);
      Point B(BottomRight.x, O.y);
      Point C(O.x, BottomRight.y);
      Point D(O.x, TopLeft.y);

      Pixel a(this, A, Black);
      Pixel b(this, B, Black);
      Pixel c(this, C, Black);
      Pixel d(this, D, Black);

      HPEN ThePen = CreatePen(PS_SOLID, Width, Color);
      HPEN OldPen = SelectPen(hDC, ThePen);
```

```
        MoveTo(hDC, a.x, a.y);          // draw axes
        LineTo(hDC, b.x, b.y);
        MoveTo(hDC, c.x, c.y);
        LineTo(hDC, d.x, d.y);

        SelectPen(hDC, OldPen);
        DeletePen(ThePen);

        FloatOut(a.x-30, a.y+5, TopLeft.x);
        FloatOut(b.x-30, a.y+5, BottomRight.x);
        FloatOut(c.x-30, c.y+5, BottomRight.y);
        FloatOut(d.x-30, d.y-20, TopLeft.y);
}
```

This function is somewhat long, but is mostly self-explanatory, especially after reference to the corresponding work in 15.2.4. It is necessary, however, to choose the manner in which the axes are drawn. We define a "pen" in the same manner as we chose a font in `TScreen::FloatOut`, deleting it before we leave the function. The extreme values are output using `FloatOut`.

The function which draws the show rectangle is simpler:

```
void  TScreen::ShowRect(HDC hDC, int Width,
                                    COLORREF color)
{
        HPEN ThePen = CreatePen(PS_SOLID, Width, color);
        HPEN OldPen = SelectPen(hDC, ThePen);
        MoveTo(hDC, showrect.left, showrect.top);
        LineTo(hDC, showrect.right, showrect.top);
        LineTo(hDC, showrect.right, showrect.bottom);
        LineTo(hDC, showrect.left, showrect.bottom);
        LineTo(hDC, showrect.left, showrect.top);
        SelectPen(hDC, OldPen);
        DeletePen(ThePen);
}
```

Then comes the function `SetScreen`, which seeks information about the coordinates in point space of the region to be shown, sets the corresponding show rectangle in pixel space, and determines the values of `tsxinc` and `tsyinc`. It makes use of some of the other `Set...` functions:

```
void  TScreen::SetScreen(RECT &rect)
{
        SetPoint(TopLeft, "Input Top Left");
        SetPoint(BottomRight, "Input Bottom Right");
```

```
    SetOffsets();
    SetShowRect(rect);
    int nox = showrect.right - showrect.left;
    int noy = showrect.bottom - showrect.top;
    tsxinc = (BottomRight.x - TopLeft.x)/nox;
    tsyinc = (TopLeft.y - BottomRight.y)/noy;
}
```

This completes the Windows equivalent of 15.2.4, but there remain two functions which will prove useful for later work. The first, `Spot`, prints a small filled circle of pixel radius `size` centred on a given pixel. We shall not explain it except to say that it uses the Windows function `Ellipse`, and defines a "brush" with which to shade the interior of the spot:

```
void  TScreen::Spot(HDC hDC, RPixel s, int size,
                                    COLORREF color)
{
    HPEN hPen = CreatePen(PS_SOLID, 1, color);
    HPEN hOldPen = SelectPen(hDC, hPen);
    HBRUSH hBrush = CreateSolidBrush(color);
    HBRUSH hOldBrush = SelectBrush(hDC, hBrush);

    Ellipse(hDC, s.x-size, s.y-size,
                    s.x+size, s.y+size);

    SelectPen(hDC, hOldPen);
    SelectBrush(hDC, hOldBrush);
    DeleteObject(hPen);
    DeleteObject(hBrush);
}
```

The second function is a little more complex. It enables us to write to the `TScreen` window an arrow, tangential to some curve, showing the direction in which the curve has been drawn—its use is to show the direction taken by a trajectory, for example. It calls the private function `Slope`, already defined. It also declares and uses instances of the Windows structure `POINT`, which somewhat confusingly gives the x and y coordinates of a *pixel*, but not the colour; and it draws an arrow head as a filled triangle obtained by calling the more general Windows function `Polygon`:

```
void  TScreen::Arrow(HDC hDC, double tangent,
                            double x, double y,
                            bool dir, COLORREF color)
{
    const int r = 8;                // blade edge length
    const double a = PI/10.0;       // half angle of blade
```

```
double ca = cos(a);
double sa = sin(a);

double t = Slope(tangent);        // tan
double c = 1.0/sqrt(1.0 + t*t);   // cos
double s = c*t;                   // sin

double cc1 = c*ca + s*sa;
double cc2 = c*ca - s*sa;
double ss1 = s*ca - c*sa;
double ss2 = s*ca + c*sa;


Point p(x, y);
Pixel S(this, p, color);
int Sx = S.x, Sy = S.y;

POINT points[3];   // Windows of Pixel, no colour
points[0].x = Sx;
points[0].y = Sy;

if (dir)
{
      points[1].x = Sx - r*cc1;
      points[1].y = Sy + r*ss1;
      points[2].x = Sx - r*cc2;
      points[2].y = Sy + r*ss2;

}     else
{
      points[1].x = Sx + r*cc1;
      points[1].y = Sy - r*ss1;
      points[2].x = Sx + r*cc2;
      points[2].y = Sy - r*ss2;
}


HPEN hPen = CreatePen(PS_SOLID, 1, color);
HPEN hOldPen = SelectPen(hDC, hPen);
HBRUSH hBrush = CreateSolidBrush(color);
HBRUSH hOldBrush = SelectBrush(hDC, hBrush);

Polygon(hDC, points, 3);
```

```
        SelectPen(hDC, hOldPen);
        SelectBrush(hDC, hOldBrush);
        DeleteObject(hPen);
        DeleteObject(hBrush);
}
```

Apart from the inline definitions, which are given in `screen.h`, all definitions are written to a file `screen.cpp` which is compiled to an object file `screen.obj`, which must be linked to any application file that is written using `TScreen`. In `screen.h`, we also include a definition of the class `TScreenApplication` precisely analogous to `TMyApp` in earlier sections.

That strenuous piece of work completes the definition of the class `TScreen`, which we shall now use as the base window for all our graphical needs. The sections which follow are illustrations of its use.

A7 Drawing a function

Each application will require a few functions peculiar to itself. In each case we therefore define a class derived from `TScreen` to contain these functions. In order to draw a function, as we did in 15.2.5, we first include the header file `screen.h`, then define a function type:

```
typedef double (*func)(double);
```

The derived class we take to be:

```
class TFuncScreen : public TScreen {
private:
        void SetFuncRange();
public:
        TFuncScreen(LPSTR ATitle) : TScreen(NULL, ATitle) {};
        void SetScreen(RECT&, func);
        void DrawFunc(HDC, func, COLORREF);
        virtual void Paint(HDC PiantDC,
                                PAINTSTRUCT &PaintInfo);
};
```

It contains a private function `SetFuncRange` which, with the aid of an input dialogue box, determines the interval `[xlo,xhi]`, over which the function is to be examined, and hence the values `TopLeft.x` and `BottomRight.x`:

```
void   TFuncScreen::SetFuncRange()
{
        char Buffer[20];
        strcpy(Buffer, "");
```

```
      float xlo, xhi;
      if (GetApplication()->ExecDialog
             (new TInputDialog(this, "Input Interval",
      "Enter XLO XHI:", Buffer,
      sizeof(Buffer))) == IDOK)
             sscanf(Buffer, "%f %f", &xlo, &xhi);
      TopLeft.x = double(xlo);
      BottomRight.x = double(xhi);
}
```

There is then the function `SetScreen`, which being of a different type from `TScreen::SetScreen` overrides the latter without the need for its being declared `virtual`. It seeks out the maximum and minimum values of the function over the chosen interval, thereby determining the y components of `TopLeft` and `BottomRight`, and the values of the increments `tsxinc` and `tsyinc`:

```
void   TFuncScreen::SetScreen(RECT &rect, func f)
{
      SetFuncRange();
      SetOffsets();
      SetShowRect(rect);
      int nox = showrect.right - showrect.left;
      tsxinc = (BottomRight.x - TopLeft.x)/nox;

      const double HUGE = 1.0E+100;
      double ylo = HUGE, yhi = -HUGE;

      for (double u = TopLeft.x; u <= BottomRight.x;
                                               u += tsxinc)
      {
             double v = f(u);
             if (v > yhi) yhi = v;
             if (v < ylo) ylo = v;
      }

      int noy = showrect.bottom - showrect.top;
      tsyinc = (yhi - ylo)/noy;

      TopLeft.y = yhi;
      BottomRight.y = ylo;
}
```

Then there is the function `DrawFunc` itself, which is notable mainly because it uses the function `SetPixel` for the first time:

```
void   TFuncScreen::DrawFunc(HDC PaintDC, func f,
                                        COLORREF color)
{
       double h = tsxinc/5.0;
       for (double u = TopLeft.x;
                    u <= BottomRight.x; u += h)
       {
              Point P(u, f(u));
              Pixel s(this, P, color);
              SetPixel(PaintDC, s);
       }
}
```

The factor 5.0 in the definition of the plotting interval h is chosen arbitrarily to give a reasonably continuous representation of the function.

Finally, there is the function `Paint` which brings together all the functions required to set the screen, to display the rectangle within which the function is to be drawn, to draw the axes, and of course the function itself:

```
#pragma argsused
void   TFuncScreen::Paint(HDC PaintDC,
                                   PAINTSTRUCT &PaintInfo)
{
       RECT rect;
       GetClientRect(HWindow, &rect);
       SetScreen(rect, F);
       ShowRect(PaintDC, 1, Green);
       Axes(PaintDC, 1, Red);
       DrawFunc(PaintDC, F, Black);
}
```

It remains only to initialize the main window, and to define a suitable function; for the latter we choose the one we used in 15.2.5:

```
void   TScreenApplication::InitMainWindow()
{
       MainWindow = new TFuncScreen(Name);
}


double F(double x)
{       return exp(-x) * sin(2.0*PI*x);
}
```

The function `WinMain` is a straightforward rewrite of the one we have used many times before. We reproduce it here for the sake of completeness (the definition of the class `TScreenApplication` is included in `screen.h`):

```
int PASCAL WinMain(HINSTANCE hInstance,
                   HINSTANCE hPrevInstance,
                   LPSTR lpCmdLine, int nCmdShow)
{
     TScreenApplication ThisApp
                   ("Graph of exp(-x)sin(2*PI*x)",
                   hInstance, hPrevInstance,
                   lpCmdLine, nCmdShow);
     ThisApp.Run();
     return ThisApp.Status;
}
```

When run, we obtain the result shown in Figure A7.

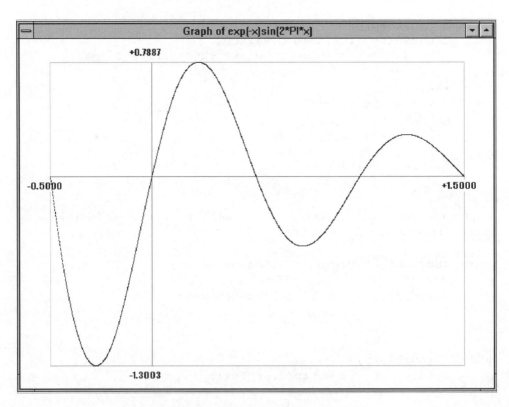

Figure A.7 Drawing a function

A8 Mapping the screen

In this section we show how to map the screen, as we did in 15.3, in order to show a quadratic Julia set. We take over unchanged the functions inside and julia from 15.3, treating them as global functions. An appropriate class is:

```
class TMapScreen : public TScreen {
public:
      TMapScreen(LPSTR ATitle) : TScreen(NULL, ATitle) {};
      void MapScreen(HDC, RECT&, complex&,
                        COLORREF, COLORREF, COLORREF);
      virtual void Paint(HDC PaintDC,
                        PAINTSTRUCT &PaintInfo);
};
```

The function MapScreen closely resembles drawblock in 15.3:

```
#pragma argsused
void   TMapScreen::MapScreen(HDC hDC, RECT &rect,
                  complex &c, COLORREF bkdcol,
                  COLORREF insetcol, COLORREF outsetcol)
{
      int nx = rect.right - rect.left;
      int ny = rect.bottom - rect.top;
      for (int j = 0; j < ny; ++j)        // scan columns
      {     for (int i = 0; i < nx; ++i)  // scan rows
            {
                  Pixel s(i, j, bkdcol);
                  if (InShowRect(s))
                  {
                        Point p(this, s);
                        complex z(p.x, p.y);
                        if (julia(z, c))
                              s.SetCol(insetcol);
                        else s.SetCol(outsetcol);
                  }
                  SetPixel(hDC, s);
            }
      }
}
```

and correspondingly the function Paint is:

```
#pragma argsused
void   TMapScreen::Paint(HDC PaintDC,
                                    PAINTSTRUCT &PaintInfo)
{
      Point p;
      SetPoint(p, "Input Julia Constant");
      RECT rect;
      GetClientRect(HWindow, &rect);
      SetScreen(rect);
      complex c(p.x, p.y);
      MapScreen(PaintDC, rect, c, Gray, Black, Yellow);
      ShowRect(PaintDC, 1, Green);
      Axes(PaintDC, 1, Red);
}
```

Main window initialization is obtained as usual by the function:

```
void   TScreenApplication::InitMainWindow()
{
      MainWindow = new TMapScreen(Name);
}
```

When `WinMain` is run, giving `maxiter` the value 300, which allows greater resolution than the value 100 used in 15.3, the result in black and white is Figure A8.

A9 Ordinary differential equations

Although `TScreen` provides a perfectly adequate base, additional functionality is again required for displaying graphically the solutions of differential equations, especially when two or more simultaneous equations are involved. In this section we construct the functions required for solving single ordinary differential equations of the form $dy/dx = f(x, y)$; in the last section we treat two simultaneous equations.

Of course, we must first include the header files:

```
#include <screen.h>
```

and then the definition of the function type:

```
typedef double (*func)(double, double);
```

We shall restrict ourselves to Runge–Kutta methods for simplicity (*cf* 18.5), although we give ourselves a choice of the order of RK approximation (the technique may easily be extended to include other possibilities). These methods are encapsulated in the class `TRKOneDim`:

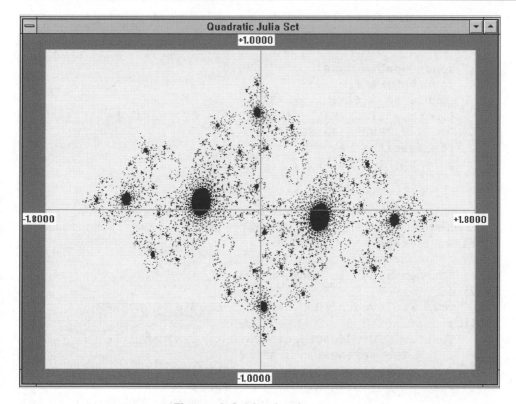

Figure A.8 Mapping the screen

```
class TRKOneDim {
public:
      typedef void (TRKOneDim::*trkf)(func, double,
                          double&, double, bool&);
      void RK1(func, double, double&, double, bool&);
      void RK2(func, double, double&, double, bool&);
      void RK4(func, double, double&, double, bool&);
};
```

Notice the appearance of a `typedef` statement *inside* the class declaration, and the form that it takes. This allows us to refer to a RK method without initially specifying which one. The RK functions are familiar to us from chapters 18 and 19. There, however, we did not take account of extreme values. If the derivative $dy/dx = f(x, y)$ becomes very large at some point, there will be a singularity in the solution which will give rise to trouble when we attempt to plot it out. In the following function we allow for that possibility in the case of the second-order function RK2:

```
void  TRKOneDim::RK2(func f, double x, double &y,
                          double h, bool &finite)
{
      const double maxslope = 1000.0;
      double K1 = f(x, y);
      double K2 = f(x+h, y+h*K1);
      finite = (fabs(K1) + fabs(K2)) < 2.0*maxslope
            ? TRUE : FALSE;
      if (!finite) return;
      y += h*(K1 + K2)/2.0;
};
```

Similar order 1 and 4 methods can readily be written down by analogy with 18.5.

The other functions required for a solution may be included as usual in a class `TOneDim`, derived form `TScreen`:

```
class TOneDim : public TScreen {
private:
      void FieldVector(HDC, RPoint, double, COLORREF);
public:
      TOneDim(PTWindowsObject AParent, LPSTR ATitle) :
            TScreen(AParent, ATitle)          {};
      void MapField(HDC, func, COLORREF);
      void SolveYXOneDim(HDC, int, func,
                              RPoint, double, COLORREF);
      virtual void Paint(HDC, PAINTSTRUCT&);
};
```

The private function `FieldVector` plots a short line of given length and of slope t ($= dy/dx$), centred on a given point p, in the expectation that solutions of equations will be tangential to such lines. It is a familiar tool in the solution of differential equations. The function is self-explanatory:

```
void  TOneDim::FieldVector(HDC hDC, RPoint p,
                          double t, COLORREF color)
{
      const int len = 10;
      t *= tsxinc/tsyinc;      // point to pixel tangent
      double c = 1.0/sqrt(1 + t*t);
      double s = c*t;
      int dx = int(round((c*len)/2.0));
      int dy = int(round((s*len)/2.0));
```

```
        Pixel S(this, p, 0);
        Pixel S1(S.GetX() - dx, S.GetY() + dy, 0);
        Pixel S2(S.GetX() + dx, S.GetY() - dy, 0);
        HPEN hPen = CreatePen(PS_SOLID, 1, color);
        HPEN hOldPen = SelectPen(hDC, hPen);
        MoveTo(hDC, S1);
        LineTo(hDC, S2);
        SelectPen(hDC, hOldPen);
        DeleteObject(hPen);
}
```

The constructor merely calls the TScreen constructor. There follows a function MapField which uses the private function FieldVector to print to window a grid of short lines 10 pixels long and 25 pixels apart, tangential to the solution at each grid point. It is otherwise self-explanatory:

```
void   TOneDim::MapField(HDC hDC, func f, COLORREF color)
{
        double hx = (BottomRight.x - TopLeft.x)/25;
        double hy = (TopLeft.y - BottomRight.y)/25;
        for (double y = TopLeft.y;
                    y >= BottomRight.y; y -= hy)
        {
               for (double x = TopLeft.x;
                           x <= BottomRight.x; x += hx)
               {
                      double tan = f(x, y);
                      Point p(x, y);
                      FieldVector(hDC, p, tan, color);
               }
        }
}
```

We then come to the function SolveYX which produces a solution passing through a given point P. The integer parameter c allows one to choose the order of RK approximation, color the colour of the curve representing the solution. The solution is sketched both forward and backkward from the point P:

```
void   TOneDim::SolveYXOneDim(HDC hDC, int c, func f,
                         RPoint P, double h, COLORREF color)
{
        TRKOneDim RK;
        TRKOneDim::trkf rk;
        switch(c) {
```

```
        case 1 : rk = &TRKOneDim::RK1; break;
        case 2 : rk = &TRKOneDim::RK2; break;
        case 4 : rk = &TRKOneDim::RK4; break;
    }

    double x, y;
    bool finite = TRUE;                    // FALSE near singularity

    y = P.y;
    for (x = P.x; x <= BottomRight.x; x += h) // forward
    {
        if (y <= TopLeft.y && y >= BottomRight.y)
        {
            Point p(x, y);
            Pixel s(this, p, color);
            SetPixel(hDC, s);
        }


        (RK.*rk)(f, x, y, h, finite);
        if (!finite) break;
    }

    y = P.y;
    for (x = P.x; x >= TopLeft.x; x -= h) // backward
    {
        if (y <= TopLeft.y && y >= BottomRight.y)
        {
            Point p(x, y);
            Pixel s(this, p, color);
            SetPixel(hDC, s);
        }


        (RK.*rk)(f, x, y, -h, finite);
        if (!finite) break;
    }
}
```

The heart of the program, as usual, is the function Paint. In this case, it sets the screen, maps the vector field, draws the axes, asks whether the parameters with which the screen was set up need to be changed in the light of the vector field, requests decisions on the order of RK approximation to be used, the number of points through which solutions are to be drawn, and what those points are, marks the points with spots, and draws the solutions:

```
#pragma argsused
void   TOneDim::Paint(HDC PaintDC, PAINTSTRUCT &PaintInfo)
{
       RECT rect;
       GetClientRect(HWindow, &rect);

       SetScreen(rect);
       MapField(PaintDC, F, Gray);
       ShowRect(PaintDC, 1, Green);
       Axes(PaintDC, 1, Red);

       if (MessageBox(HWindow, "Set New Parameters?",
                        "Adjust Field", MB_OKCANCEL) == IDOK)
              InvalidateRect(HWindow, NULL, TRUE);

       int Order;
       bool done = FALSE;
       while (!done)
       {
              SetInteger(Order,
                     "Choose Approximation: 1, 2, or 4 only");
              switch(Order) {
                     case 1 :
                     case 2 :
                     case 4 : done = TRUE; break;
                     default:

                     MessageBox(HWindow,
                                "Enter 1, 2, or 4 only",
                                "No such approximation", MB_OK);
              }
       }

       int N;
       SetInteger(N, "Number of Initial Points");
       PPoint p = new Point [N];
       double h = tsxinc/5.0;
       for (int i = 0; i < N; ++i)
       {
              SetPoint(p[i], "Initial Point");
              Pixel s(this, p[i], Black);
```

```
                    Spot(PaintDC, s, 2, Black);
                    SolveYXOneDim(PaintDC, Order, F,
                                          p[i], h, Blue);
            }
    }
```

Finally, we must initialize the main Window:

```
void  TScreenApplication::InitMainWindow()
{
        MainWindow = new TOneDim(NULL, Name);
}
```

When **WinMain** has been included, the completed program compiled, linked to
screen.obj and to **indiabox.res**, and run, the results are shown in Figure A9, where
the chosen solutions are those through the points (±0.3, ±0.3). The "separatrix" through (0.0,
0.0) is also shown; it is not itself a solution (although the program produces it without protest),
but separates solutions in which $y \to +\infty$ from those in which $y \to -\infty$, as $x \to 0$.

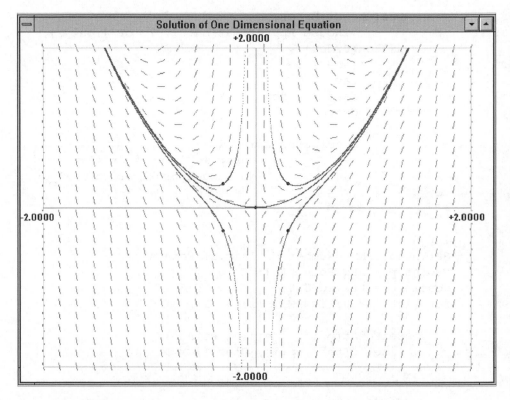

Figure A.9 Solutions of one dimensional differential equation

A10 Two simultaneous differential equations

Finally, we consider the more complex problem of two simultaneous (autonomous) differential equations $dy/dt = f(x, y)$, $dy/dt = g(x, y)$, the solutions of which pass through given points. In particular, let us study the non-linear problem:

$$dx/dt = x - y^2, \, dy/dt = x^2 - y.$$

Clearly, this system of equations has 2 fixed points, where $dx/dt = dy/dt = 0$, at $(0, 0)$, which is a saddleback, and at $(1, 1)$, which is a centre. By linearization, the stable and unstable invariant manifolds local to the origin are given respectively by:

$$W_{loc}^s(0.\,0) = \{\ (x, y) \mid x \approx y^2/3\ \}, \, W_{loc}^u(0, 0) = \{\ (x, y) \mid y \approx x^2/3\ \}.$$

That is already enough information with which to sketch by hand the essential nature of the solutions. Our concern, however, is to compute the solutions, and to display them graphically. We shall require two windows: one to show the phase diagram—the (x, y) trajectories passing through given points; the other to show the time dependence $x(t)$ and $y(t)$ of a chosen trajectory. This calls for the multiple document interface technique introduced in A5. Therefore, we must include the header file mdi.h as well as screen.h.

By analogy with the last section, we define a class to encapsulate the RK increment functions:

```
class TRKTwoDim {
public:
    typedef void (TRKTwoDim::*trkf)(func, func,
              double&, double&, double,
                  bool&, bool&);
    void RK1(func, func, double&, double&,
              double, bool&, bool&);
    void RK2(func, func, double&, double&,
              double, bool&, bool&);
    void RK4(func, func, double&, double&,
              double, bool&, bool&);
};
```

In the case of two dimensions there is the additional possibility that dx/dt and dy/dt are simultaneously zero, giving rise to a fixed point, or even a continuous curve of fixed points. We allow for this possibility in the two-dimensional RK2 function:

```
void  TRKTwoDim::RK2(func f, func g, double &x, double &y,
            double h, bool &finite,
                    bool &fixedpoint)
{
```

```
        const double maxslope = 1.0E+6;
        const double smallest = 1.0E-3;
        double s = x, t = y;
        double K1 = f(s, t), L1 = g(s, t);
        fixedpoint =
                fabs(K1) < smallest && fabs(L1) < smallest
                ? TRUE : FALSE;
        if (fixedpoint) return;
        s += h*K1; t += h*L1;
        double K2 = f(s, t), L2 = g(s, t);
        finite = (fabs(K1) + fabs(K2) < 2.0*maxslope)
                    && (fabs(L1) + fabs(L2) < 2.0*maxslope)
                ? TRUE : FALSE;
        if (!finite) return;
        x += h*(K1 + K2)/2.0;
        y += h*(L1 + L2)/2.0;
}
```

The RK functions of orders 1 and 4 may again be written down by analogy.

We then come to the derived class TDiffTwo, which contains the functions required to solve the two-dimensional problem:

```
class TDiffTwo : public TScreen {
private:
        BOOL Active;
        double tangent(func, func, double, double);
public:
        TDiffTwo(PTWindowsObject AParent, LPSTR ATitle) :
        TScreen(AParent, ATitle)        { Active = FALSE; }
        void SetActive(BOOL b)          { Active = b; }
        void SetMessage(HDC);
        void FPCurve(HDC, RECT&, func, func, COLORREF);
        void SolveYXTwoDim(HDC, int, func, func,
                                double, double, RPoint,
                                double, COLORREF);
        void SolveXYTTwoDim(HDC, int, func, func, double,
                            RPoint, double, COLORREF, COLORREF);
        virtual void WMRButtonDown(RTMessage)
                    = [ WM_FIRST + WM_RBUTTONDOWN ];
        virtual void Paint(HDC, PAINTSTRUCT&);
};
```

The private function tangent is simply the ratio $(dy/dt)/(dx/dt)$:

```
double TDiffTwo::tangent(func f, func g,
                                 double x, double y)
{
       const double toosmall = 1.0E-10;
       double F = f(x, y), G = g(x, y);
       if (fabs(F) < toosmall) F = sign(F)*toosmall;
       return G/F;
}
```

Notice that we take avoiding action in case the tangent becomes too nearly vertical.

For the two-dimensional case we include a function `SetMessage` which prints to window a reminder to activate child windows by means of the right mouse button:

```
void  TDiffTwo::SetMessage(HDC hDC)
{
       RECT rect;
       rect.top = 25;
       rect.left = 25;
       rect.bottom = 100;
       rect.right = 150;
       char String[] = "Activate each Child Window "
                   "by clicking it with Right Button";
       DrawText(hDC, String, -1, &rect, DT_WORDBREAK);
}
```

The function `DrawText` is an alternative output function which prints within the rectangle `rect`.

There is the possibility already mentioned that there may be a continuous curve of fixed points (although not in the example we have chosen). The following function displays any such curve, and otherwise isolated fixed points, by simply scanning the screen:

```
void  TDiffTwo::FPCurve(HDC hDC, RECT &rect,
                           func f, func g, COLORREF color)
{
       const double smallest = 1.0E-2;
       int nx = rect.right - rect.left;
       int ny = rect.bottom - rect.top;
       for (int j = 0; j < ny; ++j)
       {    for (int i = 0; i < nx; ++i)
            {
                   Pixel s(i, j, color);
                   if (InShowRect(s))
                   {
```

```
              Point p(this, s);
              double F = f(p.x, p.y);
              double G = g(p.x, p.y);
              if (fabs(F) < smallest
                     && fabs(G) < smallest)
                  SetPixel(hDC, s);
          }
      }
   }
}
```

The two `Solve` functions resemble `TDiffOne::SolveYXOneDim` but are sufficiently different from it, and from each other, that we must write them down explicitly. `SolveYXTwoDim` produces the phase diagram in the (x, y) plane, with an arrow in the direction of increasing t (cf A6); `SolveYXTTwoDim` produces the time behaviour x(t) and y(t):

```
void  TDiffTwo::SolveYXTwoDim(HDC hDC, int c,
                  func f, func g, double T, double A,
                  RPoint P, double h, COLORREF color)
{
    TRKTwoDim RK;
    TRKTwoDim::trkf rk;
    switch(c) {
           case 1 : rk = &TRKTwoDim::RK1; break;
           case 2 : rk = &TRKTwoDim::RK2; break;
           case 4 : rk = &TRKTwoDim::RK4; break;
    }
    bool finite = TRUE;
    bool fixedpoint = FALSE;
    double x = P.x, y = P.y;
    for (double t = 0; t <= T; t += h)
    {
           Point p(x, y);
           double oldx = x;
           Pixel s(this, p, color);

           if (InShowRect(s))
                  SetPixel(hDC, s);
           (RK.*rk)(f, g, x, y, h, finite, fixedpoint);
           if (!finite) break;
           if (fixedpoint)
           {
```

```
                Spot(hDC, s, 3, Black);
                break;
        }
        if (fabs(t - A) < h/2.0)
        {
                bool ToRight = (x >= oldx) ? TRUE : FALSE;
                double tan = tangent(f, g, x, y);
                Arrow(hDC, tan, x, y, ToRight, color);
        }
    }

    fixedpoint = FALSE;
    x = P.x; y = P.y;
    for (t = 0; t <= T; t += h)
    {
        Point p(x, y);
        Pixel s(this, p, color);
        if (InShowRect(s))
                SetPixel(hDC, s);
        (RK.*rk)(f, g, x, y, -h, finite, fixedpoint);
        if (!finite) break;
        if (fixedpoint)
        {
                Spot(hDC, s, 3, Black);
                break;
        }
    }
}

void   TDiffTwo::SolveXYTTwoDim(HDC hDC, int c,
                func f, func g, double T,
                RPoint P, double h,
                COLORREF Xcolor, COLORREF Ycolor)
{
    TRKTwoDim RK;
    TRKTwoDim::trkf rk;
    switch(c) {
        case 1 : rk = &TRKTwoDim::RK1; break;
        case 2 : rk = &TRKTwoDim::RK2; break;
        case 4 : rk = &TRKTwoDim::RK4; break;
    }
    bool finite = TRUE;
    bool fixedpoint = FALSE;
```

```
      double x = P.x, y = P.y;
      for (double t = 0; t <= T && t <= BottomRight.x;
                                              t += h)
      {
            Point px(t, x), py(t, y);
            Pixel sx(this, px, Xcolor);
            Pixel sy(this, py, Ycolor);
            if (InShowRect(sx))
                  SetPixel(hDC, sx);
            if (InShowRect(sy))
            SetPixel(hDC, sy);
            (RK.*rk)(f, g, x, y, h, finite, fixedpoint);
            if (!finite || fixedpoint) break;
      }
}
```

For the right mouse button function we take, as before:

```
void   TDiffTwo::WMRButtonDown(RTMessage)
{
      SetActive(TRUE);
      InvalidateRect(HWindow, NULL, TRUE);
}
```

The function `Paint` is reminiscent of that used in A5. It begins with a number of declarations, specified as `static` so that they may be used in the different MDI windows. The program runs from t = 0 to t = T, and the arrow is superimposed at t = A. Apart from that, the function needs no explanation:

```
#pragma argsused
void   TDiffTwo::Paint(HDC PaintDC, PAINTSTRUCT &PaintInfo)
{
      static RECT rect;
      static int Order;
      static double T;
      static double A;
      static int N;
      static PPoint P;
      static PPixel S;

      GetClientRect(HWindow, &rect);
```

```
if (Attr.Id == 1 && Active)
{
      Active = FALSE;
      MessageBox(HWindow, "Ready to Set?",
                       "Problem Parameters", MB_OK);
      bool done = FALSE;
      while (!done)
      {
            SetInteger(Order,
            "Choose Approximation: 1, 2, or 4 only");
            switch(Order) {
                  case 1 : done = TRUE; break;
                  case 2 : done = TRUE; break;
                  case 4 : done = TRUE; break;
                  default:
                  MessageBox(HWindow,
                             "Enter 1, 2, or 4 only",
                       "No such approximation", MB_OK);
            }
      }

      SetDouble(T, "Set Run Duration");
      SetDouble(A, "Set Arrow Time");
      SetInteger(N, "Number of Trajectories");
      P = new Point [N];
      S = new Pixel [N];
      SetMessage(PaintDC);
      MessageBox(HWindow, "Ready?",
                    "Delete Window", MB_OK);
      CloseWindow();
}
if (Attr.Id == 2 && Active)
{
      Active = FALSE;
      MessageBox(HWindow, "Ready to Set?",
                       "Y vs X Screen", MB_OK);
      SetScreen(rect);
      ShowRect(PaintDC, 1, Green);
      double h = tsxinc/10.0;
      if (MessageBox(HWindow, "Choose Y or N:",
            "Show Lines of Fixed Points?",
                             MB_YESNO) == IDYES)
            FPCurve(PaintDC, rect, F, G, Black);
      for (int i = 0; i < N; ++i)
```

```
          {
                    SetPoint(P[i], "Initial Values");
                    SolveYXTwoDim(PaintDC, Order,
                                  F, G, T, A, P[i], h, PaleGray);
                    S[i] = Pixel(this, P[i], Black);
                    DigitOut(PaintDC, S[i], i);    // traj num
          }
          Axes(PaintDC, 1, Red);
     }

     if (Attr.Id == 3 && Active)
     {
          Active = FALSE;
          MessageBox(HWindow, "Ready to Set?",
                              "X,Y vs T Screen", MB_OK);
          int Traj;
          SetInteger(Traj, "Choose Trajectory");
          SetScreen(rect);
          ShowRect(PaintDC, 1, Green);
          double h = tsxinc/10.0;
          SolveXYTTwoDim(PaintDC, Order,
                         F, G, T, P[Traj], h, Cyan, Magenta);

          Point PTraj;                 // show traj num
          PTraj.x = BottomRight.x - (BottomRight.x
                                    - TopLeft.x)/10.0;
          PTraj.y = TopLeft.y - (TopLeft.y
                                    - BottomRight.y)/10.0;
          Pixel STraj(this, PTraj, Black);
          DigitOut(PaintDC, STraj, Traj);
          Axes(PaintDC, 1, Red);
     }
}
```

There follows the MDI apparatus. The class `TFrame` declaration is identical with that in A5, and so are the function definitions, except that `TFrame::SetupWindow` should be given suitable captions for its child windows. We do not repeat these details. We initialize the main window in the usual way, but using `TFrame`:

```
void  TScreenApplication::InitMainWindow()
{
     MainWindow = new TFrame
          ("First Order - Two Dimensions", "MAINMENU");
}
```

WinMain may be given the name "Two First Order Equations", but is otherwise the same as usual. When compiled, there must be linking to `screen.obj` and to a suitable resource file which combines the attributes of `indiabox.res` and `owlmdi.res`. When run, the outcome is shown in black and white in Figure A.10, where, however, we have superimposed the invariant manifold obtained by calling the function `SolveYXTwoDim` twice more, through the points (`eps, eps*eps/3`) and (`eps*eps/3, eps`), where `eps` is a sufficiently small quantity that these points are effectively on the local parabolic manifolds near the origin. The phase diagrams are on the left; the time dependence on the right. The invariant manifold will be seen to divide the plane into an outer and an inner region; in the latter, solutions are near sinusoidal.

The same technique may with some effort be extended to represent the solutions of three simultaneous differential equations, including their chaotic solutions, if any. But that would take us much too far afield.

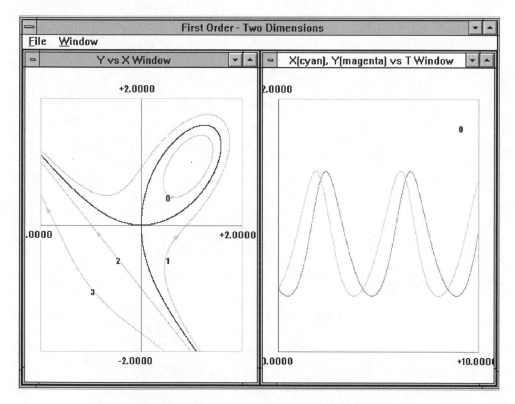

Figure A.10 Solutions of two dimensional differential equation

Bibliography

Atkinson, Kendall E., *An Introduction to Numerical Analysis*, Wiley, 1978.

Atkinson, L. V. and Harley, P. J., *An Introduction to Numerical Methods with Pascal*, Addison-Wesley, 1983.

Bird, Richard and Wadler, Philip, *Introduction to Functional Programming*, Prentice-Hall, 1988.

Borland International Inc, *TurboC++ Manual*, 4 Vols, 1990.

Boyce, William E. and DiPrima, Richard C., *Elementary Differential Equations and Boundary Value Problems*, Wiley, 5th Edition, 1992.

Burden, Richard L. and Faires, J. Douglas, *Numerical Analysis*, PWS-Kent, 4th Edition, 1989.

Copson, E. T., *An Introduction to the Theory of Functions of a Complex Variable*, Oxford University Press, 1935.

Ellis, Margaret A. and Stroustrup, Bjarne, *The Annotated C++ Reference Manual*, Addison-Wesley, 1994.

Gonnet, G. H., *Handbook of Algorithms and Data Structures*, Addison-Wesley, 1984.

Hardy, G. H. and Wright, E. M., *An Introduction to the Theory of Numbers*, Oxford University Press, 4th Edition, 1975.

Kernighan, Brian W. and Ritchie, Dennis M., *The C Programming Language*, Prentice-Hall, 1978.

Knuth, Donald E., *The Art of Computer Programming*, 3 Vols, Addison-Wesley, 2nd Edition, 1973.

Lippman, Stanley B., *C++ Primer*, Addison-Wesley, 1989.

Peitgen, H.-O. and Saupe, D. (Editors), *The Science of Fractal Images*, Springer-Verlag, 1988.

Perry, Paul, *Turbo C++ for Windows*, SAMS Publishing, 1993.

Petzold, Charles, *Programming Windows 3.1*, Microsoft Press, 1992.

Press, William H., Flannery, Brian P., Tenkolsky, Saul A. and Vetterling, William T., *Numerical Recipes in C*, Cambridge University Press, 1988.

Rudin, Walter, *Principles of Mathematical Analysis*, McGraw-Hill, 3rd Edition, 1976.

Sterling, Leon and Shapiro, Ehud, *The Art of Prolog*, MIT Press, 1986.

Stroustrup, Bjarne, *The C++ Programming Language*, Addison-Wesley, 1987.

Tenenbaum, Aaron M. and Augenstein, Moshe J., *Data Structures Using Pascal*, Prentice-Hall, 1981.

Welsh, Jim, Elder, John and Bustard, David, *Sequential Program Structures*, Prentice-Hall, 1984.

Appendix A

```
#ifndef STANDARD
#define STANDARD

                                /**************************/
                                /*       STANDARD.H       */
                                /**************************/
                                /*                        */
                                /*   Some standard things */
                                /*                        */
                                /**************************/


#include <stdlib.h>
#include <iostream.h>
#include <iomanip.h>
#include <math.h>

const char tab  = '\t';
const char bell = '\7';
const char *const sep = ", ";

#define newline      cout<<endl
#define clear        clrscr()

inline void statement(char *s) { cout << s; }

template <class T>
inline void output(char *s, T x) { cout << s << x; }

template <class T>
inline void input(char *s, T &x) { cout << s;  cin >> x; }

const    double PI       = 3.141592653589793;
const    double EXP      = 2.718281828459045;
const    double GAMMA    = 0.577215664901533;
const    double GOLDMEAN = 0.618033988749895;
const    double LN2      = 0.693147180559945;
const    double LN10     = 2.302585092994046;

typedef unsigned char byte;
typedef unsigned int  word;
typedef unsigned long longword;

enum    bool { FALSE, TRUE };

enum    state { ITERATING, SUCCESS, WONTSTOP, NEARZERO, SAMESIGN };
```

```
void    error(char*);                    // runtime error termination

bool    getyes(const char*);             // boolian choice

// roots of quadratic equation:  a*x*x + b*x + c = 0
void    quadroots(double, double, double, double&, double&);

inline  double round(double x)   { return floor(x + 0.5); }

double  mod(double, double);

template <class T>
inline  T sign(T x) { return x < 0 ? -1 : 1; }

template <class T>
inline T max(T x, T y) { return x < y ? y : x; }

template <class T>
inline T min(T x, T y) { return x > y ? y : x; }

template <class T>
inline void swap(T &x, T &y) { T t = x; x = y; y = t; }

class   format {                    // three parameter manipulator
private:
        long lb;
        int pr, wd;
        ios& iosfunc(ios&, long, int, int);       // formats stream output
public:
        format(long l, int p, int w) : lb(l), pr(p), wd(w) { }
        friend ostream& operator<<(ostream&, format&);
};

inline  ostream& operator<<(ostream &s, format &f)
{       f.iosfunc(s, f.lb, f.pr, f.wd); return s; }


#endif  //STANDARD
```

Appendix B

```cpp
#ifndef RESTRING
#define RESTRING

                               /***************************/
                               /*       RESTRING.H       */
                               /***************************/
                               /*                         */
                               /*      Classes strbuf     */
                               /*        and string       */
                               /*                         */
                               /***************************/


class    strbuf {
private:
         int range(int);                                         // index in range
protected:
         char *buf;                                              // char array
         int size;                                               // string length + 1
public:
         strbuf(int sz);
         ~strbuf()                       { delete buf; }
         char& operator[](int i)         { return buf[range(i)]; }
         void swap(int, int);
};



class    string : public strbuf {
friend   ostream& operator<<(ostream&, string&);
friend   istream& operator>>(istream&, string&);
private:
         static int inbufsize;                                   // input buffer
public:
         string()          : strbuf(1)       {}                  // string x;
         string(int n)     : strbuf(n+1)     {}                  // string x(n);
         string(char *str);                                      // string x = "...";
         string(string &x);                                      // string x = y;
         ~string()                           {}
         string& operator=(char*);                               // assignment
         string& operator=(string&);                             // assignment
         string& operator+(char*);                               // concatenation
         string& operator+(string&);                             // concatenation
};


#endif   // RESTRING
```

Appendix C

```
#ifndef RATIONAL
#define RATIONAL
                                     /***************************/
                                     /*       RATIONAL.H       */
                                     /***************************/
                                     /*                         */
                                     /*    Class of rational    */
                                     /*  numbers represented by */
                                     /*      two long integers  */
                                     /*                         */
                                     /***************************/
#include <standard.h>

class rational {
friend      rational rabs(rational&);
friend      rational operator+(rational&, rational&);
friend      rational operator-(rational&, rational&);
friend      rational operator*(rational&, rational&);
friend      rational operator/(rational&, rational&);
friend      ostream& operator<<(ostream&, rational&);
friend      istream& operator>>(istream&, rational&);
private:
            long num, den;
            long gcd(long, long);
public:
            rational()                                  { num = 0; den = 1; }
            rational(long n)                            { num = n; den = 1; }
            rational(long, long);
            rational(rational &r)                       { num = r.num; den = r.den; }
            operator double()                           { return double(num)/den; }
            rational operator+()                        { return *this; }
            rational operator-()                        { return rational(-num, den); }
            bool operator==(rational&);
            bool operator>(rational&);
            bool operator<(rational&);
            bool operator>=(rational&);
            bool operator<=(rational&);
            rational& operator=(rational&);
            rational& operator+=(rational&);
            rational& operator-=(rational&);
            rational& operator*=(rational&);
            rational& operator*=(long);
            rational& operator/=(rational&);
            rational& operator/=(long);
};

#endif     // RATIONAL
```

Appendix D

```
#ifndef RANDOM
#define RANDOM


                              /***************************/
                              /*        RANDOM.H         */
                              /***************************/
                              /*                         */
                              /*      Random number      */
                              /*        generators       */
                              /*                         */
                              /***************************/


#include <standard.h>


long    next(long&);                              // long integer generator
float   uniform(long&);                           // float uniform in (0, 1)
float   boltzmann(long&, double);                 // exponential variate
float   gaussian(long&, double);                  // gaussian variate
int     bitrand(long&);                           // slow random bit generator

//      the above should be seeded with a large negative long


//      fast sequence of about 6 million random bits
int     randbit(longword&);
//      this should be seeded with an unsigned long (longword)


#endif   //RANDOM
```

Appendix E

```
#ifndef VECMAT
#define VECMAT
                                /***************************/
                                /*       VECMAT.H         */
                                /***************************/
                                /*                         */
                                /*         Classes         */
                                /*     vector and matrix   */
                                /*                         */
                                /***************************/
#include <standard.h>

class    matrix;                                          // incomplete declaration

class    vector {
friend   matrix;
private:
        int size;
        double *vec;
        int range(int);                                  // index check
public:
        vector(int);                                     // init to zero elems
        vector(const vector&);                           // init to given vec
        vector(const double*, int);                      // init to given array
        ~vector()            {delete vec;}
        double& operator[](int i)    {return vec[range(i)];}
        vector& operator+()              { return *this; } // unary plus
        vector& operator=(const vector&);
        vector& operator+=(const vector&);
        vector& operator-=(const vector&);
        vector& operator*=(double);                      // mult by double
        vector& operator/=(double);                      // div by double
        int getsize()                    {return size;}
        void swap(int, int);
        friend vector operator-(const vector&);          // unary minus
        friend vector operator+(const vector&, const vector&);
        friend vector operator-(const vector&, const vector&);
        friend vector operator*(const vector&, double);
        friend vector operator*(double, const vector&);
        friend vector operator/(const vector&, double);
        friend double scalar(const vector&, const vector&);  // scalar product
        friend double norm(const vector&);               // euclidean norm
        friend double norminf(const vector&);            // infinite norm
        friend vector operator*(const matrix&, const vector&);  // v = m.u
        friend vector operator*(const vector&, const matrix&);  // v = u.m
        friend matrix operator*(const vector&, const vector&);  // m = u.v
        friend matrix operator*(const matrix&, const matrix&);  // m = m1.m2
```

```
            friend double norm(const matrix&);                    // euclidean norm
            friend double norminf(const matrix&);                 // infinite norm
            friend ostream& operator<<(ostream&, const vector&);
            friend istream& operator>>(istream&, vector&);
};

class   matrix {
private:
            int numrows;
            int numcols;
            vector **mat;
            int range(int);                                       // row index check
public:
            matrix(int, int);                                     // rectangular matrix
            matrix(int);                                          // square matrix
            matrix(const matrix&);                                // init to given matrix
            ~matrix();
            vector& operator[](int i)    { return *mat[range(i)]; }
            matrix& operator+()                       { return *this; }  // unary plus
            matrix& operator=(const matrix&);
            matrix& operator+=(const matrix&);
            matrix& operator-=(const matrix&);
            matrix& operator*=(double);
            matrix& operator/=(double);
            int getnumrows()                     {return numrows;}
            int getnumcols()                     {return numcols;}
            int getsize();                                        // square matrix only!
            void swap(int, int);
            matrix transpose();                                   // transposed matrix
            friend matrix operator-(const matrix&);               // unary minus
            friend matrix operator+(const matrix&, const matrix&);
            friend matrix operator-(const matrix&, const matrix&);
            friend matrix operator*(const matrix&, const matrix&);   // m = m1.m2
            friend matrix operator*(const matrix&, double);          // m = m1.d
            friend matrix operator*(double, const matrix&);          // m = d.m1
            friend matrix operator/(const matrix&, double);          // m = m1/d
            friend vector operator*(const matrix&, const vector&);   // v = m.u
            friend vector operator*(const vector&, const matrix&);   // v = u.m
            friend matrix operator*(const vector&, const vector&);   // m = u.v
            friend double norm(const matrix&);                    // euclidean norm
            friend double norminf(const matrix&);                 // infinite norm
            friend ostream& operator<<(ostream&, const matrix&);
            friend istream& operator>>(istream&, matrix&);
};

//      inline functions

inline  int vector::range(int i)
{       if (i<0 || i>=size) error("vector index out of range!");
        return i;
}

inline  int matrix::range(int i)
{       if (i<0 || i>=numrows) error("matrix row index out of range!");
        return i;
}

inline  vector operator*(double d, const vector &v)
{       return v * d;       }

inline  matrix operator*(double d, const matrix &m)
{       return m * d;       }

#endif    //VECMAT
```

Appendix F

```
#ifndef STDGRAPH
#define STDGRAPH

                                    /**************************/
                                    /*      STDGRAPH.H       */
                                    /**************************/
                                    /*                        */
                                    /*  Some Standard Graphics */
                                    /*       Maths Routines    */
                                    /*                        */
                                    /**************************/

#include <standard.h>
#include <graphics.h>
#include <complex.h>

typedef   double (*func)(double);

class     pixel;                                        // incomplete declaration

struct  point {                                         // mathematical point
        double x, y;
        point()                                         { x = y = 0; }
        point(double a, double b)                       { x = a; y = b; }
        point(point &p)                                 { x = p.x; y = p.y; }
        point(complex &z)                               { x = real(z); y = imag(z); }
        point(pixel&, point&, point&);                  // point image of pixel in box
};

ostream& operator<<(ostream&, point&);                  // point output
istream& operator>>(istream&, point&);                  // point input


void    getblock(point&, point&);
void    getblock(point[], int, point&, point&);
void    getblock(func, double, double, point&, point&);
void    getblock(complex&, complex&);

void    far plot(point&, point&, point&, int);          // plot point on screen
void    far drawfunc(func, point&, point&, int);        // draw function

class   pixel {                                         // pixel image of point
friend  point;
private:
        int x, y, c;                                    // graphics conventions!
public:
        pixel();                                // top-left of screen, default colour
        pixel(int, int, int);
```

```
            pixel(point &p, point &tl, point &br, int color);        // image of p
            pixel(pixel&);
            pixel& operator=(pixel&);
            void setcol(int);
            int getcol();
            int getx();
            int gety();
            friend bool inbox(pixel&);
            friend void far putpixel(pixel&);
            friend void far line(pixel&, pixel&);
            friend void far scalegraph(point&, point&, int, int, int);
};

inline  pixel::pixel()
{ x = y = c = 0; }

inline    pixel::pixel(int a, int b, int col)
{ x = a; y = b; c = col; }

inline    pixel::pixel(pixel &p)
{ x = p.x; y = p.y; c = p.c; }

inline  pixel& pixel::operator=(pixel &p)
{ x = p.x; y = p.y; c = p.c; return *this; }

inline    void pixel::setcol(int color)
{ c = color; }

inline    int pixel::getcol()
{ return c; }

inline    int pixel::getx()
{ return x; }

inline    int pixel::gety()
{ return y; }

inline  void far putpixel(pixel &p)
{ putpixel(p.x, p.y, p.c); }                                         // TurboC++

inline    void far line(pixel &p, pixel &q)
{ line(p.x, p.y, q.x, q.y); }                                       // TurboC++

double    getaspect();
bool      inbox(pixel&);

void      far opengraph();                                          // set offsets and open graph-
ics
void      far scalegraph(point&, point&, int, int, int); // box and axes
void      far scalegraph(complex&, complex&, int, int, int);// complex ditto
void      far shutgraph(int);                                      // pause and close graphics

#endif    // STDGRAPH
```

Appendix G

```
#ifndef LIST
#define LIST

                                    /****************************/
                                    /*          LIST.H          */
                                    /****************************/
                                    /*                          */
                                    /*      Base Class list     */
                                    /*    and class template    */
                                    /*    for arbitrary type    */
                                    /****************************/


#include <standard.h>

class    node;                      // incomplete declarations
class    list;                      // for defining pointer types
typedef node *nptr;                 // node pointer
typedef list *lptr;                 // list pointer
typedef void *vptr;                 // void pointer for base class

// abstract list implemented as cyclic singly linked list of nodes

class    node {
friend   class list;
friend   class listscan;
private:
        vptr data;                  // void data pointer
        nptr next;                  // pointer to next node
        node(vptr, nptr=0);         // null default pointer
};

class    list {
friend   class listscan;
protected:
        nptr last;                  // last->next is head of cyclic list
        int  len;                   // length of list, defined dynamically
private:
        void clearlist();           // remove all nodes
public:
        list();                     // empty list constructor
        list(list&);                // copy list constructor
        ~list();                    // destructor
        list& operator=(list&);     // assignment
        list& operator+(list&);     // concatenation
        void prepend(vptr);         // add at head of list
        void append(vptr);          // add at foot of list
```

```
        vptr get();             // return and remove head
        int  length();          // current length of list
        vptr operator[](int);   // data pointer of given node
        void swap(int, int);    // swap given items in list
        vptr foot();            // data pointer at foot
        vptr head();            // data pointer at head
};

class   listscan {
private:
        nptr np;
        lptr lp;
public:
        listscan(list&);        // constructor
        vptr operator()();      // scanner
};

inline  list::list()
{       last = 0; len = 0; }

inline  list::~list()
{       clearlist(); }

inline  node::node(vptr vp, nptr np)
{       data = vp; next = np; }

inline  int list::length()
{       return len; }

inline  listscan::listscan(list &vl)
{       lp = &vl; np = lp->last; }


template <class T> class Tlist : public list {
public:
        Tlist() {}
        Tlist(Tlist<T> &tl) : list(tl) {}
        ~Tlist() {}
        void prepend(T* tp) { list::prepend((void*)tp); }
        void append(T* tp) { list::append((void*)tp); }
        Tlist& operator=(Tlist<T> &tl) \
                { return *(Tlist<T>*)&list::operator=(tl); }
        Tlist& operator+(Tlist<T> &tl) \
                { return *(Tlist<T>*)&list::operator+(tl); }
        T* get() { return (T*)list::get(); }
        T* operator[](int i) { return (T*)list::operator[](i); }
        T* foot() { return (T*)list::foot(); }
        T* head() { return (T*)list::head(); }
};


template <class T> class Tlistscan : public listscan {
public:
        Tlistscan(Tlist<T> &tl) : listscan(*(list*)&tl) {}
        T* operator()() { return (T*)listscan::operator()(); }
};

#endif // LIST
```

Appendix H

```
#ifndef SCREEN
#define SCREEN

                              /***************************/
                              /*      SCREEN.H         */
                              /***************************/
                              /*                       */
                              /*  Classes Point, Pixel, */
                              /*      and TScreen       */
                              /*                       */
                              /***************************/

#define WIN31
#define STRICT

#include <owl.h>
#include <windowsx.h>
#include <inputdia.h>
#include <string.h>
#include <standard.h>
#include <stdio.h>

// colour constants

const    COLORREF Black     = RGB(0,    0,    0  );
const    COLORREF White     = RGB(255, 255, 255);
const    COLORREF Gray      = RGB(127, 127, 127);
const    COLORREF PaleGray  = RGB(192, 192, 192);
const    COLORREF Red       = RGB(255, 0,    0  );
const    COLORREF Brown     = RGB(127, 0,    0  );
const    COLORREF Green     = RGB(0,   255, 0  );
const    COLORREF DarkGreen = RGB(0,   127, 0  );
const    COLORREF Blue      = RGB(0,    0,   255);
const    COLORREF Indigo    = RGB(0,    0,   127);
const    COLORREF Yellow    = RGB(255, 255, 0  );
const    COLORREF Khaki     = RGB(127, 127, 0  );
const    COLORREF Magenta   = RGB(255, 0,   255);
const    COLORREF Purple    = RGB(127, 0,   127);
const    COLORREF Cyan      = RGB(0,   255, 255);
const    COLORREF BlueGreen = RGB(0,   127, 127);

// forward class declarations

_CLASSDEF(Point)
class    Point;
```

```
_CLASSDEF(Pixel)
class    Pixel;

_CLASSDEF(TScreen)
class    TScreen;

_CLASSDEF(TScreenApplication)
class TScreenApplication;

// class declarations

class    Point {
public:
        double x, y;
        Point()                         { x = y = 0; }
        Point(double a, double b)       { x = a; y = b; }
        Point(RPoint p)                 { x = p.x; y = p.y; }
        Point(PTScreen, RPixel);  // image of pixel in screen
        RPoint operator=(RPoint);  // assignment
};

class    Pixel {
friend   class Point;
friend   class TScreen;
private:
        int x, y;
        COLORREF c;
public:
        Pixel(); // top left client rect, default color
        Pixel(int, int, COLORREF);
        Pixel(PTScreen, RPoint p, COLORREF color);// image of point in screen
        Pixel(RPixel);
        RPixel operator=(RPixel);
        int GetX();
        int GetY();
        COLORREF GetCol();
        void SetCol(COLORREF);
        friend COLORREF SetPixel(HDC, RPixel);
        friend void MoveTo(HDC, RPixel);
        friend void LineTo(HDC, RPixel);
};

class    TScreen : public TWindow {
friend   class Point;
friend   class Pixel;
private:
        int tsox, tsoy;
        double Slope(double);      // convert tangent from Point to Pixel

protected:
        double tsxinc, tsyinc;
        Point TopLeft, BottomRight;
        RECT showrect;

        void SetPoint(RPoint, LPSTR);
        void SetOffsets();
        void SetShowRect(RECT&);
        void SetInteger(int&, LPSTR);
        void SetLong(long&, LPSTR);
        void SetDouble(double&, LPSTR);
        double GetAspect();

public:
```

```
        TScreen(PTWindowsObject, LPSTR);   // maximized window
        BOOL InShowRect(RPixel);
        void FloatOut(int, int, float);
        void DigitOut(HDC, RPixel, int);
        void Axes(HDC, int, COLORREF);
        void ShowRect(HDC, int, COLORREF);
        void SetScreen(RECT&);
        void Spot(HDC, RPixel, int, COLORREF);
        void Arrow(HDC, double, double, double, BOOL, COLORREF);
};


class   TScreenApplication : public TApplication {
public:
        TScreenApplication(LPSTR AName, HINSTANCE hInstance,
                              HINSTANCE hPrevInstance,
                              LPSTR lpCmdLine, int nCmdShow) :
        TApplication(AName, hInstance, hPrevInstance,
                              lpCmdLine, nCmdShow) {};
        virtual void InitMainWindow();
};

// class member functions

inline   RPoint Point::operator=(RPoint p)
{ x = p.x; y = p.y; return *this; }

inline   Pixel::Pixel()
{ x = y = 0; c = Black; }

inline   Pixel::Pixel(int a, int b, COLORREF color)
{ x = a; y = b; c = color; }

inline   Pixel::Pixel(RPixel s)
{ x = s.x; y = s.y; c = s.c; }

inline   RPixel Pixel::operator=(RPixel s)
{ x = s.x; y = s.y; c = s.c; return *this; }

inline   int Pixel::GetX() { return x; }

inline   int Pixel::GetY() { return y; }

inline   COLORREF Pixel::GetCol() { return c; }

inline   void Pixel::SetCol(COLORREF color)
{ c = color; }

inline   double TScreen::Slope(double t)
{ return t*tsxinc/tsyinc; }

// other inline functions

inline   COLORREF SetPixel(HDC hDC, RPixel s)
{ return SetPixel(hDC, s.x, s.y, s.c); }

inline   void MoveTo(HDC hDC, RPixel s)
{        MoveTo(hDC, s.x, s.y); }

inline   void LineTo(HDC hDC, RPixel s)
{        LineTo(hDC, s.x, s.y); }

#endif   // SCREEN
```

Index

In this index, references to C++ identifiers, including file, function, and program names, and keywords, are given in `monotype`.

NB: *All C++ operators and their precedences are listed on p. 24.*